Biometric Systems

James Wayman, Anil Jain, Davide Maltoni
and Dario Maio (Eds)

Biometric Systems

Technology, Design and Performance Evaluation

Springer

James Wayman
San Jose State University, USA

Anil Jain
Michigan State University, USA

Davide Maltoni
University of Bologna, Italy

Dario Maio
University of Bologna, Italy

British Library Cataloguing in Publication Data
Biometric Systems : technology, design and performance
evaluation
1.Biometric identification 2.Electronic security systems
I.Wayman, James
621.3'8928
ISBN 1852335963
A catalog record for this book is available from the Library of Congress

ISBN 1-85233-596-3 Springer-Verlag London Berlin Heidelberg
Springer Science+Business Media
springeronline.com

Wayman et. al.: Biometric Systems

First Indian Reprint 2011

ISBN 978-81-322-0252-3

This edition is published by Springer (India) Private Limited, (a part of Springer Science+Business Media), Registered Office: 3rd Floor, Gandharva Mahavidyalaya, 212, Deen Dayal Upadhyaya Marg, New Delhi – 110 002, India

Printed in India by Rakmo Press (P) Ltd., New Delhi.

Preface

The use of computers to recognize humans from physical and behavioral traits dates back to the digital computer evolution of the 1960s. But even after decades of research and hundreds of major deployments, the field of biometrics remains fresh and exciting as new technologies are developed and old technologies are improved and fielded in new applications. Worldwide over the past few years, there has been a marked increase in both government and private sector interest in large-scale biometric deployments for accelerating human–machine processes, efficiently delivering human services, fighting identity fraud and even combating terrorism. The purpose of this book is to explore the current state of the art in biometric *systems* and it is the system aspect that we have wished to emphasize.

By their nature, biometric technologies sit at the exact boundary of the human–machine interface. But like all technologies, by themselves they can provide no value until deployed in a *system* with support hardware, network connections, computers, policies and procedures, all tuned together to work with *people* to improve some real business process within a social structure.

In this book, we bring together some of the most respected and experienced international researchers and practitioners in the field to look closely at biometric systems from many disciplinary angles. We focus on the technologies of fingerprint, iris, face and speaker recognition, how those technologies have evolved, how they work, and how well they work as determined in recent test programs. We look at the challenges of designing and deploying biometrics in people-centered systems, particularly when those systems become large. We conclude with discussions on the legal and privacy issues of biometric deployments from both European and US perspectives. We hope you find this book valuable in understanding both the historical accomplishments and remaining challenges in this fascinating field.

James Wayman
Anil Jain
Davide Maltoni
Dario Maio
31 July 2004

Contents

List of Contributors

Robert J. Allen
Allen Consultants LLC
Robert.Allen7@att.net

Julian Ashbourn
International Biometric Foundation
ibf@1to1.org

J. Mike Bone
NAVSEA Crane Division
Bone_Mike@crane.navy.mil

Duane Blackburn
Federal Bureau of Investigation
Duane.Blackburn@ic.fbi.gov

Joseph P. Campbell, Jr.
Massachusetts Institute of Technology, Lincoln Laboratory
j.campbell@ieee.org

Raffaele Cappelli
University of Bologna
cappelli@csr.unibo.it

Jean-Christophe Fondeur
Sagem
jean-christophe.fondeur@sagem.com

Herbert Gish
BBN Technologies
hgish@bbn.com

Patrick Grother
National Institute of Standards and Technology
pgrother@NIST.gov

Herve Jarosz
Sagem
herve.jarosz@sagem.com

Anil K. Jain
Michigan State University
jain@cse.msu.edu

Chengjun Liu
New Jersey Institute of Technology
liu@cs.njit.edu

Davide Maltoni
University of Bologna
maltoni@csr.unibo.it

Dario Maio
University of Bologna
dmaio@deis.unibo.it

Alvin Martin
National Institute of Standards and Technology
alvin.martin@nist.gov

Elaine Newton
Rand Corporation
enewton@cmu.edu

Kenneth P. Nuger
San Jose State University
kpnuger@email.sjsu.edu

P. Jonathon Phillips
National Institute of Standards and Technology
Jonathon@nist.gov

Salil Prabhakar
Digital Persona, Inc.
SalilP@digitalpersona.com

Mark Przybocki
National Institute of Standards and Technology
mark.przybocki@nist.gov

Marek Rejman-Greene
BTExact Technologies
marek.rejman-greene@bt.com

Pat Sankar
U.S. Naval Postgraduate School
patsankar@yahoo.com

James L. Wayman
San Jose State University
jlwayman@aol.com

Harry Wechsler
George Mason University
Wechsler@cs.gmu.edu

Richard Wildes
York University
wildes@cs.yorku.ca

An Introduction to Biometric Authentication Systems 1

James Wayman, Anil Jain, Davide Maltoni and Dario Maio

1.1 Introduction

Immigration cards holding both passport number and measures of the user's hand [1]; fingerprints taken as a legal requirement for a driver license, but not stored anywhere on the license [2]; automatic facial recognition systems searching for known card cheats in a casino [3]; season tickets to an amusement park linked to the shape of the purchaser's fingers [4]; home incarceration programs supervised by automatic voice recognition systems [5]; and confidential delivery of health care through iris recognition [6]: these systems seem completely different in terms of purpose, procedures, and technologies, but each uses "biometric authentication" in some way. In this book, we will be exploring many of the technologies and applications that make up the field of "biometric authentication" – what unites them and what differentiates them from each other. In this chapter, we want to present a systematic approach to understanding in a unified way the multitude of technologies and applications of the field.

We start with a narrow definition, designed as much to limit the scope of our inquiry as to determine it.

> "Biometric technologies" are automated methods of verifying or recognizing the identity of a living person based on a physiological or behavioral characteristic [7, 8].

There are two key words in this definition: "automated" and "person". The word "automated" differentiates biometrics from the larger field of human identification science. Biometric authentication techniques are done completely by machine, generally (but not always) a digital computer. Forensic laboratory techniques, such as latent fingerprint, DNA, hair and fiber analysis, are not considered part of this field. Although automated identification techniques can be used on animals, fruits and vegetables [9], manufactured goods and the deceased, the subjects of biometric authentication are living humans. For this reason, the field should perhaps be more accurately called "anthropometric authentication".

The second key word is "person". Statistical techniques, particularly using fingerprint patterns, have been used to differentiate or connect groups of people [10, 11] or to probabilistically link persons to groups, but biometrics is interested only in recognizing people as individuals. All of the measures used contain both physiological and behavioral components, both of which can vary widely or be quite similar across a population of individuals. No technology is purely one or the other, although some measures seem to be more behaviorally influenced and some more physiologically influenced. The behavioral component of all biometric measures introduces a "human factors" or "psychological" aspect to biometric authentication as well.

In practice, we often abbreviate the term "biometric authentication" as "biometrics", although the latter term has been historically used to mean the branch of biology that deals with its data statistically and by quantitative analysis [12].

So "biometrics", in this context, is the use of computers to recognize people, despite all of the across-individual similarities and within-individual variations. Determining "true" identity is beyond the scope of any biometric technology. Rather, biometric technology can only link a person to a biometric pattern and any identity data (common name) and personal attributes (age, gender, profession, residence, nationality) presented at the time of enrollment in the system. Biometric systems inherently require no identity data, thus allowing anonymous recognition [4].

Ultimately, the performance of a biometric authentication system, and its suitability for any particular task, will depend upon the interaction of individuals with the automated mechanism. It is this interaction of technology with human physiology and psychology that makes "biometrics" such a fascinating subject.

1.2 A Quick Historical Overview

The scientific literature on quantitative measurement of humans for the purpose of identification dates back to the 1870s and the measurement system of Alphonse Bertillon [13–17]. Bertillon's system of body measurements, including such measures as skull diameter and arm and foot length, was used in the USA to identify prisoners until the 1920s. Henry Faulds, William Herschel and Sir Francis Galton proposed quantitative identification through fingerprint and facial measurements in the 1880s [18–20]. The development of digital signal processing techniques in the 1960s led immediately to work in automating human identification. Speaker [21–26] and fingerprint recognition [27] systems were among the first to be explored. The potential for application of this technology to high-security access control, personal locks and financial transactions was recognized in the early 1960s [28]. The 1970s saw development and deployment of hand geometry systems [29], the start of large-scale testing [30] and increasing interest in government use of these "automated personal identification" technologies [31]. Retinal [32, 33] and signature verification [34, 35] systems came in the 1980s, followed by face [36–42] systems. Iris recognition [43, 44] systems were developed in the 1990s.

1.3 The "Best" Biometric Characteristic

Examples of physiological and behavioral characteristics currently used for automatic identification include fingerprints, voice, iris, retina, hand, face, handwriting, keystroke, and finger shape. But this is only a partial list as new measures (such as gait, ear shape, head resonance, optical skin reflectance and body odor) are being developed all of the time. Because of the broad range of characteristics used, the imaging requirements for the technology vary greatly. Systems might measure a single one-dimensional signal (voice); several simultaneous one-dimensional signals (handwriting); a single two-dimensional image (fingerprint); multiple two-dimensional measures (hand geometry); a time series of two-dimensional images (face and iris); or a three-dimensional image (some facial recognition systems).

Which biometric characteristic is best? The ideal biometric characteristic has five qualities: robustness, distinctiveness, availability, accessibility and acceptability [45, 46]. By "robust", we mean unchanging on an individual over time. By "distinctive", we mean showing great variation over the population. By "available", we mean that the entire population should ideally have this measure in multiples. By "accessible", we mean easy to image using electronic sensors. By "acceptable", we mean that people do not object to having this measurement taken from them.

Quantitative measures of these five qualities have been developed [47–50]. Robustness is measured by the "false non-match rate" (also known as "Type I error"), the probability that a submitted sample will not match the enrollment image. Distinctiveness is measured by the "false match rate" (also known as "Type II error") – the probability that a submitted sample will match the enrollment image of another user. Availability is measured by the "failure to enroll" rate, the probability that a user will not be able to supply a readable measure to the system upon enrollment. Accessibility can be quantified by the "throughput rate" of the system, the number of individuals that can be processed in a unit time, such as a minute or an hour. Acceptability is measured by polling the device users. The first four qualities are inversely related to their above measures, a higher "false non-match rate", for instance, indicating a lower level of robustness.

Having identified the required qualities and measures for each quality, it would seem a straightforward problem to simply run some experiments, determine the measures, and set a weighting value for the importance of each, thereby determining the "best" biometric characteristic. Unfortunately, for all biometric characteristics, all of the desired qualities have been found to be highly dependent on the specifics of the application, the population (both their physiological and psychological states), and the hardware/software system used [51–54]. We cannot predict performance metrics for one application from tests on another. Further, the five metrics, which are correlated in a highly complex way, can be manipulated to some extent by administration policy.

System administrators might ultimately be concerned with: (1) the "false rejection rate", which is the probability that a true user identity claim will be falsely rejected, thus causing inconvenience; (2) the "false acceptance rate", which is the probability that a false identity claim will be accepted, thus allowing fraud; (3) the system throughput rate, measuring the number of users that can be processed in a time period; (4) the user acceptance of the system, which may be highly dependent upon the way the system is "packaged" and marketed; and (5) the ultimate total cost savings realized from implementing the system [55]. These latter, more practical, measures depend upon the basic system qualities in highly complex and competitive ways that are not at all well understood, and can be controlled only to a limited extent through administrative decisions [56, 57]. Predicting the "false acceptance" and "false rejection" rates, and system throughput, user acceptance and cost savings for operational systems from test data, is a surprisingly difficult task.

For the users, the questions are simple: "Is this system easier, faster, friendlier and more convenient than the alternatives?". These issues, too, are highly application-, technology- and marketing-specific.

Consequently, it is impossible to state that a single biometric characteristic is "best" for all applications, populations, technologies and administration policies. Yet some biometric characteristics are clearly more appropriate than others for any particular application. System administrators wishing to employ biometric authentication need to articulate clearly the specifics of their application. In the following sections, we look more carefully at the distinctions between applications.

1.4 The Applications

The operational goals of biometric applications are just as variable as the technologies: some systems search for known individuals; some search for unknown individuals; some verify a claimed identity; some verify an unclaimed identity; and some verify that the individual has no identity in the system at all. Some systems search one or multiple submitted samples against a large database of millions of previously stored "templates" – the biometric data given at the time of enrollment. Some systems search one or multiple samples against a database of a few "models" – mathematical representations of the signal generation process created at the time of enrollment. Some systems compare submitted samples against models of both the claimed identity and impostor identities. Some systems search one or multiple samples against only one "template" or "model".

And the application environments can vary greatly – outdoors or indoors, supervised or unsupervised, with people trained or not trained in the use of the acquisition device.

To make sense out of all of the technologies, application goals and environments, we need a systematic method of approach – taxonomies of uses and applications.

1.5 A Taxonomy of Uses

A biometric system can be designed to test one of only two possible hypotheses: (1) that the submitted samples are from an individual known to the system; or (2) that the submitted samples are from an individual not known to the system. Applications to test the first hypothesis are called "positive identification" systems (verifying a positive claim of enrollment), while applications testing the latter are "negative identification" systems (verifying a claim of no enrollment). All biometric systems are of one type or the other. This is the most important distinction between systems, and controls potential architectures, vulnerabilities and system error rates.

"Positive" and "negative" identification are "duals" of each other. Positive identification systems generally[1] serve to prevent multiple users of a single identity, while negative identification systems serve to prevent multiple identities of a single user. In positive identification systems, enrolled template or model storage can be centralized or decentralized in manner, including placement on optically read, magnetic stripe or smart cards. Negative identification systems demand centralized storage. Positive identification systems reject a user's claim to identity if no match between submitted samples and enrolled templates is found. Negative identification systems reject a user's claim to no identity if a match is found. Regardless of type of system, false rejections are a nuisance to users and false acceptances allow fraud.

An example of a positive identification system is the use of biometrics for employee access control at San Francisco International Airport. Hand geometry has been used since the early 1990s to control access by employees to secured airport areas. There are currently 180 readers used by about 18,000 enrolled users. Employees activate the system by swiping a magnetic stripe identity card through a reader. The purpose of the system is to limit use of the identification card to the enrolled owner, thereby prohibiting use of the card by multiple users. Although the 9-byte template could be stored on the magnetic stripe, in this case it is stored centrally to allow updating upon successful use. The stored hand shape template indexed to the card is transmitted from the central server to the access control device. The user then places the right hand in the hand geometry reader, making the implicit claim, "I am the user who is enrolled to use this card". If the submitted hand sample is found to be "close enough" to the stored template, the user's claim is accepted.

Santa Clara County, located in California near the San Francisco International Airport, requires the fingerprints of both left and right index fingers

---

1 Surveillance systems are also "positive" and "negative", but do not seek to prevent either multiple users of a single identity or multiple identities of a single user. A surveillance system for positive identification tests the hypothesis that all persons are on a list of authorized personnel. A negative system tests the hypothesis that no person is on the list of forbidden personnel.

from all applicants for social service benefits. Citizens are only eligible for benefits under a single identity and must attest upon enrollment that they are not already enrolled in the system. Consequently, this biometric system is for "negative identification". When an applicant applies for benefits, he or she places the index fingers on an electronic scanner with the implicit claim, "I am not known to this system". The submitted fingerprints are searched against the entire centralized database of enrolled persons – although to facilitate the search, the prints in the database might be partitioned by gender. If no match is found, the claim of non-identity in the system is accepted.

Use of biometrics in positive identification systems can be voluntary because alternative methods for verifying a claimed identity exist. Those electing not to use biometrics can have their identity verified in other ways, such as by presentation of a passport or driver's license. Use of biometrics in negative identification systems must be mandatory for all users because no alternative methods exist for verifying a claim of no known identity.

Those wishing to circumvent a positive identification system need to create a false match by impersonating an enrolled user. The possibility of biometric mimicry and forgery has been recognized since the 1970s [47, 58, 59]. Those wishing to circumvent a negative identification system need to submit altered samples not matching a previous enrollment. Table 1.1 summarizes these differences.

Historically, a distinction has been made between systems that verify a claimed identity and those that identify users without a claim of identity, perhaps returning a result that no identity was found. Some systems compare a single input sample to a single stored template or model to produce a "verification", or compare a single input sample to many stored templates to produce an "identification". Identification systems are said to compare

Table 1.1 Identification: "positive" and "negative".

Positive	Negative
To prove I am someone known to the system	To prove I am not someone known to the system
To prevent multiple users of a single identity	To prevent multiple identities of a single user
Comparison of submitted sample to single claimed template – "one-to-one" under the most common system design	Comparison of submitted sample to all enrolled templates – "one-to-many"
A "false match" leads to "false acceptance"	A "false match" or a "failure to acquire" leads to a "false rejection"
A "false non-match" or a "failure to acquire" leads to a "false rejection"	A "false non-match" leads to a "false acceptance"
Alternative identification methods exist	No alternative methods exist
Can be voluntary	Must be mandatory for all
Spoofed by submitting someone else's biometric measures	Spoofed by submitting no or altered measures

samples from one person to templates from many persons, with verification being the degenerate case of "many" equal to one. In the mid-1990s, several companies began to promote "PIN-less verification" systems, in which verification was accomplished without a claim to identity. The "verification/identification" dichotomy has been further clouded by the development of surveillance and modern "few-to-many" access control systems, which cannot be consistently classified as either "verification" or "identification". The uses and search strategies of biometric systems have expanded to the point where these distinctions of "verification/identification" and "one-to-one/one-to-many" are no longer fully informative.

Ultimately, a biometric system can only link a submitted sample to an enrolled template or model: that record created upon first use of the system by a person. That enrollment template/model need not be connected with any identifying information, such as a name or registration number. In fact, biometric measures and the enrollment templates/models derived from them contain no information about name, age, nationality, race or gender. Consequently, use of a biometric system without linkages of stored data to common identifiers allows for anonymous authentication. If system administrators have a need to connect the stored biometric data to other information, such as a name, that must be done by the presentation and human certification of trusted identifying credentials at the time of enrollment. Subsequent identification by the biometric system is no more reliable than this source documentation. But once that link has been made, subsequent identifications can be made without reference to the original source documents.

1.6 A Taxonomy of Application Environments

In the early 1990s, as we gained experience with the use of biometric devices, it became apparent that variations in the application environment had a significant impact on the way the devices performed. In fact, accurate characterization of the operational environment is primary in selecting the best biometric technology and in predicting the system's operational characteristics. In this section, we will present a method for analyzing a proposed operational environment by differentiating applications based on partitioning into six categories beyond the "positive" and "negative" applications already discussed.

1.6.1 Overt Versus Covert

The first partition is "overt/covert". If the user is aware that a biometric identifier is being measured, the use is overt. If unaware, the use is covert. Almost all conceivable access control and non-forensic applications are overt. Forensic applications can be covert.

1.6.2 Habituated Versus Non-Habituated

The second partition, "habituated/non-habituated", applies to the intended users of the application. Users presenting a biometric trait on a daily basis can be considered habituated after a short period of time. Users who have not presented the trait recently can be considered "non-habituated". A more precise definition will be possible after we have better information relating system performance to frequency of use for a wide population over a wide field of devices. If all the intended users are "habituated", the application is considered a "habituated" application. If all the intended users are "non-habituated", the application is considered "non-habituated". In general, all applications will be "non-habituated" during the first week of operation, and can have a mixture of habituated and non-habituated users at any time thereafter. Access control to a secure work area is generally "habituated". Access control to a sporting event is generally "non-habituated".

1.6.3 Attended Versus Non-Attended

A third partition is "attended/unattended", and refers to whether the use of the biometric device during operation will be observed and guided by system management. Non-cooperative applications will generally require supervised operation, while cooperative operation may or may not. Nearly all systems supervise the enrollment process, although some do not [4].

1.6.4 Standard Versus Non-Standard Environment

A fourth partition is "standard/non-standard operating environment". If the application will take place indoors at standard temperature (20 °C), pressure (1 atm), and other environmental conditions, particularly where lighting conditions can be controlled, it is considered a "standard environment" application. Outdoor systems, and perhaps some unusual indoor systems, are considered "non-standard environment" applications.

1.6.5 Public Versus Private

A fifth partition is "public/private". Will the users of the system be customers of the system management (public) or employees (private)? Clearly, attitudes toward usage of the devices, which will directly affect performance, vary depending upon the relationship between the end-users and system management.

1.6.6 Open Versus Closed

A sixth partition is "open/closed". Will the system be required, now or in the future, to exchange data with other biometric systems run by other management? For instance, some US state social services agencies want to be able to exchange biometric information with other states. If a system is to be open, data collection, compression and format standards are required. A closed system can operate perfectly well on completely proprietary formats.

This list is open, meaning that additional partitions might also be appropriate. We could also argue that not all possible partition permutations are equally likely or even permissible.

1.6.7 Examples of the Classification of Applications

Every application can be classified according to the above partitions. For instance, the positive biometric identification of users of the Immigration and Naturalization Service's Passenger Accelerated Service System (INSPASS) [1, 60], currently in place at Kennedy, Newark, Los Angeles, Miami, Detroit, Washington Dulles, Vancouver and Toronto airports for rapidly admitting frequent travelers into the USA, can be classified as a cooperative, overt, non-attended, non-habituated, standard environment, public, closed application. The system is cooperative because those wishing to defeat the system will attempt to be identified as someone already holding a pass. It will be overt because all will be aware that they are required to give a biometric measure as a condition of enrollment into this system. It will be non-attended and in a standard environment because collection of the biometric will occur near the passport inspection counter inside the airports, but not under the direct observation of an INS employee. It will be non-habituated because most international travelers use the system less than once per month. The system is public because enrollment is open to any frequent traveler into the USA. It is closed because INSPASS does not exchange biometric information with any other system.

The negative identification systems for preventing multiple identities of social service recipients can be classified as non-cooperative, overt, attended, non-habituated, open, standard environment systems.

Clearly, the latter application is more difficult than the former. Therefore we cannot directly compare hand geometry and facial recognition technologies based on the error rates across these very different applications.

1.7 A System Model

Although these devices rely on widely different technologies, much can be said about them in general. Figure 1.1 shows a generic biometric authentication system divided into five subsystems: data collection, transmission, signal processing, decision and data storage. We will consider these subsystems one at a time.

1.7.1 Data Collection

Biometric systems begin with the measurement of a behavioral/physiological characteristic. Key to all systems is the underlying assumption that the

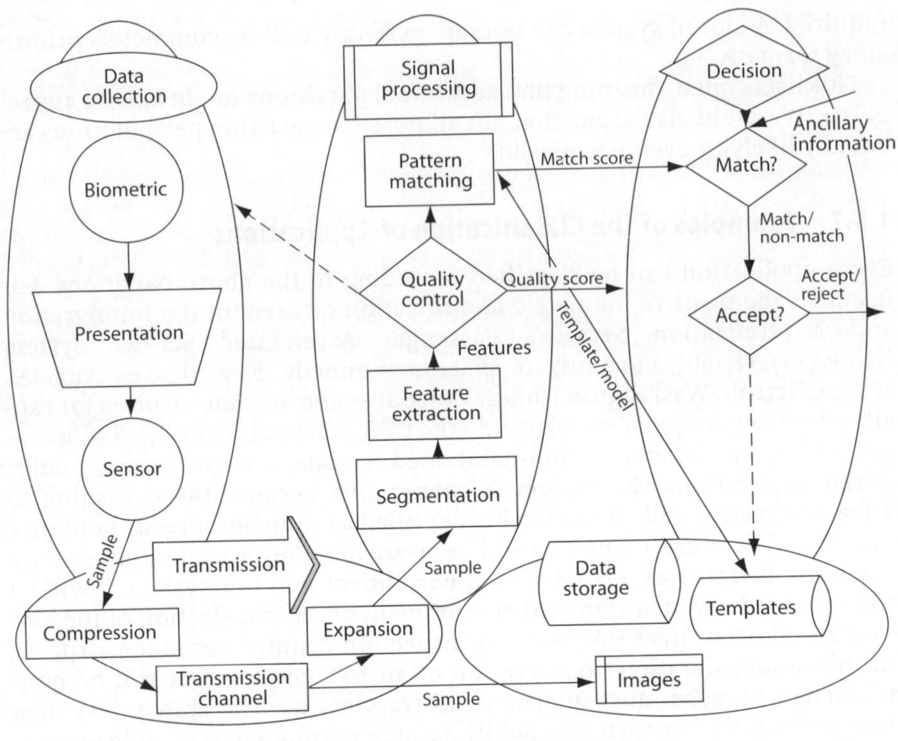

Figure 1.1 A generic biometric system.

measured biometric characteristic is both distinctive between individuals and repeatable over time for the same individual. The problems in measuring and controlling these variations begin in the data collection subsystem.

The user's characteristic must be presented to a sensor. The presentation of any biometric characteristic to the sensor introduces a behavioral (and, consequently, psychological) component to every biometric method. This behavioral component may vary widely between users, between applications, and between the test laboratory and the operational environment. The output of the sensor, which is the input data upon which the system is built, is the convolution of: (1) the biometric measure; (2) the way the measure is presented; and (3) the technical characteristics of the sensor. Both the repeatability and the distinctiveness of the measurement are negatively impacted by changes in any of these factors. If a system is to be open, the presentation and sensor characteristics must be standardized to ensure that biometric characteristics collected with one system will match those collected on the same individual by another system. If a system is to be used in an overt, non-cooperative application, the user must not be able to willfully change the biometric or its presentation sufficiently to avoid being matched to previous records.

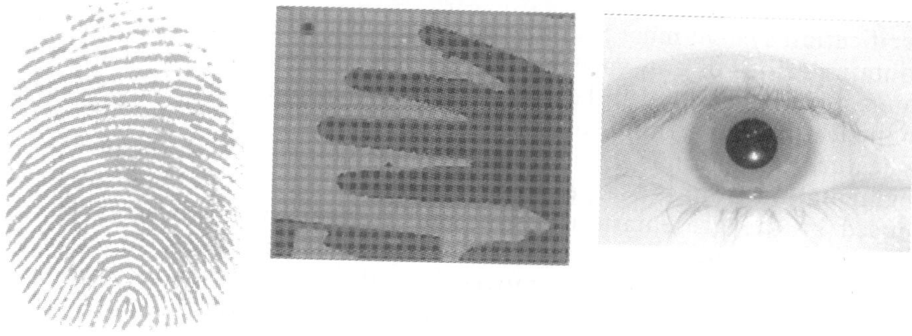

Figure 1.2 Fingerprint, hand and iris system input images.

Figure 1.2 shows input images from fingerprint, hand geometry and iris recognition systems.

1.7.2 Transmission

Some, but not all, biometric systems collect data at one location but store and/or process it at another. Such systems require data transmission. If a great amount of data is involved, compression may be required before transmission or storage to conserve bandwidth and storage space. Figure 1.1 shows compression and transmission occurring before the signal processing and image storage. In such cases, the transmitted or stored compressed data must be expanded before further use. The process of compression and expansion generally causes quality loss in the restored signal, with loss increasing with increasing compression ratio. The compression technique used will depend upon the biometric signal. An interesting area of research is in finding, for a given biometric technique, compression methods with minimum impact on the signal-processing subsystem.

If a system is to be open, compression and transmission protocols must be standardized so that every user of the data can reconstruct the original signal. Standards currently exist for the compression of fingerprints (Wavelet Scalar Quantization), facial images (JPEG), and voice data (Code Excited Linear Prediction).

1.7.3 Signal Processing

Having acquired and possibly transmitted a biometric characteristic, we must prepare it for matching with other like measures. Figure 1.1 divides the signal-processing subsystem into four tasks: segmentation, feature extraction, quality control, and pattern matching.

Segmentation is the process of finding the biometric pattern within the transmitted signal. For example, a facial recognition system must first find the boundaries of the face or faces in the transmitted image. A speaker verification system must find the speech activity within a signal that may contain periods of non-speech sounds. Once the raw biometric pattern of interest has been found and extracted from larger signal, the pattern is sent to the feature extraction process.

Feature extraction is fascinating. The raw biometric pattern, even after segmentation from the larger signal, contains non-repeatable distortions caused by the presentation, sensor and transmission processes of the system. These non-controllable distortions and any non-distinctive or redundant elements must be removed from the biometric pattern, while at the same time preserving those qualities that are both distinctive and repeatable. These qualities expressed in mathematical form are called "features". In a text-independent speaker recognition system, for instance, we may want to find the features, such as the mathematical frequency relationships in the vowels, that depend only upon the speaker and not upon the words being spoken, the health status of the speaker, or the speed, volume and pitch of the speech. There are as many wonderfully creative mathematical approaches to feature extraction as there are scientists and engineers in the biometrics industry. You can understand why such algorithms are always considered proprietary. Consequently, in an open system, the "open" stops here.

In general, feature extraction is a form of non-reversible compression, meaning that the original biometric image cannot be reconstructed from the extracted features. In some systems, transmission occurs after feature extraction to reduce the requirement for bandwidth.

After feature extraction, or maybe even before, we will want to check to see if the signal received from the data collection subsystem is of good quality. If the features "don't make sense" or are insufficient in some way, we can conclude quickly that the received signal was defective and request a new sample from the data collection subsystem while the user is still at the sensor. The development of this "quality control" process has greatly improved the performance of biometric systems in the last few short years. On the other hand, some people seem never to be able to present an acceptable signal to the system. If a negative decision by the quality control module cannot be overridden, a "failure to enroll" error results.

The feature "sample", now of very small size compared to the original signal, will be sent to the pattern matching process for comparison with one or more previously identified and stored feature templates or models. We use the term "template" to indicate stored features. The features in the template are of the same type as those of a sample. For instance, if the sample features are a "vector" in the mathematical sense, then the stored template will also be a "vector". The term "model" is used to indicate the construction of a more complex mathematical representation capable of generating features characteristic of a particular user. Models and features will be of different mathematical types and structures. Models are used in some speaker and facial recognition systems. Templates are used in fingerprint, iris, and hand geometry recognition systems.

The term "enrollment" refers to the placing of a template or model into the database for the very first time. Once in the database and associated with an identity by external information (provided by the enrollee or others), the enrollment biometric data is referred to as the template or model for the individual to which it refers.

The purpose of the pattern matching process is to compare a presented feature sample to the stored data, and to send to the decision subsystem a quantitative measure of the comparison. An exception is enrollment in systems allowing multiple enrollments. In this application, the pattern matching process can be skipped. In the cooperative case where the user has claimed an identity or where there is but a single record in the current database (which might be a magnetic stripe card), the pattern matching process might only make a comparison against a single stored template. In all other cases, such as large-scale identification, the pattern matching process compares the present sample to multiple templates or models from the database one at a time, as instructed by the decision subsystem, sending on a quantitative "distance" measure for each comparison. In place of a "distance" measure, some systems use "similarity" measures, such as maximum likelihood values.

The signal processing subsystem is designed with the goal of yielding small distances between enrolled models/templates and later samples from the same individual and large distances between enrolled models/templates and samples of different individuals. Even for models and samples from the same individual, however, distances will rarely, if ever, be zero, as there will always be some non-repeatable biometric-, presentation-, sensor- or transmission-related variation remaining after processing.

1.7.4 Storage

The remaining subsystem to be considered is that of storage. There will be one or more forms of storage used, depending upon the biometric system. Templates or models from enrolled users will be stored in a database for comparison by the pattern matcher to incoming feature samples. For systems only performing "one-to-one" matching, the database may be distributed on smart cards, optically read cards or magnetic stripe cards carried by each enrolled user. Depending upon system policy, no central database need exist, although in this application a centralized database can be used to detect counterfeit cards or to reissue lost cards without re-collecting the biometric pattern.

The database will be centralized if the system performs one-to-N matching with N greater than one, as in the case of identification or "PIN-less verification" systems. As N gets very large, system speed requirements dictate that the database be partitioned into smaller subsets such that any feature sample need only be matched to the templates or models stored in one partition, or indexed by using an appropriate data structure which allows the templates to be visited in an advantageous order during the retrieval [61]. These strategies have the effect of increasing system speed and decreasing false matches, at the expense of increasing the false non-match rate owing to partitioning errors. This means that system error rates do not remain constant with increasing database size and identification systems do not scale linearly. Consequently, database partitioning/indexing strategies represent a complex policy decision [56].

If it may be necessary to reconstruct the biometric patterns from stored data, raw (although possibly compressed) data storage will be required. The biometric pattern is generally not reconstructable from the stored templates or models, although some methods [41] do allow a coarse reconstruction of patterns from templates. Further, the templates themselves are created using the proprietary feature extraction algorithms of the system vendor. The storage of raw data allows changes in the system or system vendor to be made without the need to re-collect data from all enrolled users.

1.7.5 Decision

The decision subsystem implements system policy by directing the database search, determines "matches" or "non-matches" based on the distance or similarity measures received from the pattern matcher, and ultimately makes an "accept/reject" decision based on the system policy. Such a decision policy could be to reject the identity claim (either positive or negative) of any user whose pattern could not be acquired. For an acquired pattern, the policy might declare a match for any distance lower than a fixed threshold and "accept" a user identity claim on the basis of this single match, or the policy could be to declare a match for any distance lower than a user-dependent, time-variant, or environmentally linked threshold and require matches from multiple measures for an "accept" decision. The policy could be to give all users, good guys and bad guys alike, three tries to return a low distance measure and be "accepted" as matching a claimed template. Or, in the absence of a claimed template, the system policy could be to direct the search of all, or only a portion, of the database and return a single match or multiple "candidate" matches. The decision policy employed is a management decision that is specific to the operational and security requirements of the system. In general, lowering the number of false non-matches can be traded against raising the number of false matches. The optimal system policy in this regard depends both upon the statistical characteristics of the comparison distances coming from the pattern matcher, the relative penalties for false match and false non-match within the system, and the *a priori* (guessed in advance) probabilities that a user is, in fact, an impostor. In any case, in the testing of biometric devices, it is necessary to decouple the performance of the signal processing subsystem from the policies implemented by the decision subsystem.

1.8 Biometrics and Privacy

Whenever biometric identification is discussed, people always want to know about the implications for personal privacy. If a biometric system is used, will the government, or some other group, be able to get personal information about the users? Biometric measures themselves contain no personal information. Hand shape, fingerprints or eye scans do not reveal name, age, race, gender, health or immigration status. Although voice patterns can give a good estimation of gender, no other biometric identification technology currently used reveals anything about the person being measured. More common identification methods, such as a driver's license, reveal name, address, age, gender, vision impairment, height and even weight! Driver's licenses, however, may be easier to steal or counterfeit than biometric measures.

Biometric measures can be used in place of a name, Social Security number or other form of identification to secure anonymous transactions. Walt Disney World sells season passes to buyers anonymously, then uses finger geometry to verify that the passes are not being transferred. Use of iris or fingerprint recognition for anonymous health care screening has also been proposed. A patient would use an anonymous biometric measure, not a name or Social Security number, when registering at a clinic. All records held at the clinic for that patient would be identified, linked and retrieved only by the measure. No one at the clinic, not even the doctors, would know the patient's "real" (publicly recognized) identity.

The real fear is that biometric measures will link people to personal data, or allow movements to be tracked. After all, credit card and phone records can be used in court to establish a person's activities and movements. There are several important points to be made on this issue.

Phone books are public databases linking people to their phone number. These databases are even accessible on the Internet. Because phone numbers are unique to phone lines[2], "reverse" phone books also exist, allowing a name to be determined from a phone number. Even if a number is unlisted, all information on calls made from that number may be available to law enforcement agencies through the subpoena process. There are no public databases, however, containing biometric identifiers, and there are only a few limited-access government databases. Five US states have electronic fingerprint records of social service recipients (Arizona, California, Connecticut, New York and Texas); six states (California, Colorado, Georgia, Hawaii, Oklahoma and Texas) maintain electronic fingerprints of all licensed drivers[3]; nearly all states maintain copies of driver's license and social service recipient photos; the FBI and state governments maintain fingerprint databases on convicted felons and sex offenders; and the federal government maintains hand geometry records on those who have voluntarily requested border crossing cards [62]. General access to this data is limited to the agencies that collected it,

2 In the days of multi-user "party lines" this was not true, and phone numbers did not uniquely map to phone lines and households. Such "party lines" are now mostly gone, allowing phone numbers to indicate a user household or business uniquely.

3 West Virginia maintains a voluntary fingerprint database on drivers who wish to use biometric identification.

but like credit card and phone "toll records", this information can be released or searched by law enforcement groups acting under court order.

Unlike phone books, however, databases of biometric measures cannot generally be reversed to reveal names from measures because biometric measures, although distinctive, are not unique. Fingerprint, retinal and iris databases may be exceptions, allowing reversal if the biometric data was carefully collected. But general biometric measures do not serve as useful pointers to other types of data. The linking of records is always done by unique identifiers such as Social Security and credit card numbers. Biometric measures are not generally useful in this regard, even if databases linking information to measures were to exist. For these reasons, biometric measures are not useful for tracking the movements of people, as is already possible using telephone and credit card numbers.

Databases of biometric images, and the numerical models or templates derived from them, are often encrypted with the intention of inhibiting their compromise in bulk. But compromise of individual measures cannot always be prevented by protecting databases and transmission channels because biometric measures, although privately owned, are sometimes publicly observable (e.g. a photo of a person's face can be taken with a camera or downloaded from a web page). In general, biometric measures are not secret, even if it might be quite complicated to acquire usable copies (e.g. a retinal map) without the cooperation of the owner. When used for security, biometric characteristics are more like public keys than private keys. Unlike public keys, however, biometric measures cannot be revoked if stolen or mimicked. The industry is currently working on methods for "live-ness testing" and revocation, hoping to ameliorate these problems [63–65].

Table 1.2 summarizes the privacy issues raised by the use of biometrics.

Table 1.2 Biometrics and privacy.

1. Unlike more common forms of identification, biometric measures contain no personal information and are more difficult to forge or steal.
2. Biometric measures can be used in place of a name or Social Security number to secure anonymous transactions.
3. Some biometric measures (face images, voice signals and "latent" fingerprints left on surfaces) can be taken without a person's knowledge, but cannot be linked to an identity without a pre-existing invertible database.
4. A Social Security or credit card number, and sometimes even a legal name, can identify a person in a large population. This capability has not been demonstrated using any single biometric measure.
5. Like telephone and credit card information, biometric databases can be searched outside of their intended purpose by court order.
6. Unlike credit card, telephone or Social Security numbers, biometric characteristics change from one measurement to the next.
7. Searching for personal data based on biometric measures is not as reliable or efficient as using better identifiers, like legal name or Social Security number.
8. Biometric measures are not always secret, but are sometimes publicly observable and cannot be revoked if compromised.

1.9 The Road Ahead

Market estimates put the total hardware sales for the industry at US$6.6 million in 1990 and nearly US$200 million in 2000 [66]. Whether the next decade will result in a similar 2500% increase will depend upon user demand for positive identification biometrics. That demand will be created by imaginatively created systems designed for convenience, friendliness, cost-effectiveness and ease of use.

The use of negative identification biometrics will be fueled by government requirements to limit citizens to a single identity in driver licensing, social service and other civil applications [67, 68]. That demand will require the development of stronger criteria for cost/benefit assessment, security assurance, and privacy protection. Although we cannot predict the future rate of growth of the industry with any certainty, we do know that long-term growth is inevitable. With this book, we hope to stimulate further inquiry into the technologies, applications and issues that will shape this industry in the years to come.

References

[1] B. Wing, Overview of all INS biometrics projects. *Proc. CTST'98*, pp. 543–552.

[2] G. Slagle, Standards for the driver's license. *Proc. CTST'99*, pp. 891–902.

[3] J. Walters, Casinos must tell customers that police are scanning faces. *Toronto Star*, February 27, 2001, Edition 1.

[4] G. Levin, Real world, most demanding biometric system usage. *Proc. Biometrics Consortium, 2001/02*, Crystal City, VA, February 14–15, 2002.

[5] J. Markowitz, Voice biometrics: speaker recognition applications and markets 1999. *Voice Europe 1999: European Symposium on Voice Technologies*, London.

[6] J. Perkins, FT-IT: New services will keep eye on security: biometrics. *Financial Times* (London), February 21, 2001, Wednesday Surveys ITC1.

[7] B. Miller, Everything you need to know about biometric identification. *Personal Identification News 1988 Biometric Industry Directory*, Warfel & Miller, Inc., Washington DC, January 1988.

[8] J. Wayman, A definition of biometrics *National Biometric Test Center Collected Works 1997–2000*, San Jose State University, 2000.

[9] R. M. Bolle, J. H. Connell, N. Haas, R. Mohan and G. Taubin, VeggieVision: a produce recognition system. *Workshop on Automatic Identification Advanced Technologies*, November 1997, pp. 35–38.

[10] R. Jantz, Anthropological dermatoglyphic research. *Ann. Rev. Anthropol.*, **16**, 161–177, 1987.

[11] R. Jantz, Variation among European populations in summary finger ridge-count variables. *Ann. Human Biol.*, **24**(2), 97–108, 1997.

[12] *Webster's New World Dictionary of the American Language*, College Edition. World Publishing Co., New York, 1966.

[13] C. Beavan, *Fingerprints: The Origins of Crime Detection and the Murder Case that Launched Forensic Science*. Hyperion, New York, 2001.

[14] S. Cole, What counts for identity?: the historical origins of the methodology of latent fingerprint identification. *Fingerprint Whorld*, **27**, 103, January 2001.

[15] S. Cole, *Suspect Identities: A History of Fingerprinting and Criminal Identification*. Harvard University Press, 2001.
[16] C. Reedman, Biometrics and law enforcement. Available from http://www.dss.state.ct.us/digital/biometrics%20and%20law%20enforcement.htm (accessed May 31, 2004).
[17] http://www.cimm.jcu.edu.au/hist/stats/bert/index.htm
[18] H. Faulds, On the skin furrows of the hand. *Nature*, **22**, 605, October 28, 1880.
[19] W. Herschel, Skin furrows of the hand. *Nature*, **23**, 76, November 25, 1880.
[20] F. Galton, Personal identification and description. *Nature*, June 21 and 28, 1888, pp. 173–177, 201–202.
[21] S. Pruzansky, Pattern-matching procedure for automatic talker recognition. *J. Acoust. Soc. Am.*, **35**, 354–358, 1963.
[22] K. P. Li, J. E. Dammann and W. D. Chapman, Experimental studies in speaker verification using an adaptive system. *J. Acoust. Soc. Am.*, **40**, 966–978, 1966.
[23] J. Luck, Automatic speaker verification using cepstral measurements. *J. Acoust. Soc. Am.*, **46**, 1026–1031, 1969.
[24] K. Stevens, C. Williams, J. Carbonell and B. Woods, Speaker authentication and identification: a comparison of spectrographic and auditory presentation of speech material. *J. Acoust. Soc. Am.*, **44**, 596–607, 1968.
[25] B. Atal, Automatic recognition of speakers from their voices. *Proc. IEEE*, **64**(4), 460–474, 1976.
[26] A. Rosenberg, Automatic speaker recognition: a review. *Proc. IEEE*, **64**(4), 475–487, 1976.
[27] M. Trauring, Automatic comparison of finger-ridge patterns. *Nature*, **197**, 938–940, 1963.
[28] M. Trauring, On the automatic comparison of finger-ridge patterns. *Hughes Laboratory Research Report No. 190*, 1961.
[29] R. Zunkel, Hand geometry based verifications, in A. Jain, *et al.* (eds) *Biometrics: Personal Identification in Networked Society*. Kluwer Academic Press, 1999.
[30] A. Fejfar and J. Myers, The testing of 3 automatic ID verification techniques for entry control. *2nd Int. Conf. on Crime Countermeasures*, Oxford, 25–29 July, 1977.
[31] National Bureau of Standards, Guidelines on the evaluation of techniques for automated personal identification. *Federal Information Processing Standards Publication 48*, April 1, 1977.
[32] H. D. Crane and J. S. Ostrem, Automatic signature verification using a three-axis force-sensitive pen. *IEEE Trans. on Systems, Man and Cybernetics*, **SMC-13**(3), 329–337, 1983.
[33] V. S. Nalwa, Automatic on-line signature verification. *Proc. IEEE*, **85**(2), 215–239, 1997.
[34] J. R. Samples and R. V. Hill, Use of infrared fundus reflection for an identification device. *Am. J. Ophthalmol.*, **98**(5), 636–640, 1984.
[35] R. H. Hill, Retina identification, in A. Jain, *et al.* (eds) *Biometrics: Personal Identification in Networked Society*. Kluwer Academic Press, 1999.
[36] L. D. Harmon, M. K. Khan, R. Lasch and P. F. Ramig, Machine recognition of human faces. *Pattern Recognition*, **31**(2), 97–110, 1981.
[37] A. Samal and P. Iyengar, Automatic recognition and analysis of human faces and facial expressions: a survey. *Pattern Recognition*, **25**, 65–77, 1992.
[38] R. Chellappa, C. L. Wilson and S. Sirohey, Human and machine recognition of faces: a survey. *Proc. IEEE*, **83**(5), 705–740, 1995.
[39] L. Sirovich and M. Kirby, Low-dimensional procedure for the characterization of human faces. *J. Optical Soc. Am.*, **4**, 519–524, 1987.

[40] M. Turk and A. Pentland, Eigenfaces for recognition. *J. Cognitive Neuroscience*, **3**(1), 71–86, 1991.

[41] J. Zhang, Y. Yan and M. Lades, Face recognition: eigenface, elastic matching and neural nets. *Proc. IEEE*, **85**(9), 1423–1436, 1997.

[42] J. D. Daugman, High confidence visual recognition of persons by a test of statistical independence, *IEEE Trans. Pattern Analysis and Machine Intelligence*, **15**(11), 1148–1161, 1993.

[43] R. P. Wildes, Iris recognition: an emerging biometric technology, *Proc. IEEE*, **85**(9), 1348–1364, 1997.

[44] A. Jain, R. Bolle and S. Pankati, Introduction to biometrics, in A. Jain, *et al.* (eds) *Biometrics: Personal Identification in Networked Society*. Kluwer Academic Press, 1999.

[45] J. Wayman, Fundamentals of biometric authentication technologies. *Int. J. Imaging and Graphics*, **1**(1), 2001.

[46] J. L. Wayman, Technical testing and evaluation of biometric identification devices, in A. Jain, *et al.* (eds) *Biometrics: Personal Identification in Networked Society*. Kluwer Academic Press, 1999.

[47] D. E. Raphael and J. R. Young, *Automated Personal Identification*. SRI International, 1974.

[48] W. Haberman and A. Fejfar, Automatic identification of personnel through speaker and signature verification – system description and testing. *Proc. 1976 Carnahan Conference on Crime Countermeasures*, Lexington, KY, May 1976, pp. 23–30.

[49] R. L. Maxwell, General comparison of six different personnel identity verifiers. *Sandia National Laboratories, Organization 5252 Report*, June 20, 1984.

[50] A. J. Mansfield and J. L. Wayman, *Best Practices in Testing and Reporting Biometric Device Performance*, version 2.0. U.K. Biometrics Working Group. Available online at `http://www.cesg.gov.uk/site/ast/biometrics/media/BiometricTestReportpt1.pdf`

[51] J. P. Phillips, A. Martin, C. Wilson and M. Przybocki, An introduction to evaluating biometric systems. *IEEE Computer*, February 2000, p. 56–63.

[52] D. Maio, D. Maltoni, J. Wayman and A. Jain, FVC2000: Fingerprint verification competition 2000, *Proc. 15th International Conference on Pattern Recognition*, Barcelona, September 2000. Available online at `http://www.csr.unibo.it/research/biolab/`.

[53] A. Mansfield, G. Kelly, D. Chandler and J. Kane, *Biometric Product Testing Final Report*. National Physical Laboratory, London, March 19, 2001. Available online at `http://www.cesg.gov.uk/site/ast/biometrics/media/BestPractice.pdf`.

[54] D. Blackburn, M. Bone, P. Grother and J. Phillips, *Facial Recognition Vendor Test 2000: Evaluation Report*, January 2001. Available online at `http://www.dodcounterdrug.com/facialrecognition/FRVT2000/documents.htm`.

[55] W. Wilson, Establishing the business case for biometrics. *Proc. Biometric Consortium 2000*, Gaithersburg, MD, September, 2000.

[56] J. L. Wayman, Error rate equations for the general biometric system. *IEEE Automation and Robotics*, **6**(1), 35–48, 1999.

[57] J. Ashbourne, *Biometrics: Advanced Identification Technology*. Springer, 2000.

[58] R. C. Lummis and A. Rosenberg, Test of an automatic speaker verification method with intensively trained mimics. *J. Acoust. Soc. Am.*, **51**, 131(A), 1972.

[59] G. Warfel, *Identification Technologies: Computer, Optical, and Chemical Aids to Personal ID*. Charles C. Thomas, Springfield, IL, 1979.

[60] J. L. Wayman, Report on the evaluation of the INSPASS hand geometry system. In *National Biometric Test Center Collected Works 1997–2000*, San Jose State University, 2000.

[61] R. Cappelli, D. Maio and D. Maltoni, Indexing fingerprint databases for efficient 1:*N* matching. *Int. Conf. (6th) on Control, Automation, Robotics and Vision (ICARCV2000)*, Singapore, December 2000.

[62] J. Wayman, Federal biometric technology legislation. *IEEE Computer*, **33**(2), 76–80, 2000.

[63] R. Derakhshani, S. Schuckers, L. Hornak and L. O'Gorman, Determination of vitality from a non-invasive biomedical measurement for use in fingerprint scanners. *Pattern Recognition*, **17**(2), 2003.

[64] N. Ratha, J. Connell and R. Bolle, Cancelable biometrics. *Proc. Biometrics Consortium 2000*, Gaithersburg, MD, September 13–14, 2000.

[65] J. Cambier, U.C. von Seelen, R. Glass, R. Moore, I. Scott, M. Braithwaite and J. Daugman, Application-specific biometric templates. *Proc. Third Workshop on Automatic Identification and Advanced Technologies*, Tarrytown, New York, March 14–15, 2002.

[66] E. Bowman, Identifying trends: the evolving biometrics market. *ID World*, **1**(5), 7, 1999.

[67] *National Standard for Driver's License/Identification Card*, AAMVA June 30, 2000. Available online at `http://www.aamva.org/`.

[68] D. Mintie, Biometrics for state identification applications – operational experiences. *Proc. CTST'98*, **1**, 299–312.

Fingerprint Identification Technology 2

Robert Allen, Pat Sankar and Salil Prabhakar

2.1 History

Archaeological evidence of the use of fingerprints to associate a person with an event or transaction has been reported from ancient China, Babylonia, and Assyria as early as 6,000 BC [1]. In 1686, Marcello Malpighi, an anatomy professor at the University of Bologna, wrote in a paper that fingerprints contained ridges, spirals and loops, but did not refer to their possible use for identification [2]. In 1823, John Purkinji, an anatomy professor at Breslau University published a thesis in which he discussed nine types of fingerprint patterns; he also did not refer to the possibility of their use for identification [3]. However, this early evidence and documentation led researchers and practitioners to explore many uses of fingerprints, including human identification.

2.1.1 Early Biometric Efforts

2.1.1.1 Alphonse Bertillon

Alphonse Bertillon, a Paris police department employee and son of an anthropologist, developed a system of anthropometry in 1880 as a means for classifying criminals and used this system to identify recidivists [4]. Anthropometry (a system of cataloging an individual's body measurements such as height, weight, lengths of arm, leg, index finger etc.) was shown to fail in a famous case at Leavenworth Prison, where two prisoners, both named William West, were found to have nearly identical measurements even though they claimed not to be biologically related [5]. Bertillon had theoretically calculated the probability of occurrence of such an event as one in four million.

2.1.1.2 William Herschel

In 1856 Sir William Herschel, a British Chief Magistrate in Jungipoor, India, used fingerprints (actually palmprints) to certify native contracts, playing more on the superstitious beliefs of the natives than on science [6]. As his fingerprint collection grew, Herschel came to the conclusion that fingerprints could be used to prove or disprove identity.

2.1.1.3 Henry Faulds

During the 1870s, Dr Henry Faulds, a British surgeon in Japan, after noticing finger marks on ancient pottery, studied fingerprints, recognized the potential for identification, and devised a method for classifying fingerprint patterns [7]. Faulds is also credited with the first fingerprint identification of a greasy fingerprint left on an alcohol bottle. Finally, in 1880, Faulds forwarded his classification system and method for recording fingerprints to Charles Darwin, who in turn forwarded this material to his cousin, Sir Francis Galton.

2.1.1.4 Francis Galton

Francis Galton, an anthropologist, began a systematic study of fingerprints as a means of identification in the 1880s [8]. In 1892, he published the first book on fingerprints, entitled *Fingerprints* [9]. He added scientific support to what Herschel and Faulds suspected: that fingerprints are permanent throughout life, and that no two fingerprints are identical (by calculating the odds of two individual fingerprints being identical to be 1 in 64 billion). Galton devised a system of classifying fingerprints into what are now called "Galton pattern types". He also identified minute fingerprint characteristics (called "minutiae" and often referred to as "Galton details") that are used to determine whether two fingerprints match.

2.1.1.5 Juan Vucetich

In 1891, Juan Vucetich, an Argentinian police officer, began the first systematic filing of fingerprints using the Galton pattern types. In 1892, Vucetich made his first criminal fingerprint identification in a murder investigation, using a bloody fingerprint to prove the identity of the murderer [1].

2.1.1.6 Edward Henry

In 1897, Sir Edward Henry, a British police officer in India, established a modified fingerprint classification system using Galton's observations. This system was ultimately adopted by Scotland Yard in 1901 and is still used in many English-speaking countries [10].

2.2 Applications of Fingerprints

2.2.1 Forensics

Fingerprints are perhaps the most widely used identifier in the field of forensics. They are used not only to link suspects to crime scenes, but also to link persons arrested under another name to previous arrests, identify deceased persons (both criminal and non-criminal), and to associate persons with questioned documents. For over a century, forensic applications have been the primary focus of fingerprint identification techniques. Manual paper or card files have, for most of that time, provided the source of the fingerprint data, and several classification techniques have been used to organize these records in order to divide the search process into sizes manageable by humans. Additional manual filing systems have been developed to organize fingerprints lifted from crime scenes (called "latents") into manageable groups that may have to be manually searched for known suspects. The cumbersome and time-consuming nature of filing, searching and matching fingerprints manually led to efforts in automating parts of the process as computer technology became more readily available to law enforcement agencies.

2.2.2 Genetics

As indicated in the brief historical references above, some of the early work and observations involving fingerprints were derived from the work of early researchers in genetics, such as Sir Francis Galton. There is a rather large body of work tracing the genetic history of population groups through the study of fingerprint pattern characteristics [11]. There is also evidence of work to associate certain fingerprint characteristics with certain birth defects and diseases in an attempt to study the correlation between these unique characteristics and a predisposition to such defects [12]. Such studies triggered much of the work to establish the unique fingerprint features (patterns, minutiae, pores) that led to the use of fingerprints as the most reliable form of identification.

2.2.3 Civil and Commercial

The brief history of fingerprints above also indicates the early use of fingerprints in associating an individual to an item or event. It is not clear whether the very early innovators realized that fingerprints could be used for identification or only took advantage of the fact that people readily believed that to be so. This is not much different from the requirement over the last 60 years for fingerprinting as a prerequisite for obtaining a driver license in some parts of the USA, despite the inability to use the fingerprints, except in extreme situations where a driver license number might be known.

In more recent times, fingerprints have been applied to application/registration forms in an attempt to associate applicants with certain benefits (welfare [13], voting, banking [14] etc.). In most of these uses, a key component has been missing: the ability to effectively search any large fingerprint database in a reasonable amount of time. To further complicate matters, fingerprint image capture is done by people with little or no training in the proper collection methods. Examination of sample fingerprint images from such collections has shown that captured fingerprints do not contain consistent areas of the finger. It is not uncommon to find as much as three-quarters of the captured fingerprints to be from the tip, right or left side of the finger rather than from the preferred "core" region at the center.

2.2.4 Government

The most common application of fingerprint identification has been, and still is, to associate an individual with a criminal record. Individuals inducted into the military and those applying for a position with the government or a government contractor have been required to submit fingerprints for comparison with criminal records on file at the Federal Bureau of Investigation (FBI). Indeed, there are well over 200 million sets of such fingerprint records stored in Fairmont, West Virginia (at one time in file cabinets occupying several floors of an office building in Washington, DC). In this case, all 10 fingerprints (called a "tenprint") have been recorded for comparison.

In many countries outside North America and Western Europe, it has been, and still is, a common practice to capture fingerprints for all individuals when they reach a certain age (e.g. 16 years) in order to issue a national identity card. Re-registration of these individuals is required often (e.g. every 5 years), at which time a search in the fingerprint database is made to confirm that the individual is indeed the original applicant (for fraud prevention). Traditionally, such systems have usually employed automated searches of a name/number file and a manual comparison of the fingerprints to confirm the identity.

2.3 Early Systems

Most of the early fingerprint identification systems were put into place in major metropolitan areas or as national repositories. Juan Vucetich established a fingerprint file system in Argentina in 1891, followed by Sir Edward Henry in 1901 at Scotland Yard in England. The first fingerprint identification system in the USA was put in place by the New York Civil Service Commission in 1902, followed by a system in the New York State prison system in 1903. At the national level in the USA, the Federal Penitentiary in Kansas instituted a fingerprint system (with the assistance of Scotland Yard) in 1904. From 1905 to the early 1920s, the US military and many state and local law enforcement agencies inaugurated fingerprint identification systems.

It was not until 1924, when the FBI established an Identification Division by an act of Congress, that a criminal file based on the work done by Sir Edward Henry was created. Over the next 47 years, the FBI manually arranged over 200 million fingerprint records into files using the Henry system of classification.

2.3.1 Manual Card Files

Manual fingerprint card files were usually organized by a pattern classification system based on combination of the patterns on each of the ten fingers of individuals. Two similar classification systems were developed, one by Sir Edward Henry in the UK, and one by Juan Vucetich in Argentina. The

Henry system became a standard for many countries outside South America, while the Vucetich system was used primarily in South America. In the Henry classification system, numerical weights are assigned to fingers with a whorl pattern [15]. A bin number, based on the sum of the weights for the right hand and sum of the weights for the left hand is computed to generate 1,024 possible bins. Letter symbols are assigned to fingers: capital letters to the index fingers and lower-case letters to other fingers. These are combined with the numeric code to further subdivide the 1,024 bins. Each of these pattern groupings defines bins into which fingerprint cards with the same pattern group are placed. A bin might be a folder in a file drawer or several file drawers if it contains a common pattern group and the file is large.

Two problems existed in manual files: first, the patterns assigned to each finger might not be exactly the same on each occurrence of the same card; and second, if a pattern type error was made, the search might not reach the correct bin. In the early stages of automation, when more sophisticated means of searching the fingerprint database became possible, the accuracy of the manual fingerprint system was estimated to be only 75%. To further complicate matters, the distribution of pattern types is not uniform; thus there were a few bins that contained most of the fingerprint cards. For example, nearly 65% of fingers have loop patterns, 30% have whorl patterns and the remaining 5% have arch patterns. To overcome this difficulty, it was necessary to devise secondary (and tertiary) breakdowns of the bin numbers to subdivide the large bins. Although this aggravated the error possibilities alluded to above, it made the search more tractable for large files.

While the reliability of the search could be compromised by the factors mentioned above, the selectivity of a search using the binning system developed by Henry and Vucetich was a big improvement over not using bins at all. However, in the case of arch patterns, the bin was not subdivided through any of the breakdown schemes used for whorls and loops. As a result, this bin could become arbitrarily large as the file grew. Thus, in a large file, a search for a record containing arch patterns on all fingers could be very difficult. As with many such indexing schemes, there is a trade-off between selectivity and reliability in a system that reduces manual searching effort (which has its own inherent errors). The method of indexing fingerprints using the Henry/Vucetich method introduces many errors through the often complicated rules for assigning primary, secondary and tertiary labels to fingerprints. In practice, these errors can be minimized through rigorous training of human indexers in order to benefit from the increased search efficiency.

2.3.2 Classification

Fingerprint classification systems based upon the work of Vucetich and Henry generally recognize fingerprint patterns as being loops (left or right), whorls and arches. Figures 2.1(a)–(d) show these four basic fingerprint pattern types, which form the basis of most fingerprint classification systems. Broadly defined, the patterns are differentiated based on the

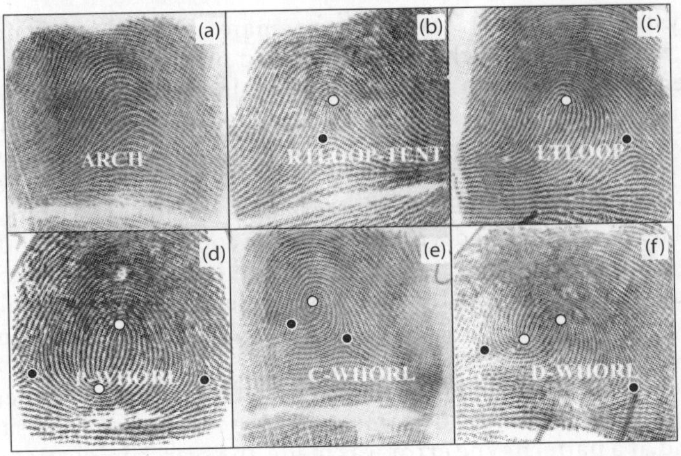

Figure 2.1 Some of the common fingerprint types. The core points are marked with solid white circles while the delta points are marked with solid black circles.

presence of zero, one or two delta regions. A delta region is defined by a triradial ridge direction at a point. There are transitional patterns between these three that have been used to define further subdivision of fingerprint records. Arch patterns have no delta, loops have one delta, and whorls have two deltas. Transitions from arches to loops occur for small loops and give rise to the so-called tented arch pattern that appears as a notable cusp to the arch. Loop patterns have a single delta and tend to transition to the plain whorl through patterns such as the central pocket loop (C-Whorl) and the twinned loop (D-Whorl), as seen in Figures 2.1(e) and (f). Whorl patterns are characterized by two deltas. This is clearly evident in the plain whorl and double loop whorl, but not so evident in the central pocket whorl pattern. In many instances, the classification decision requires the application of a topological test to determine whether there is a true re-curve, as necessary for a loop, or a delta present to support a whorl pattern. As the files become larger, it becomes necessary to further subdivide these basic patterns using ridge tracings and/or ridge counts in order to maintain reasonably sized bins. The details of these further subdivisions are interesting to understand, but beyond the scope of this chapter.

As computers were introduced into the fingerprint identification process [16], it became desirable to revise some of the pattern subdivisions to make the search more effective. Initially, the alphanumeric Henry system of classification was converted to a numeric model that encoded alphanumeric pattern symbols to numeric form, for both primary and subordinate classifications. Also, since computers could quickly search a database indexed by fingerprint classification and return record identification numbers, the manual files could be organized by identification number rather than pattern classifications. In this way, some of the search error could be minimized. However, the reliability of the manual classifications that were assigned over many years of operation remained questionable. Simply transcribing the manual codes to more computer-friendly codes had the potential to eliminate the manual search and filing errors (probably the more significant component), but did not offset the original pattern type assignment errors.

2.3.3 Searching

Searching of a fingerprint database typically involved the use of the finger patterns, subordinate breakdown (ridge counts/tracings) for loops and whorls, and alternate patterns for those fingers that had patterns that might be seen as one of several possibilities. Further, if the subordinate features might assume a range of possible values, the various combinations would need to be searched as well. Many agencies established policies limiting the possible alternate values allowed (for labor considerations), and the alternate patterns themselves might be suspect, leading to inaccuracy. At some level, if the comparison of fingerprints was to be done manually, the resulting search needed to be restrictive enough to ensure an accurate fingerprint feature comparison. If the number of search candidates were too large, then the potential for human error in comparison would rise. A compromise had to be reached that allowed for searches to be conducted in a reasonable time with a reasonable level of accuracy. As the search parameter became indexed by computers, searching became less error-prone, but not completely error-free [17].

2.3.4 Matching

Fingerprint matching prior to automation involved the manual examination of the so-called Galton details (ridge endings, bifurcations, lakes, islands, pores etc., collectively known as "minutiae"). Prior to the late 1960s, neither the available computer systems that could display fingerprint images for comparison were affordable, nor a significant number of digital fingerprint images were available for display. Consequently, the comparison process was manual, requiring a magnification glass for comparing the features of each search print to each of the many candidate prints manually retrieved from the database files. If the correct record could be retrieved from the files, it was fairly likely that the comparison process would identify the individual. It was very unlikely that an incorrect identification would be made because the fingerprint-matching technicians were rigorously trained and faced sanctions/penalties if they made any errors. However, it was possible that the correctly matching file record would not be retrieved (and thus not identified) because of incorrect classification of the search record, or an incorrectly filed original record.

2.4 Early Automation Efforts

By the mid-1960s, it become apparent to agencies with large fingerprint files and increasing workloads that some form of automation was required to keep labor costs within reasonable bounds. It was also apparent (although to a lesser extent) that the inaccuracy of the manual systems was an important performance limitation, especially in large fingerprint files. It was becoming clear to those agencies that measured the identification error rates of their systems that classification assignment and search were significant sources of error that increased as the file size and search workload grew.

2.4.1 US NBS/NIST Research

In the mid-1960s, the National Institute of Standards and Technology (NIST) (known as the National Bureau of Standards at the time) initiated several research projects to automate the fingerprint identification process. These efforts were supported by the Federal Bureau of Investigation (FBI) as part of an initiative to automate many of the processes in the Bureau. The NIST looked at automatic methods of digitization of inked fingerprint images, the effect of image compression on fingerprint image quality, classification, extraction of Galton features, and matching of fingerprints (tenprint-to-tenprint and latent-to-tenprint) [17]. The results of these efforts, together with a collaboration/interaction with private industry and national agencies (e.g. the Scientific Research and Development Branch and the Home Office in the UK), led to important implementation of this core technology.

2.4.2 Royal Canadian Police

By the mid-1960s, the fingerprint collection of the Royal Canadian Mounted Police (RCMP) had grown to over a million tenprint records. Their identification workload was increasing, as was the need to keep labor costs in check. They investigated the technology available to automate the search process in their tenprint identification section and ultimately selected an automated video tape-based filing system that was already in use in other industries (e.g. railroads and insurance).

The video file system, manufactured by Ampex Corporation, featured a digital track on a two-inch commercial video tape that could be searched by a minicomputer. A complex video disk buffering of search data and display stations capable of presenting single fingerprint images for comparison made it possible to display the record images selected by a classification-based search of the video tapes. The video-file system was operational until the mid-1970s, when the RCMP installed the first automated fingerprint identification system (AFIS).

2.4.3 FBI

In the USA, at about the same time that the RCMP and the UK Home Office were looking for automation technologies, the FBI was investigating the possibilities for automating various Identification Division processes, including the fingerprint identification operations. The identification operations received requests to match about 25,000 tenprints daily, had a criminal fingerprint file of over 20 million tenprint records, and employed several thousand people to handle the daily workload.

In the mid-1960s, the FBI signed research contracts with three companies to build a working prototype for scanning FBI fingerprint cards. By the early 1970s, the prototypes were completed and evaluated for their capabilities. A small-scale test resulted in the selection of one design approach and a contract was awarded to build five production card readers to begin the digitization of the tenprint files.

Over the next five or six years, the FBI worked with the computer industry to build other core technologies, including fingerprint matching hardware and automated classification software and hardware, and began a broad-based program to automate several manual functions within the fingerprint identification section. This plan ultimately resulted in an automated tenprint card-handling system, with functional workstations to support the manual data entry, classification, validation and completion of result response.

Study of the automation process continued and system designs were developed for what would become the next stage of fingerprint automation at the FBI. By the end of 1994, competition for the Integrated Automated Fingerprint Identification System (IAFIS) basic demonstration model was completed. Model systems had to demonstrate the ability to meet the performance requirements defined for the system. By the end of 1995, Lockheed Martin Corporation was selected to build the IAFIS, and by 1999 the major components were operational.

2.4.4 United Kingdom

In the UK, over about the same time-scale as the FBI, the Home Office was working within its own Scientific Research and Development Branch (SRDB), and in cooperation with the computer industry to develop technology to automate fingerprint processing. The Home Office directed its attention to the design of the algorithms and processes that were needed by the National Police Service (NPS) to automate tenprint processing nationally. At the same time, the Metropolitan Police Service, located at New Scotland Yard, directed attention to latent fingerprint processing, and initially working with Ferranti Ltd, developed a prototype latent encoding and search system. The resulting system provided the fingerprint data entry capability, the encoding of minutiae data for selected tenprint file entries, and the search and matching of fingerprints to identify candidates.

The technology developed through these efforts was eventually incorporated into the model for the National Automated Fingerprint Identification System (NAFIS) constructed by the SRDB. Several technologies for the core algorithms were investigated, including not only simulation, but also implementation into a specific hardware realization. To support these efforts, SRDB carried out substantial development internally and contracted with several software houses in the UK to carry out the programming and utilizing. For example, the Transputer parallel processing array was evaluated to serve as the execution engine for a variety of tasks. As a result of the simulation, research and implementation studies, when the NAFIS project was ready for tender there already existed a great deal of quantitative data on the requirements of performance and response times and the desired characteristics of the human–machine interfaces.

2.4.5 Japan

At about the same time as interest began to build in the USA and the UK for the automation of the manual fingerprint processes in the national repositories, the Japanese National Police (JNP), who had a fingerprint file of over six million records, also initiated study of the automation possibilities. JNP posted its technical people to collaborate with researchers at the FBI and the Home Office on the details of the processing required. The system ultimately developed for the national police contained many of the concepts included in both the USA and UK efforts.

2.5 The Technology

Certain essential components that are required for the automation of fingerprint searching and matching were already employed by the US military as a part of both defensive and offensive weapons systems. For example, imaging devices capable of resolving small objects in a large field were already in use to identify and classify potential targets. Methods had been developed for recognizing different signatures of potential targets using both spatial and frequency domain techniques for separating the signal from the background clutter. It was natural to look to the organizations developing these image processing-based military applications for solutions in the application of fingerprint matching. A catalyst for this process was found in the Law Enforcement Assistance Administration (LEAA), which was formed in the early 1970s to infuse automation technology into law enforcement organizations. Through grants and contracts, LEAA (1968–1977) funded a number of research programs to investigate techniques that could be applied to fingerprint automation projects together with other operational aspects of the law enforcement.

2.5.1 Scanning and Digitizing

One of the most important elements needed for fingerprint automation was a method for scanning inked fingerprint forms/cards that would provide images of sufficient quality for subsequent enhancement, feature extraction and matching. The FBI initiated a research program to build an engineering model of a scanner that could sample an object area of 1.5 × 1.5 in at 500 pixels per inch (often called DPI), with an effective sampling spot size of 0.0015 in, with a signal-to-noise (S/N) ratio in excess of 100:1, and digitized to at least 6 bits (64 gray levels). In the late 1960s, these requirements could only be met by a system that used a cathode ray tube and a precision deflection system, an array of photomultiplier tubes to measure the incident and reflected light, and an amplifier–digitizer to convert the electrical signal into a digital value for each pixel. Further, this system needed to scan the objects (i.e. fingerprint cards) in complete darkness (a light-tight enclosure). Three companies were selected to build these engineering models to demonstrate the capability to scan, process and extract minutiae features from an example set of fingerprint cards.

Based on an evaluation of these three competing engineering models, the supplier of the best model was selected to build a prototype of a fingerprint card scanner capable of transporting fingerprint cards to and through the scanner and of processing and extracting features from the captured digital images. This device was successfully built, delivered and tested by the FBI in the early 1970s. This prototype formed the basis of a contract to build at least five such scanning and data capture systems, each capable of scanning and extracting data from 250 tenprint cards per hour. These five scanning systems were operational by 1978 and began digitizing the 22 million record tenprint files that existed at the FBI at that time. One of the major drawbacks of this system was that the fingerprint cards had to be transported into a light-tight area during scanning, and in the event of a problem manual intervention was not possible.

There were relatively few scanning devices by the late 1970s that met the technical characteristics requirements of 500 DPI, a 0.0015 inch effective sample size, greater than 100 S/N ratio and 6 bit dynamic range. Further, almost all of these scanners were analog devices. A few digital cameras were available that used analog sensors to produce a digital output, but it was not until the early 1980s that relatively low cost digital scanners capable of scanning fingerprints at the required resolution and quality became available. It was another ten years before the scan quality standards were clearly articulated in the IAFIS Appendix F specification, which is the current benchmark for scanning and digital capture of fingerprint images [18]. The most notable change brought about by the Appendix F specification is the gray-scale range requirement of 200 or more gray levels, with no missing values, for a broad range of input inked fingerprint card samples (extremely light to extremely dark). In fact, this requires that a scanner is able to scan and digitize some fingerprint cards at greater than 8 bits and then compress to an 8-bit dynamic range to satisfy this specification.

Today, there are reasonably priced scanners (under US$10,000) that are capable of scanning a 1.5 × 1.5 inch fingerprint area at 1,000 (or more) DPI with a digitizing range of 10 or 12 bits, S/N ratio in excess of 100:1, and digital controls to extract the desired data range. However, these days there is a trend to move away from capturing fingerprints on paper using ink or other such media. Most of the fingerprint input devices now used in both criminal and civil fingerprint systems directly scan the fingerprint from the finger. These scanners are called "live-scan" fingerprint capture devices. The most common types of live-scan fingerprint devices either directly digitize the fingerprint image (by electronically scanning a planar array of samples) or digitize the fingerprint image created through optical means (frustrated total internal reflection – FTIR). Many of these live-scan fingerprint devices are capable of capturing digital data at the quality levels established by the IAFIS Appendix F specification. If a paper copy of the fingerprints is required, the fingerprint images can be printed. The FBI specifies fingerprint image quality for paper copy submissions to them.

The most recent American National Standards Institute (ANSI) standard for fingerprint data interchange recommends 1,000 DPI resolution to yield greater definition of minute fingerprint features [19]. Apart from the obvious issue of the large size of the resulting image files, there are many other issues related to such high-resolution imaging of fingerprints (e.g. how to design subsequent image processing algorithms to take advantage of this high-resolution data).

For many civil and commercial applications, there is no mandate for a set of ten fingerprints for each individual to be recorded by the system. Often, it is sufficient for the scanning device to capture a fingerprint from a single finger. Moreover, it is not necessary to capture a fingerprint image rolled across the finger from one side of the fingernail to the other. In fact, capturing a "dab" (or "flat") impression of 1 in^2 (or even lesser) area of the finger is quite acceptable. There are many fingerprint sensors available today that capture areas substantially less than 1 in^2 (even down to 0.25 in^2) of a single finger. Many of these scanners sample fingerprints at less than the 500 DPI that is mandated for forensic uses by the IAFIS specifications. Indeed, there is not yet a set of specifications articulated for finger scanners in civil or commercial applications as there is for criminal applications. It may well be that the device manufacturers will ultimately be required to address the issue of a common specification for fingerprint image quality for live-scan devices in these applications. A more likely possibility is that the environment of these applications will drive the manufacturers to meet certain minimum scanner quality specifications in order to meet the performance and interchangeability goals.

In many cases, the live-scan finger scanning devices are implemented using optical (FTIR and scattering) techniques, using planar fabrication techniques to build capacitor arrays (various semiconductor charge transfer techniques) and ultrasound transducers. The optical and ultrasound devices generally capture larger areas of the fingerprint, but the optical devices often suffer more from inconsistent contact (dry finger) problems over a large sample of people than do the ultrasound devices. The planar devices capture a substantially smaller area and have more resistance to contact problems, but may have environmental/usability problems (wear or damage) due to surface fabrication issues that are not yet fully understood. Only extensive use and testing will resolve the advantages and disadvantages of these competing designs. However, there is growing evidence that live-scan fingerprint scanning devices will play a leading role in many automatic civil and commercial applications as a way to certify an individual's identity.

Figure 2.2 The core is within the octagon, a ridge bifurcation within the circle, a delta within the triangle, and a ridge ending within the square.

2.5.2 Enhancement

A fingerprint image is shown in Figure 2.2. A ridge bifurcation minutia is marked with a circle, and a ridge ending minutia is marked with a square. Ridge bifurcations and endings represent the two types of local fingerprint singularities used by modern AFIS implementations. To show the fingerprint singularities at the global level in this fingerprint image, the core point is marked with an octagon and the delta is marked with a triangle. Traditionally, ridge count is described as the number of ridges between the core and the delta, that is, along the solid line marked on the fingerprint image. More recently, the definition of the ridge count has been extended to the number of ridges between any two points (typically minutiae) in the fingerprint images.

A crucial element in the processing of digital fingerprint data is the enhancement of the ridge structure to allow accurate extraction of minute features. Figure 2.3(a) is an example of an inked fingerprint that has both good and poor quality areas that must be dealt with during the enhancement process. The histogram of gray levels for this image is shown in Figure 2.4(a) (gray value 0 is at the left and gray value 255 is at the right in this histogram). It is apparent that the predominant gray range is somewhat more than 128 gray values. Initial enhancement may involve the normalization of the inherent intensity variation in a digitized fingerprint caused either by the inking (as in this case) or the live-scan device. One such process – local area contrast enhancement (LACE) – is useful to provide such normalization through the scaling of local neighborhood pixels in relation to a calculated global mean. The form of LACE used here calculates a global pixel mean (*GlobalMean*) for the entire image, and then computes a local mean and variance for a 15 × 15 neighborhood about each pixel (building a table of statistics over the entire image). A pixel gain is calculated as indicated below, and subjected to the constraints $1.0 \leq PixelGain \leq 5.0$:

$$PixelGain = GlobalGain \times (1/\sqrt{LocalVariance}) \tag{2.1}$$

The *GlobalGain* factor is calculated using the *GlobalMean* and a *GlobalCorrection* factor which is determined for fingerprint images empirically (a typical value could be 0.5). This calculation is as follows:

$$GlobalGain = GlobalCorrection \times GlobalMean \tag{2.2}$$

A new intensity for each original pixel (*RawPixel*) of the image is calculated using the *PixelGain* and *LocalMean* as follows:

$$NewIntensity = PixelGain \times (RawPixel - LocalMean) + LocalMean \tag{2.3}$$

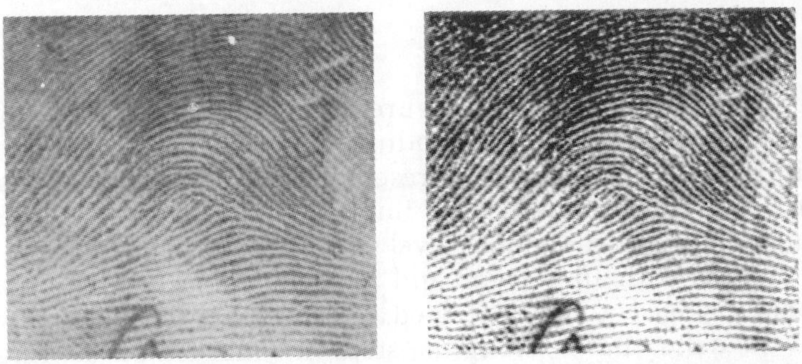

(a) (b)

Figure 2.3 (a) An inked fingerprint image; (b) the results of the local area contrast enhancement algorithm on (a).

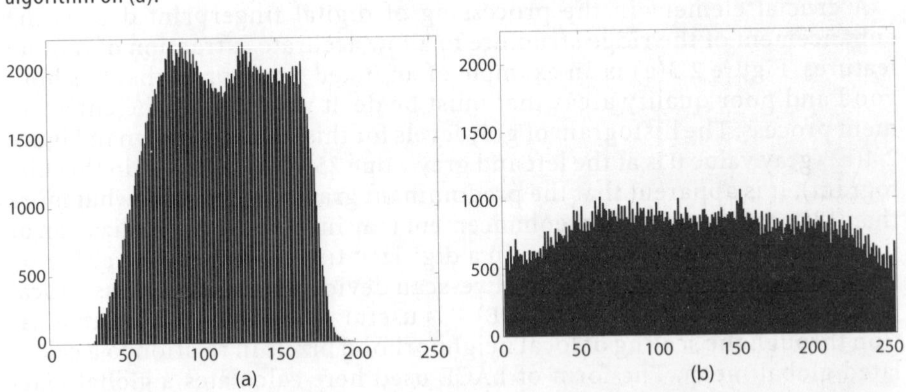

Figure 2.4 (a), (b): Histograms of fingerprint images in Figures 2.3(a) and (b), respectively.

Figure 2.3(b) shows an example of the application of LACE to the image in Figure 2.3(a). The histogram for the image after the enhancement is shown in Figure 2.4(b). Note that the gray range now spans the full 256 gray values, with a mean value near 128. By comparing Figure 2.3(b) with Figure 2.3(a), one can clearly see a marked increase in contrast over the entire image.

Another type of enhancement that can be used to preprocess fingerprint images is contextual filtering [20, 21]. This type of enhancement has the following objectives: (1) provide a low-pass (averaging) effect along the ridge direction with the aim of linking small gaps and filling impurities due to pores or noise; and (2) perform a bandpass (differentiating) effect in a direction orthogonal to the ridges to increase the discrimination between ridges and valleys and to separate parallel linked ridges. Often, Gabor filters [22] are used for this type of contextual filtering. Gabor filters have both frequency-selective and orientation-selective properties and have optimal joint resolution in both spatial and frequency domains. Consequently, these filters have the ability to minimize the bandwidth required while maximizing the spatial resolution. An even symmetric two-dimensional Gabor filter has the following form [20] (see Figure 2.5):

$$g(x,y{:}\theta,f)=\exp\left\{-\frac{1}{2}\left[\frac{x_\theta^2}{\sigma_x^2}+\frac{y_\theta^2}{\sigma_y^2}\right]\right\}\cos(2\pi f\cdot x_\theta) \tag{2.4}$$

where θ is the orientation of the filter and $[x_\theta, y_\theta]$ are the coordinates of $[x, y]$ after a clockwise rotation of the Cartesian axes by an angle of $(90° - \theta)$:

$$\begin{bmatrix} x_\theta \\ y_\theta \end{bmatrix}=\begin{bmatrix} \cos(90°-\theta) & \sin(90°-\theta) \\ -\sin(90°-\theta) & \cos(90°-\theta) \end{bmatrix}\begin{bmatrix} x \\ y \end{bmatrix}=\begin{bmatrix} \sin\theta & \cos\theta \\ -\cos\theta & \sin\theta \end{bmatrix}\begin{bmatrix} x \\ y \end{bmatrix} \tag{2.5}$$

In the above expression, f is the frequency of a two-dimensional sinusoidal surface of the fingerprint, and σ_x and σ_y are the standard deviations of the Gaussian envelope along the x- and y-axes respectively. As shown in Figure

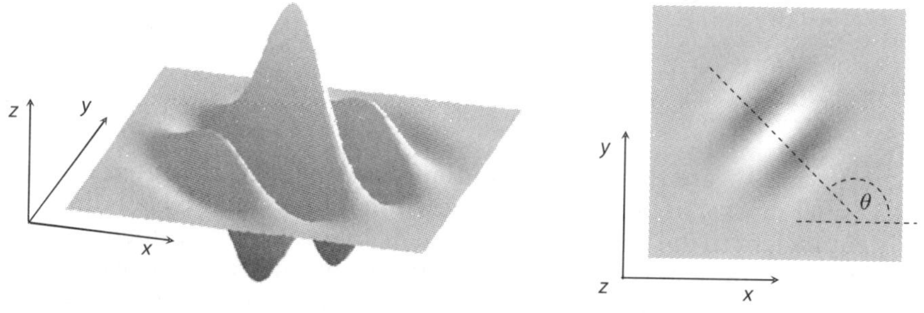

Figure 2.5 Graphical representation (lateral view and top view) of the Gabor filter defined by the parameters $\theta = 135°, f = 1/5, \sigma_x = \sigma_y = 3$ [21].

2.5, a Gabor filter is defined by a sinusoidal wave (the second term of Equation (2.4)) tapered by a Gaussian (the first term in Equation (2.4)).

To contextually filter a fingerprint image using Gabor filters, the four parameters $(\theta, f, \sigma_x, \sigma_y)$ must be specified. Obviously, the frequency, f, of the filter is completely determined by the local ridge frequency, and the orientation, θ, is determined by the local ridge orientation. The selection of the values σ_x and σ_y involves a trade-off. The larger the values, the more robust the filters are to the noise in the fingerprint image, but are also the more likely they are to create spurious ridges and valleys. Conversely, the smaller the values, the less likely the filters are to introduce spurious ridges and valleys, but they are then less effective in removing the noise. In fact, from the Modulation Transfer Function (MTF) of the Gabor filter, it can be shown that increasing σ_x, σ_y decreases the bandwidth of the filter and vice versa.

The simplest and most natural approach for extracting the local ridge orientation field image, **D**, containing elements θ_{ij}, in a fingerprint image is based on the computation of gradients in the fingerprint image. The gradient $\nabla(x_i, y_j)$ at point $[x_i, y_j]$ of fingerprint image **I**, is a two-dimensional vector $[\nabla_x(x_i, y_j), \nabla_y(x_i, y_j)]$, where ∇_x and ∇_y components are the derivatives of **I** in $[x_i, y_j]$ with respect to the x and y directions, respectively. It is well known that the gradient phase angle denotes the direction of the maximum pixel-intensity change. Therefore, the direction θ_{ij} of a hypothetical edge which crosses the region centered at $[x_i, y_j]$ is orthogonal to the gradient phase angle at $[x_i, y_j]$. This method, although simple and efficient, does not provide robust estimates of local ridge orientation in fingerprint images. As a result, an alternative method is used to compute the local ridge orientations as the dominant ridge orientation θ_{ij} by combining multiple gradient estimates within a $W \times W$ window centered at $[x_i, y_j]$ [20]:

$$\begin{aligned}
\theta_{ij} &= 90° + \frac{1}{2}\arctan\left(\frac{2G_{xy}}{G_{xx} - G_{yy}}\right), \\
G_{xy} &= \sum_{h=-W/2}^{W/2} \sum_{k=-W/2}^{W/2} \nabla_x(x_i + h, y_j + k) \cdot \nabla_y(x_i + h, y_j + k), \\
G_{xx} &= \sum_{h=-W/2}^{W/2} \sum_{k=-W/2}^{W/2} \nabla_x(x_i + h, y_j + k)^2, \\
G_{yy} &= \sum_{h=-W/2}^{W/2} \sum_{k=-W/2}^{W/2} \nabla_y(x_i + h, y_j + k)^2
\end{aligned} \tag{2.6}$$

where ∇_x and ∇_y are the x and y gradient components computed through 3×3 Sobel masks. In fact, it can be shown that this method is mathematically equivalent to the principal component analysis of the autocorrelation matrix of the gradient vectors [23]. Usually, the orientation image is further smoothed (low-pass filtered) to eliminate any false local estimates of fingerprint ridge orientations. Figure 2.7(a) shows the estimated orientation field of the fingerprint image shown in Figure 2.3(b).

The local ridge frequency (or density) f_{xy} at point $[x, y]$ is the inverse of the number of ridges per unit length along a hypothetical segment centered at $[x, y]$ and orthogonal to the local ridge orientation θ_{xy}. A frequency image **F**, analogous to the orientation image **D**, can be defined if the frequency is estimated at discrete positions and arranged into a matrix [21]. The local ridge frequency varies across different fingers, and may also vary noticeably across different regions in the same fingerprint. The ridge pattern can be locally modeled as a sinusoidal-shaped surface, and the variation theorem can be exploited to estimate the unknown frequency [24]. The variation V of a function h in the interval $[x_1, x_2]$ is the amount of "vertical" change in h:

$$V(h) = \int_{x_1}^{x_2} \left| \frac{\mathrm{d}h(x)}{\mathrm{d}x} \right| \mathrm{d}x \tag{2.7}$$

If the function h is periodic at $[x_1, x_2]$ or the amplitude changes within the interval $[x_1, x_2]$ are small, the variation may be expressed as a function of the average amplitude α_m and the average frequency f (see Figure 2.6):

$$V(h) = (x_2 - x_1) \cdot 2\alpha_m \cdot f \tag{2.8}$$

Therefore the unknown frequency can be estimated as:

$$f = \frac{V(h)}{2 \cdot (x_2 - x_1) \cdot \alpha_m} \tag{2.9}$$

In a practical method based on the above theory, the variation and the average amplitude of a two-dimensional ridge pattern can be estimated from the first- and second-order partial derivatives and the local ridge frequency can be computed from Equation (2.9).

Once the local orientation image, **D**, and the local ridge frequency image, **F**, have been estimated, the fingerprint image (shown in Figure 2.2(b)) can

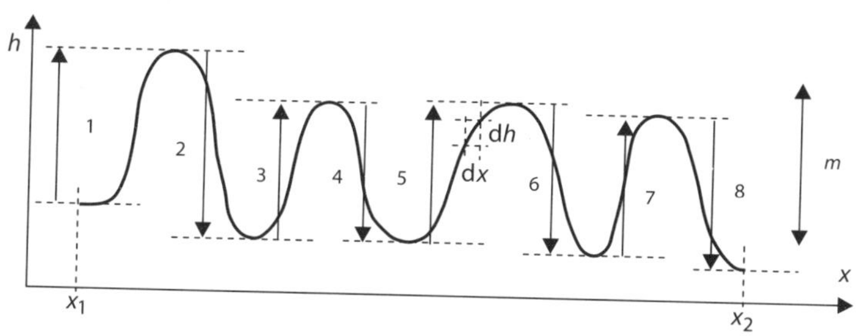

Figure 2.6 The variation of the function h in the interval $[x_1, x_2]$ is the sum of amplitudes $\alpha_1, \alpha_2, \ldots \alpha_8$ [24]. If the function is periodic or the function amplitude does not change significantly within the interval of interest, the average amplitude α_m can be used to approximate the individual α. Then the variation may be expressed as $2\alpha_m$ multiplied by the number of periods of the function over the interval [21].

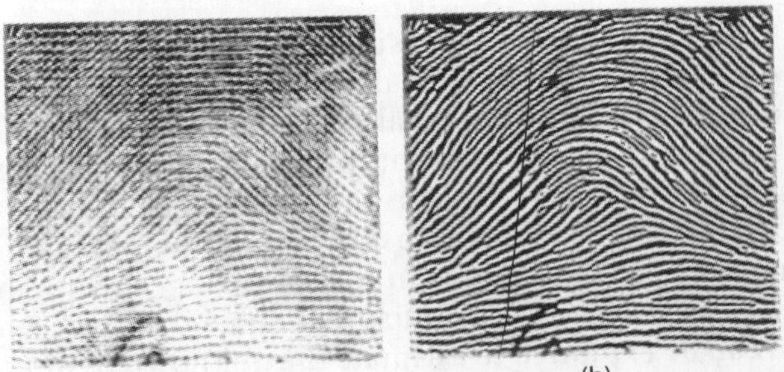

(a) (b)

Figure 2.7 The orientation field is superimposed on the fingerprint image in (a). In (b), the result of the Gabor filters-based contextual filtering of the fingerprint image in Figure 2.3(a) is shown.

be contextually filtered using the Gabor filters as in Equations (2.4) and (2.5), resulting in an enhanced image (shown in Figure 2.7(b)).

An enhancement algorithm built upon a model of fingerprint structure can be useful for the enhancement of fingerprints, but it is also important to understand that there are some non-fingerprint properties of digitized fingerprint images that need to be dealt with. The first is that inked fingerprint images may have variations in intensity due to the inking process. The ink may be absent or less dense in some areas or over the entire print. The fingerprint image may also have smudges (blobs) in some areas or over the entire print due to excessive ink. Similarly, live-scan fingerprint images may not always contain an impression of a real finger, but may contain a latent image of a previous impression (e.g. oils left on the surface of the scanner) or a three-dimensional artificial replica of a fingerprint. Consequently, inked fingerprints may need different enhancement schemes than live-scanned images. The enhancement process must neither be so aggressive that any variation of gray-scale is assumed to be caused by the finger ridge structure (e.g. at extremely low S/N ratios) nor too weak to handle the imaging non-uniformity often found in inked fingerprints. The goal is always to produce an enhanced image that does not contain artificially generated ridge structure that might later generate false minutiae features, while capturing the maximum available ridge structure to allow detection of true minutiae. Adapting the enhancement process to the fingerprint capture method will yield the optimal matching performance over a large collection of fingerprints.

A fingerprint may contain such poor quality areas that the local ridge orientation and frequency estimates are completely wrong. An algorithm that can reliably locate (and mask) these extremely poor quality areas is very useful for the feature detection and recognition stages by preventing false or unreliable features from being created [20].

2.5.3 Feature Extraction

Enhancement generally results in an image that is ready to be converted to a binary value at each pixel. For inked or optically live-scanned prints, the ridge width will likely be several black pixels and the valleys several white

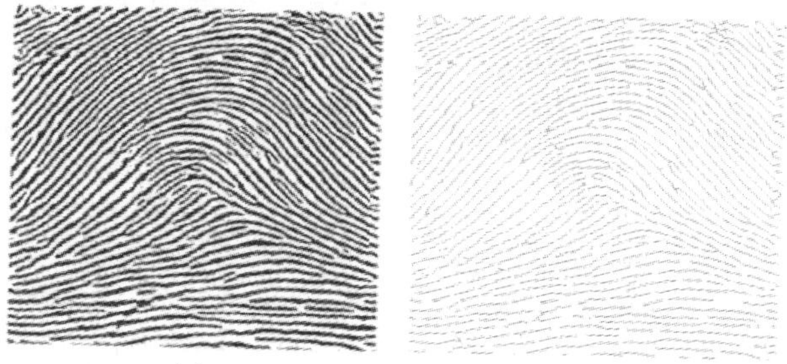

(a) (b)

Figure 2.8 (a) shows the result of binarization (through the ridge location algorithm of [25]) of the enhanced fingerprint image in Figure 2.7(b). (b) shows the results of thinning the image in (a) to a single pixel width.

pixels. Some solid state live-scan devices reverse the "polarity" of the image, yielding white ridges and black valleys.

In feature extractors found in the early systems, different types of minutiae features (e.g. ridge ending, bifurcation, and additional types such as island and dot) were often characterized by the use of models. This required an assignment of all possible minutiae features to a model or models, and required subsequent manual editing to resolve the multiple possibilities for a single minutia feature. Feature extraction that depends upon models of all possible configurations for minutiae must necessarily be a compromise, since as the number of fingerprints becomes large, the possible minutiae shapes and details become ever more complex. The compromises result in many true minutiae missed and false minutiae detected. Instead of working directly with the enhanced image, a better approach may be to deal with the fingerprint image after ridges have been symmetrically thinned about the ridge centerline.

The enhanced image shown in Figure 2.7(b) is first binarized (i.e. setting ridge pixels to 1 and valley pixels to 0; see Figure 2.8(a)) and then uniformly thinned to a single pixel width about the ridge centerline (see Figure 2.8(b)). Binarization can either be achieved simply by thresholding the image in Figure 7b or by using more sophisticated ridge location algorithms [25]. The central idea of the thinning process is to perform successive (iterative) erosions of the outer-most layers of a shape until a connected unit-width set of lines (or skeletons) is obtained. No mathematical definitions exist for the thinning process, leading to a wide range of approaches proposed in the literature. The skeletal image shown in Figure 2.8(b) was created using a method defined by Rosenfeld [26]. Rosenfeld's method examines a 3 × 3 pixel neighborhood to decide whether the center pixel (P1) should be black (indicating a ridge) or white (indicating a valley). Figure 2.9 illustrates the four conditions under which P1 would be considered as sitting on a ridge. The shading in Figure 2.9 indicates the location of black pixels in the image.

Neighborhoods for ridge end points are also defined (see Figure 2.10). There are also definitions for isolated points where none of the neighbors

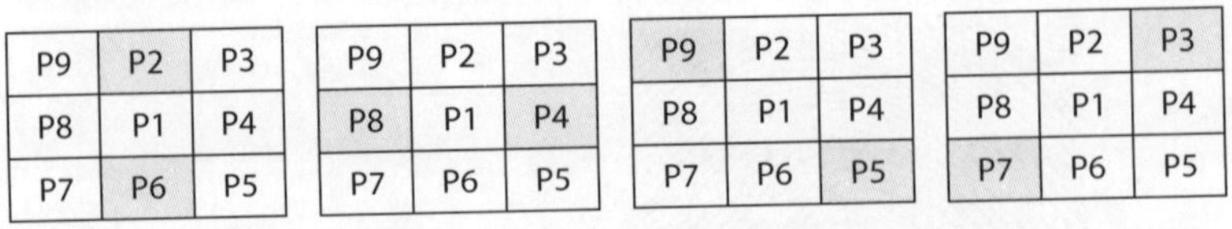

Figure 2.9 Center pixel (P1) is determined to be on a ridge during thinning.

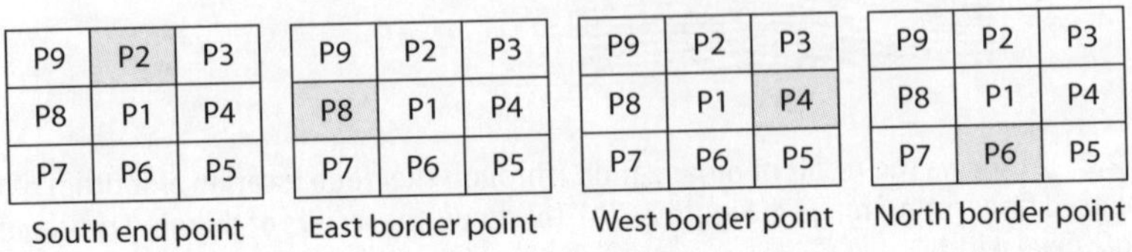

Figure 2.10 Center pixel (P1) is determined to be at the end of a ridge during thinning.

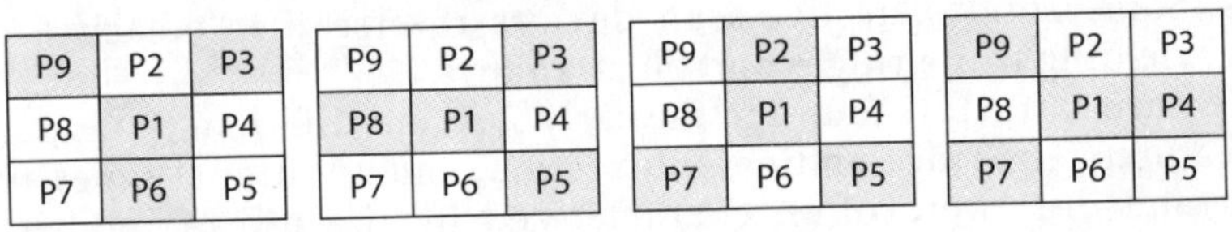

Figure 2.11 Center pixel (P1) is determined to be a ridge bifurcation minutia during minutiae detection.

of P1 is black. The 3 × 3 analysis kernels are applied to a binary image along rows and by columns (in a raster-scan fashion), and a decision is made whether to change the center pixel (P1) from black to white for simply connected, not isolated nor end point, P1s. By systematically applying these region definitions to a binary image to decide whether a pixel should be white or black, a thinned, single-pixel-width ridge structure is created. The order of application needs to be consistent, alternating from top to bottom to left to right to produce a thinned image centered on the original image. Usually some editing is required to remove short "whiskers" generated by the thinning process on certain two-dimensional structures. Additionally there is often a need to introduce some filling definitions to handle the existence of pores in fingerprint images. Pores are background white pixels surrounded by black ridge pixels in a binary image. When eroded by thinning, these pores can produce enclosed white regions of significant size such that they may be (falsely) detected as minutiae. Elimination of as many of these pore regions as possible before minutiae detection makes it easier to edit the final minutiae set.

Once the skeletal image is created, a minutiae detection stage can analyze this image using another set of 3 × 3 kernels to identify ridge ending or bifurcations points as minutiae. A ridge-ending minutia is indicated if only one pixel element, P2–P9, is black. A bifurcation minutia has three neighboring black pixels, as illustrated in Figure 2.11.

Although the process is simple in principle, false structures, resulting from imperfect restoration of ridge detail, pores or other anomalies created

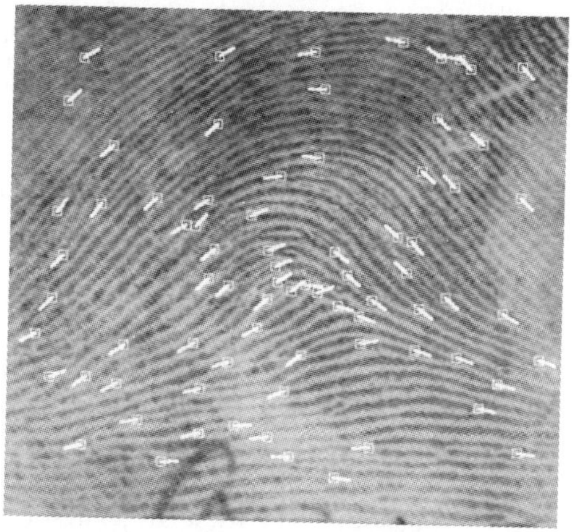

Figure 2.12 The detected minutiae features are superimposed on the original inked fingerprint of Figure 2.3(a) for display.

by aggressive enhancement, must be detected and eliminated. This may be accomplished as part of the initial validation of the detected minutia or in post processing [21]. Initial validation considers, for example, whether the ridge length running away from the minutia point is sufficient, or if the ridge direction at the point is within acceptable limits. Post processing might include an examination of the local image quality, neighboring detections or other indicators of non-fingerprint structure in the area. Further, the image can be inverted in gray-scale, converting white to black and black to white. Reprocessing of this inverted image should yield minutiae endings in place of bifurcations and vice versa, allowing a validity check on the previously detected minutiae. The final detected minutiae are those that meet all of the validity checks. Figure 2.12 shows the minutiae detected on the fingerprint image shown in Figure 2.3(a) by using the techniques described in this section.

Certain minutiae extraction algorithms work differently and detect the minutiae directly in the gray-scale fingerprint image [21, 27]. Once a validated set of minutiae is determined, additional feature data such as minutiae confidence, ridge counts between minutiae, ridge count confidence, core and delta locations, etc. can be determined. These additional features may be useful to achieve added selectivity from a minutiae matching process. Their usefulness for this purpose may be mediated by the confidence associated with each such feature. Therefore it is important to collect confidence data as a part of the image enhancement and feature extraction process to be able to properly qualify detected minutiae and associated features.

2.5.4 Classification

In some system implementations, fingerprint pattern classification (such as loop, arch, whorl) is done automatically, and is used as a selector for candidate records in a search. Submitted samples need only be compared to database records with the same classification. Such automatic classification of fingerprint patterns is not perfect (the state-of-the-art classification algorithms have an accuracy of about 99% [28]), and sometimes requires manual confirmation. Errors occur when truly matching prints are given different classifications. Such errors increase the system's false non-match rate.

Fingerprint pattern classification can be determined in several ways: explicitly characterizing regions of a fingerprint as belonging to a particular shape; or through implementation of one of many possible generalized classifiers (e.g. artificial neural networks) trained to recognize the specified patterns. The singular shapes (e.g. cores and deltas) in a fingerprint image are typically detected using the Poincaré method [29, 30] on the fingerprint orientation image, **D**, as follows.

Let **G** be the vector field associated with a fingerprint orientation image **D** (note that a fingerprint orientation image is not a true vector field since its elements are unoriented directions, but it can be converted to a pseudo-vector field by multiplying the orientation values by 2 [31]) and let $[i,j]$ be the position of the element θ_{ij} in the orientation image; then the Poincaré index $P_{\mathbf{G},C}(i,j)$ at $[i,j]$ is computed as follows:

- the curve C is a closed path defined as an ordered sequence of some elements of **D**, such that $[i,j]$ is an internal point
- $P_{\mathbf{G},C}(i,j)$ is computed by algebraically summing the orientation differences between adjacent elements of C. Summing orientation differences requires a direction (among the two possible) to be associated at each orientation. A solution to this problem is to randomly select the direction of the first element and assign the direction closest to that of the previous element to each successive element. It is well known and can be easily shown that, on closed curves, the Poincaré index assumes only one of the discrete values: 0°, ±180° and ±360°. In the case of fingerprint singularities:

$$P_{\mathbf{G},C}(i,j) = \begin{cases} 0^\circ & \text{if } [i,j] \text{ does not belong to any singular region} \\ 360^\circ & \text{if } [i,j] \text{ belongs to a whorl-type singular region} \\ 180^\circ & \text{if } [i,j] \text{ belongs to a loop-type singular region} \\ -180^\circ & \text{if } [i,j] \text{ belongs to a delta-type singular region} \end{cases} \tag{2.10}$$

Figure 2.13 shows three portions of orientation images. The path defining C is the ordered sequence of the 8 elements $\mathbf{d}_k$ $(k = 0..7)$ surrounding $[i,j]$. The direction of the elements $\mathbf{d}_k$ is chosen as follows: $\mathbf{d}_0$ is directed upward; $\mathbf{d}_k$ $(k = 1..7)$ is directed so that the absolute value of the angle between $\mathbf{d}_k$ and $\mathbf{d}_{k-1}$ is less than or equal to 90°. The Poincaré index is then computed as:

$$P_{\mathbf{G},C}(i,j) = \sum_{k=0..7} angle(\mathbf{d}_k, \mathbf{d}_{(k+1) \bmod 8}) \tag{2.11}$$

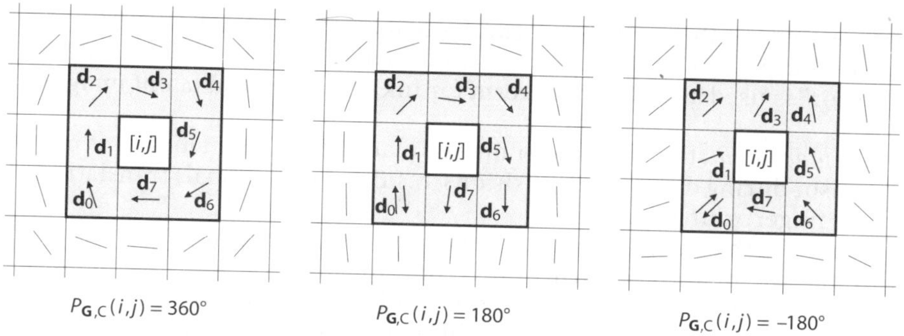

Figure 2.13 Example of computation of the Poincaré index in the 8-neighborhood of points belonging (from the left to the right) to a whorl, loop and delta singularity, respectively. Note that for the loop and delta examples (center and right), the direction of $\mathbf{d}_0$ is first chosen upward (to compute the angle between $\mathbf{d}_0$ and $\mathbf{d}_1$) and then successively downward (when computing the angle between $\mathbf{d}_7$ and $\mathbf{d}_0$) [21].

The explicit (rule-based) fingerprint classification systems first detect the fingerprint singularities using the above method and then apply a set of rules (e.g. arches have no loops and deltas, tented arches and loops have one core and one delta, and whorls have two loops and two deltas etc.) to determine the pattern type of the fingerprint image. The most successful generalized (e.g. artificial neural network-based) fingerprint classification systems utilize a combination of a number of different classifiers [21].

The effort to validate patterns during fingerprint image capture, and automate their entry into the system, is substantially less than the procedures required for manual systems. The use of fingerprint pattern information can be an effective means to limit the volume of data sent to the matching engine resulting in benefits in both the system response time and the false-match error rate.

2.5.5 Matching

Automatic fingerprint matching has been in operational use in criminal AFIS applications since the late 1970s. Running routine searches to identify criminals associated with a particular crime scene without known suspects is now possible with the help of AFIS. In many ways, automatic fingerprint matching only accomplishes what a fingerprint technician would do, but accomplishes it much faster and more effectively. Prior to the development of AFIS technology, many police agencies did not routinely search crime scene fingerprints because of the labor required. Only the most serious crimes were subjected to a search for candidates from the criminal fingerprint file. With automatic matching, it is possible to search for records without a suspect in mind, and to do this with far less labor in a matter of minutes rather than days or weeks.

Automatically matching fingerprint minutiae sets is a difficult pattern recognition problem, mainly due to the large variability in different impressions of the same finger (i.e. large intra-class variations). The main factors responsible for the intra-class variations are as follows (note that "sensor" is used as a collective term for ink, live-scan and latent imaging):

- *Displacement*: the same finger may be placed at different locations on the sensor during different acquisitions, resulting in a (global) translation of the fingerprint area.
- *Rotation*: the same finger may be rotated at different angles with respect to the sensor surface during different acquisitions.
- *Partial overlap*: finger displacement and rotation often cause part of the search fingerprint area to fall outside the file fingerprint's "field of view", resulting in a smaller overlap between the foreground areas of the search and the file fingerprints. This problem is particularly serious in latent fingerprints.
- *Nonlinear distortion*: the act of sensing maps the three-dimensional shape of a finger onto a two-dimensional surface. This mapping results in a nonlinear distortion in successive acquisitions of the same finger due to skin plasticity.
- *Pressure and skin condition*: the ridge structure of a finger would be accurately captured if ridges of the part of the finger being imaged were in uniform contact with the sensor surface. However, the finger pressure, dryness of the skin, skin disease, sweat, dirt, grease and humidity in the air all confound the situation, resulting in a non-uniform contact. As a consequence, the acquired fingerprint images are very noisy and the noise varies strongly in successive acquisitions of the same finger, depending on the magnitude of the above-cited causes.
- *Noise*: this is mainly introduced by the fingerprint sensing system; for example, excessive or too little ink causes noise in inked fingerprints, residues are left over on the glass platen from the previous fingerprint capture; and latents may be lifted from rough surfaces.
- *Feature extraction errors*: the feature extraction algorithms are imperfect and often introduce measurement errors. Errors may be made during any of the feature extraction stages (e.g. segmentation of the fingerprint area from the background; estimation of orientation and frequency images; detection of the number, type and position of the singularities; detection of minutiae; and post processing). Aggressive enhancement algorithms may introduce inconsistent biases that perturb the location and the orientation of the reported minutiae from their gray-scale counterparts. In low-quality fingerprint images, the minutiae extraction process may introduce a large number of spurious minutiae and may not be able to detect all the true minutiae.

Mathematically, the fingerprint minutiae matching problem can be described as follows [21]. Let **T** and **I** be the representation of the file and search fingerprint, respectively. Each fingerprint minutia may be described by a number of attributes, including its location in the fingerprint image, orientation, type (e.g. ridge ending or ridge bifurcation), a weight based on the quality of the fingerprint image in the neighborhood of the minutia, etc. Most common minutiae matching algorithms consider each minutia as a triplet $\mathbf{m} = \{x, y, \theta\}$ which indicates the x, y minutia location coordinates and the minutia angle θ:

$$\mathbf{T} = \{\mathbf{m}_1, \mathbf{m}_2, \ldots, \mathbf{m}_m\}, \ \ \mathbf{m}_i = \{x_i, y_i, \theta_i\}, \ \ i = 1..m \tag{2.12}$$

$$\mathbf{I} = \{\mathbf{m}'_1, \mathbf{m}'_2, \ldots, \mathbf{m}'_m\}, \ \ \mathbf{m}'_j = \{x'_i, y'_i, \theta'_i\}, \ \ j = 1..n \tag{2.13}$$

where m and n denote the number of minutiae in **T** and **I**, respectively. A minutia $\mathbf{m}'_j$ in **I** and a minutia $\mathbf{m}_i$ in **T** are considered "mating" if the spatial distance between them is smaller than a given tolerance d_0:

$$|x'_j - x_i| \le d_0 \quad \text{and} \quad |y'_j - y_i| \le d_0 \tag{2.14}$$

and the direction difference between them is smaller than an angular tolerance θ_0:

$$\min(|\theta'_j - \theta_i|, \ 360 - |\theta'_j - \theta_i|) \le \theta_0 \tag{2.15}$$

Equation (2.14) takes the minimum of $|\theta'_j - \theta_i|$ and $360° - |\theta'_j - \theta_i|$ because of the circularity of angles (the difference between angles of 2° and 358° is only 4°). The "tolerance boxes" defined by d_0 and θ_0 are necessary to compensate for the unavoidable errors made by the feature extraction algorithm and to account for the small plastic distortions that cause the minutiae positions to change (see Figure 2.14).

Aligning the two fingerprints is a mandatory step in order to maximize the number of matching minutiae. Correctly aligning two fingerprints certainly requires *displacement* (in x and y) and *rotation* (θ) to be recovered and likely involves other geometrical transformations, such as scale and nonlinear distortion.

Let *map*(.) be the function which maps a minutia $\mathbf{m}'_j$ (from **I**) into $\mathbf{m}''_j$ according to a given geometrical transformation; for example by considering

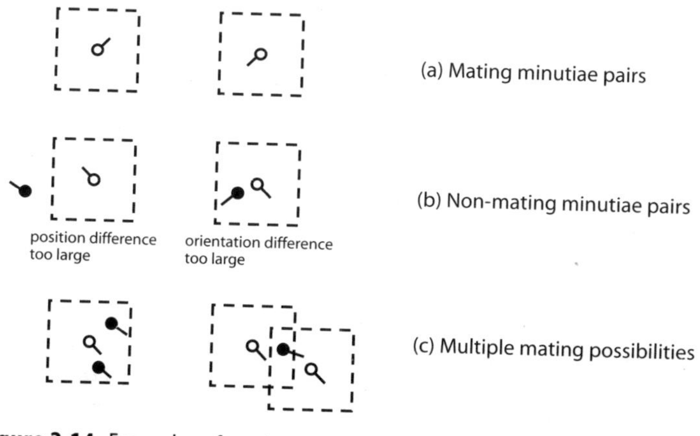

Figure 2.14 Examples of mating, non-mating and multiple mating minutiae.

a displacement of $[\Delta x, \Delta y]$ and a counterclockwise rotation θ around the origin (the origin is usually selected as the minutiae centroid, i.e. the average point; before the matching step, minutiae coordinates are adjusted by subtracting the centroid coordinates):

$$map_{\Delta x,\Delta y,\theta}(\mathbf{m}'_j = \{x'_j, y'_j, \theta'_j\}) = \mathbf{m}''_j = \{x''_j, y''_j, \theta'_j + \theta\}$$
$$\text{where} \begin{bmatrix} x''_j \\ y''_j \end{bmatrix} = \begin{bmatrix} \cos\theta & -\sin\theta \\ \sin\theta & \cos\theta \end{bmatrix} \begin{bmatrix} x'_j \\ y'_j \end{bmatrix} + \begin{bmatrix} \Delta x \\ \Delta y \end{bmatrix} \tag{2.16}$$

Let $mm(.)$ be an indicator function which returns 1 in case the minutiae $\mathbf{m}''_j$ and $\mathbf{m}_i$ match according to expressions (2.14) and (2.15):

$$mm(\mathbf{m}''_j, \mathbf{m}_i) = \begin{cases} 1 & \text{if expressions 14 and 15 are true} \\ 0 & \text{otherwise} \end{cases} \tag{2.17}$$

Then the matching problem can be formulated as:

$$\underset{\Delta x,\Delta y,\theta,P}{\text{maximize}} \sum_{i=1}^{m} mm(map_{\Delta x,\Delta y,\theta}(\mathbf{m}'_{P(i)}), \mathbf{m}_i) \tag{2.18}$$

where $P(i)$ is an unknown function which determines the "correspondence" or "pairing" between **I** and **T** minutiae.

The maximization in Equation (2.18) can be easily solved if the function P (minutiae correspondence) is known; in this case, the unknown alignment $(\Delta x, \Delta y, \theta)$ can be determined in the least square sense. Unfortunately, in practice neither the alignment parameters nor the correspondence function, P, are known, and therefore solving the matching problem is very hard. A brute force approach, i.e. evaluating all the possible solutions (correspondences and alignments), is prohibitive since the number of possible solutions is exponential in the number of minutiae (the function P is more than a permutation due to the possible *null* values). In pattern recognition literature, the minutiae matching problem has been generally addressed as a *point pattern matching* problem, and a family of approaches known as relaxation methods, algebraic and operational research solutions, tree-pruning approaches, energy-minimization methods, Hough transforms etc. are available.

Fingerprint matching can be best visualized by taking a paper copy of a file fingerprint image with its minutiae marked and a transparency of a search fingerprint with its minutiae marked. By placing the transparency of the search print over the paper copy of the file fingerprint and translating and rotating the transparency, one can locate the common minutiae that exist in both prints. From the number of common minutiae found and their closeness of fit, it is possible to assess the similarity of the two prints. Figure 2.15 shows an example of a comparison of a fingerprint pair.

Of course, to make these minutiae comparisons manually would be prohibitive in terms of time and would therefore seriously limit the

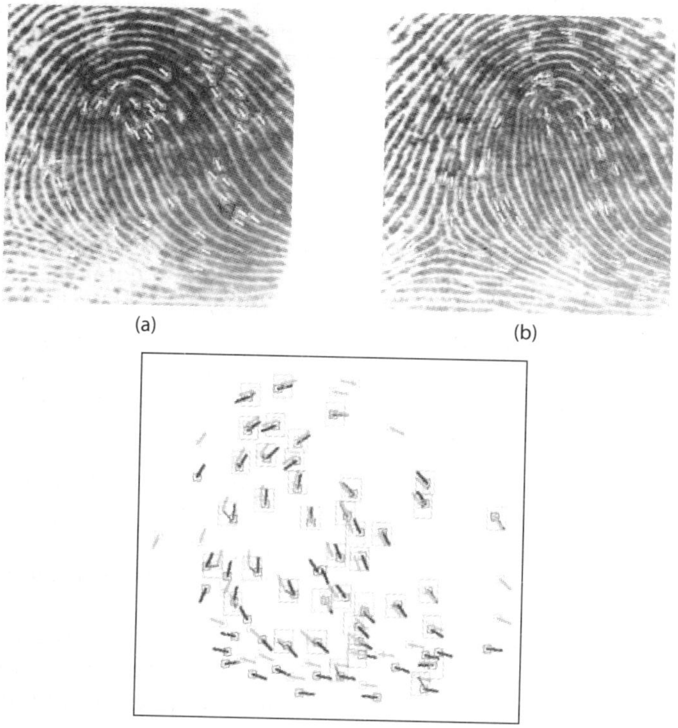

(a) (b) (c)

Figure 2.15 An example of matching the search minutiae set in (a) with the file minutiae set in (b) is shown in (c).

effectiveness of the procedure. With automatic matching, the comparisons can be done tens of thousands of times each second, and the results can be sorted according to their degree of similarity and combined with any other criteria that may be available to further restrict the candidates, all without human intervention.

Ultimately, the difficulty in carrying out an identification lies in the volume of fingerprint data that needs to be matched and the quality of the fingerprints from which that data is extracted. In many thousands or millions of fingerprints, there is always a broad range of quality present, due to varying skin conditions or the techniques used to capture the prints, or a combination of both. The extracted data from a print to be searched may be excellent, but the data for its mate on file may either be of poor quality or improperly taken (e.g. only partial capture). There may also be other poor-quality data not related to the fingerprint in question, but whose presence may produce false comparison results to mask the true match. If more than one finger is available to match, it is often possible to overcome the most pathological set of circumstances that would otherwise make finding the matching person impossible. It is an important policy to allow for sufficient fingerprint data (multiple fingers from the same individual) to be collected and stored for use when the file of fingerprints reaches its maximum size. In this way it will always be possible to achieve a specified level of accuracy almost independent of the volume of data on file.

2.5.6 Searching

Searches in a criminal AFIS are conducted with the same objective as in manual criminal systems, but are preferred because of their fast speed. AFIS can rapidly access records, filter the fingerprint database on a variety of data elements (such as gender or age, if known), and match the filtered candidates at a very high speed. Fingerprint pattern classifications can be specified in terms of primary and secondary pattern types and tolerances can be applied to account for expected coding error in the candidate selection process. A large number of possible candidates can be rapidly compared and ranked by their likelihood of being a mate and presented to a trained fingerprint technician. Such possibilities do not exist in manual systems, not only because of the labor involved in comparing a large number of records, but also because manual methods generally do not allow efficient access to the records based on the ranges of filter parameters.

Fingerprint databases in AFIS are often very large, consisting of hundreds of thousands or millions of records. The database holds not only the feature (minutiae) data for each finger but also the gray-scale fingerprint images that may be required for manual inspection and printing. The database may additionally hold the thinned ridge images that may be used for added matching selectivity and reliability in certain unattended modes of operation. Unattended "lights-out" applications require automatic decisions to be made without the immediate possibility of human intervention. Such cases arise in criminal identification operations that are not staffed around the clock. If an arrest is made during non-office hours, a decision needs to be made whether to hold (based on outstanding charges) or release the individual in custody based on fully automatic matching of his/her fingerprints. In a "lights-out" application, since the matching system needs to make an unattended decision, more detailed comparisons are necessary to provide the required level of decisional confidence. Eventually, the automatic matching results would be confirmed by a fingerprint technician, but the automated decision process must be reliable enough to ensure that initial errors are very rare.

The search process also needs to cope with a continual growth in the database size as new arrest information is added. The processing of searches must scale effectively as the number of records grows, since the response time for searches is directly proportional to the volume of data and computational resources. To achieve this scalability, the search is usually divided among a number of computers on a network that can access the same database. At the same time, efficient methods are required to maintain the system database as additions, updates and deletions are concurrently made on various computers. A reasonable way to satisfy both these needs is to provide a bi-level database architecture: the first level maintains the complete database with a general-purpose database management system (DBMS), and the second level maintains discrete data partitions resident on the matching engines. The first level provides general-purpose data access, statistics, and reporting, while the second level provides rapid access to data required by minutiae matching and filtering.

2.5.7 Manual Verification

In every large-scale identification application, whether criminal or civil, a fingerprint technician manually confirms the identification made by the AFIS. The AFIS accomplishes this by assembling a collection of the possible candidates, ordered by the search and matching process, and delivering these candidate images to a display station for review by the technician. In some cases, all the identifications are submitted to a second level of review by a supervisor to confirm the decisions made by a more junior technician. Typically, the verification workstation provides side-by-side display of the search fingerprint(s) sequentially with each of the file candidates ordered by the matching process. The individual images are presented at a size and resolution that allows the technician to examine details of the fingerprints so as to make a decision whether the fingerprints match. Images are displayed at 8 bit gray-scale (256 levels) and magnified to at least five times their normal size. It is a common practice to use large, high-resolution (e.g. 1,280 × 1,024 pixel) computer monitors to allow simultaneous display of the images being reviewed and a set of controls to manage the process of sequencing through the series of images.

2.6 Criminal Applications

Criminal identification systems that use manual techniques to file and search records in a fingerprint collection have been in productive use for over a century. Different systems operated at the national level, at the metropolitan level, and in very local populations (e.g. prisons and jails). There was seldom any interaction between these systems, not only because of geographic or operational separation, but also because there were often significant differences in the cataloging and search procedures of different systems that made the interaction difficult. Even the introduction of AFIS into law enforcement did not immediately stimulate interest in providing interoperability among such systems.

2.6.1 National Systems

Apart from the system installed by the national police of Japan in the mid-1970s, there was no rush internationally to install AFIS at national levels. This was probably the result of two important factors: first, the databases of inked tenprint files were large and the effort required to digitize them was high; and secondly, the technical risks of implementing the complex workflow, the personnel task automation and the training procedures were high. There was, however, a great deal of effort expended by national organizations (e.g. the FBI in the USA and the Home Office in the UK) to research the technology and implement pilot systems. In fact, it is likely that the commercialization of AFIS and the deployment of state and local systems would not have progressed at the pace it did were it not for the investment made by these national agencies.

In the USA, the FBI implemented several pilot systems that operated on portions of the daily workload or criminal fingerprint database. The initial implementation that started in the mid-1970s provided a means to test research results in a small but operational environment. In this way, research into image processing, automated pattern type classification, automated search strategies and minutiae matching could be evaluated and tuned to optimize performance metrics. During the 1970s, the FBI contracted with a number of organizations as well as developed their own research organization to manage the numerous projects that lead the way to the Integrated Automated Fingerprint Identification System (IAFIS) in place today. Among the incremental steps to IAFIS, in the mid-1980s the FBI implemented a comprehensive automation of the workflow of the identification section, including specialized workstations to automate many of the complex human tasks involved in processing a search. This automation also included the complex media routing dictated by the workflow to minimize one of the most serious problems in any large data processing system – the lengthy delays associated with holding paper records in manual in-baskets.

The transition to a large-scale imaging application environment provided enormous challenges for everyone at that time, but it was especially challenging for the FBI to implement a system to manage up to 35,000 image-based transactions per day. Massive amounts of storage were required to hold the transient search image data as well as the active file image data. Imaging standards were needed for fingerprint scanning, compression [44] and display to ensure that suitable data would be captured and stored. Since an overwhelming amount of FBI work is submitted from state and local agencies, the standardization needed to include these agencies as well. At the same time as the FBI was laying the plans for IAFIS , live-scan fingerprint capture devices were beginning to be deployed by many state and local agencies to eliminate the inked fingerprint collection process and the scanning of inked tenprint cards. It was necessary for the FBI to extend imaging standards originally designed for scanned inked fingerprints to include live-scan devices manufactured by several different vendors. Although the FBI had no direct control over these live-scan vendors or the agencies using live-scan equipment, they worked with both to the benefit of the identification community at large. Ultimately, the FBI efforts gained support from the international identification community and the agencies involved cooperated through Interpol to develop standards similar to those articulated by the FBI.

The IAFIS system currently in use at the identification division of the FBI in Clarksburg, VA, has been designed, built and installed by the Lockheed

Martin Corporation. This system handles in excess of 35,000 record submissions daily and provides far more rapid response to identification requests than was ever possible with the earlier manual systems. The remaining challenge is to transition the agencies requesting identification from submission of paper tenprint cards to submission of electronic fingerprint images. This is a tremendous undertaking, given the wide geographic distribution of these agencies, the variation in operational procedures, the non-interoperability of existing AFIS, and the secure electronic communications links that must be established between these agencies and the FBI. A great deal of effort is under way to facilitate this transition and provide the required technical, physical and hardware support.

Outside the USA, several national AFIS were put in place in Europe and the Middle East from 1984 onwards. In particular, systems were installed in Switzerland, Norway and Holland. Thereafter, AFISs were installed in the UK, Denmark, Sweden, Finland, Belgium, France, Germany, Spain, Portugal, Austria, Hungary, the Czech Republic, Slovenia and Slovakia. Since then, many other countries have already installed or are in the process of installing AFIS.

2.6.2 Local Systems

In the USA, local agencies, primarily cities and counties, were among the first adopters of AFIS. This was probably because many metropolises were at the forefront of criminal investigations and viewed their needs as more immediate than did the state agencies. By the late 1970s, these local agencies saw an opportunity to dramatically improve their capability to search latent prints lifted from the crime scenes through AFIS and committed to implementing fingerprint automation within their identification bureaus. For metropolises, latent print investigation was a local process rather than a state process. State crime labs focused on supporting smaller jurisdictions that did not have the trained staff necessary to handle all the details of crime scene investigations. Local agencies implementing AFIS envisioned dramatic improvements in their ability to catch criminals and to use their trained staff more effectively.

Another factor that may have led the early AFIS implementations to be at the local level was the generally smaller size of the fingerprint databases held in these bureaus. The states maintained fingerprint databases that were often several orders of magnitude larger than those of the localities within the state, and conversion to AFIS represented a formidable undertaking in both time and money. Regardless of the reasons, local agencies led the way in AFIS adoption, and by the end of the 1970s there were at least half a dozen cities in the USA with an AFIS in place. During this time, some state agencies were also evaluating the use of AFIS for latent print matching. However, this number remained rather small until the mid-1980s, when many US cities and some European agencies began AFIS procurement activities following the success of the AFIS reported by the early adopters.

In the mid-1980s, the AFIS implementations began to store the full fingerprint images in the computerized databases in addition to the minutiae data. This made it possible to manually review the candidate matches on a display screen, which was significantly faster and less laborious than the inked card processes.

2.6.3 Interoperability

As more states implemented AFIS, the problem of interoperation among the local agencies gained importance. Further, it was common for the states to receive fingerprints from the local agencies to be searched against the state criminal database before submission to the FBI. This provided some filtering of identification requests to the FBI. Therefore it became desirable to have standardized protocols for electronic submissions of fingerprint matching requests from the local agency to the state and, finally, to the national AFIS.

The efforts put into AFIS interoperability by NIST under the FBI sponsorship resulted in an ANSI/NIST standard for data interchange. This standard was initially crafted in mid-1980, is updated every 5 years, and defines data formats for images, features and text [19]. Since different AFIS vendors have different implementations and definitions of minutiae feature details, and since record data elements vary between agencies, the ANSI/NIST standard provides a common basis for interoperability by encapsulating data in defined formats.

To handle the differences in record data elements, the standard requires that the sending agency use the element definitions of the receiving agency. In practice, most agencies provide their definitions for other agencies to use. However, it is difficult to get the vendors involved in the standards process and to actively provide support for the formats required to exchange data. It can be a lengthy process to get all the definitions created, agreed upon, and then implemented, so that the data can actually be exchanged. The FBI has provided not only the forum and impetus for the creation of the interchange standard, but has also assisted their users in both technical and material ways to facilitate data exchange. However, with so many separate agencies, each with its own resources and schedules, meeting specific implementation dates has been a challenging task for all involved.

Outside North America, under the auspices of the Interpol AFIS Expert Working Group (IAEG), there is a similar effort toward interchange standardization following the basic format of the ANSI/NIST standard. There are at least 11 countries (Brazil, Canada, France, Germany, Japan, Mexico, Norway, South Africa, Spain, UK and USA), and an observer from Europol that participate in this international effort. IAEG has defined their current standard as INT-I Version 4 [32] and is committed to periodic updates. IAEG clearly demonstrates the level of international cooperation necessary in an era when criminal activity is not contained by national boundaries.

2.6.4 "Daubert" Questions

Much has been written about the theoretical infallibility of fingerprints in criminal identification. However, unlike DNA matching research that emerged from a laboratory environment with a quantitative foundation and a robust estimate of the likelihood of occurrence of the same (or similar) protein strings in nature [33], fingerprint matching has not had such rigorous quantitative development. Fingerprint collection and examination began in the field rather than in the laboratory. There is now well over 100 years of accumulated and established knowledge, observation and experimental investigation coupled with theoretical explanation of the origin and development of fingerprints and their individuality. These qualifications, as a science, have been successfully used to demonstrate the validity of fingerprints for a wide range of identification functions. Finally, theoretical quantifications of the amount of discriminatory information present in the fingerprints have now been provided [34].

Nevertheless, recently, there have been a number of "Daubert" challenges in courts of law to the validity of latent fingerprint identification in particular and to the scientific foundations of fingerprint analysis and comparison in general.

The US Supreme Court's "Daubert" decision originated in the 1993 civil case of Daubert vs. Merrill Dow Pharmaceuticals and generated an opinion regarding the admissibility of scientific evidence in Federal court [35]. Many state and local courts have also adopted this ruling. The opinion rendered was that a trial judge must screen scientific evidence to ensure that it is relevant and reliable, and the focus must be on principles and methods, not just on the conclusions. Courts must consider testing, validation, peer review of the processes, error rates and "general acceptance" of the practice.

The application of the "Daubert criteria" to fingerprint identification in criminal cases began in 1999 with the case of US vs. Byron C. Mitchell [36]. The admissibility of fingerprint evidence was challenged on the premises that, although fingerprints have been used for over 100 years, there is no scientific foundation for reliability (e.g. uniqueness and permanence), for estimating error rates, or for the uniformity and level of training of latent fingerprint examiners.

In defense of the general forensic admissibility of fingerprint evidence in the Mitchell case, the US Department of Justice expended great efforts to collect information on: fingerprint uniqueness and permanence based on the formation of friction ridges during embryonic development; the forensic techniques of fingerprint identification and the peer review process; the statistics related to the errors in identification; and the training programs in place for latent fingerprint collection, examination and comparison. The ruling in this landmark case concluded that (1) "...human friction ridges are unique and permanent throughout the area of friction ridge skin, including small friction ridge areas..." and (2) "...friction ridge skin arrangements are unique and permanent..." [36]. Notwithstanding this ruling, there have been at least 36 more Daubert challenges to the admissibility of fingerprint evidence in criminal (and one civil) proceedings till date.

The significance of these challenges to fingerprint identification is that they have forced the law enforcement community in general, and the latent fingerprint examiners in particular, to consider the processes and procedures more carefully and understand the scientific foundation for the many techniques and rules previously taken for granted. Latent examiners are now more aware of the details of the techniques used to conclude that a partial latent print from a crime scene actually matches a rolled, inked fingerprint taken perhaps many years earlier. The fact that these challenges continue to be filed is convincing evidence that fingerprint identification community must continue to explore and understand the firm scientific foundation on which it stands [37, 45]. There must also be a continuing effort in the teaching of the science (well documented in the Daubert hearings) to ensure that all technicians in the field understand and apply these well-developed techniques in their daily work. It is also important to continue a high level of peer review to ensure that there is consistency and quality necessary to maintain confidence in the error rates established in application.

2.7 Civil Applications

As criminal AFISs have matured, it has become apparent that there are many opportunities in the civil sector where fingerprints might be useful to verify a person's identity or identify a person who may try to obtain an identity fraudulently. In some instances it may be desirable to determine whether an applicant for a civil document or a sensitive employment position has a criminal record by matching the applicant's fingerprints against a criminal AFIS.

Although, there is relatively little desire within North America for a general civil identification document, it is a common practice in many parts of the world to require submission of applicant's fingerprints when applying for an identity card or registering to vote. With strictly manual fingerprint searches and comparison, these systems become intractable as the number of records grows, and a civil AFIS can supply a more reliable and practical solution.

Civil AFIS differ from criminal AFIS in several ways. Civil AFIS have no capability for the input of latent prints, the database contains only dab live-scanned fingerprints of limited finger surface area instead of the "nail-to-nail" rolled fingerprints of criminal AFIS, and there is no capability for interoperability with FBI or other national criminal systems.

2.7.1 Welfare Fraud Reduction

In the USA, there is a growing interest to control the abuse of Benefits Transfer Systems (BTS) by fraudulent application for multiple benefits through multiple registrations. Such fraud has been documented in BTS that do not use any form of identification.

In the early 1990s, a system was put into place in Los Angeles County (AFIRM) to register the two index fingerprints of all applicants for benefits transfer, and to create a database that could be searched to detect and prevent duplicate enrollments. The system also allowed identity verification of persons already in the system and demonstrated very significant savings [38]. With the implementation of the State Fingerprint Imaging System (SFIS) within California (using the AFIRM database as the starting point), these functions became statewide.

SFIS incorporates many more counties into a comprehensive network of fingerprint imaging, data entry, and processing with the goal of controlling the level of fraud within the state Health and Human Services network. Currently, several states implement fingerprint systems similar to SFIS to maintain control over fraud [39]. It seems clear that this introduction of fingerprint identification/verification technology into BTS will expand since it has been proven effective in reducing costs due to fraud in the benefit application process. As the number of these applications grows, they will demonstrate how AFIS technology may be applied to other types of benefit delivery systems.

2.7.2 Border Control

Another important application of fingerprint technology was introduced as a part of the Operation Gatekeeper program of the Border Patrol, a division of the US Immigration and Naturalization Service (INS). This program, named IDENT, was initiated in the mid-1990s in the area around San Diego, California, and its purpose is to identify individuals making repeated illegal border crossings. Records maintained in the fingerprint identification system provide support for legal action against the most severe recidivists (e.g. those with 15 or 20 recorded attempts).

Since fingerprint identification was only one of many aspects of Operation Gatekeeper, its effectiveness alone has been difficult to assess. However, it is clear that Operation Gatekeeper, in its entirety, had the effect of shifting apprehensions of illegal border crossers from the San Diego area eastward. It also established credibility for the use of fingerprint identification at the borders in other countries. Many countries, including Australia, Singapore, Hong Kong, Malaysia, Netherlands, Germany and the UK, are developing biometric systems to aid the identification of persons seeking admission to the country. In addition to fingerprints, hand geometry, iris and facial recognition are also being used. It is clear that the verification of a person's identity will become an important part of international travel.

2.7.3 Driver registration

There is a long history of use of fingerprints with driver licenses dating back at least 50 years. In California, the fingerprinting requirement was eliminated for many years but then reintroduced in the 1980s when it became possible to electronically capture fingerprint images. While the use of fingerprints in the driver registration process has great potential in fraud protection, there are serious privacy concerns over the possibility of "function creep", i.e. the potential for the system to be used for purposes other than driver identification. Immediately after September 11th 2001, at the mere suggestion of the potential for use of the driver license database to identify terrorists, there was an immediate response by groups such as the American Civil Liberties Union (ACLU) [40], the Electronic Privacy Information Center (EPIC) [40], and the National Research Council [42], pointing out the potential problems with any attempt to standardize the state driver license systems into a national ID card. However, it seems clear that there are many legitimate ways in which the driver license program can collect and use fingerprints without privacy implications. For example, fingerprints stored on the driver license could be used to verify the identity of a person renewing a license (e.g. to combat identity theft). At the same time, it also seems clear that there will be very earnest debate about any widespread application of fingerprinting to driver licensing.

2.8 Commercial Applications

After over 30 years of development, implementation, and operational use of AFIS for criminal identification, and more limited use in civil and high-security access control applications, it seems that the use of fingerprint matching in other civil and commercial applications is rapidly increasing. The past 20 years have seen the introduction of a variety of personal computer-friendly fingerprint scanners with ever decreasing price points, making them more and more affordable for use in a wide variety of commercial applications [43]. Fingerprint matching technology is now found in shrink-wrapped software marketed by companies focused on the commercial application of fingerprint technology. The question is no longer whether fingerprint matching will become an integral part of secure commercial applications, but when and in what volume.

2.8.1 Miniaturized Sensors

One of the important changes that have occurred within the last 5 to 10 years has been the miniaturization and cost reduction of the live-scan fingerprint sensors. Many vendors now manufacture optical sensors that are small and low-cost. Further, a number of solid state fingerprint sensors have been developed that are fabricated by processes quite similar to those used to manufacture semiconductor devices. These solid state sensors take advantage of sub-micron fabrication methods to provide fingerprint images of high resolution. Initially, these solid state sensors were constrained by the die size that could be fabricated with acceptable yield. The problem was not the creation of a large enough sensor area, but rather the number of good, flaw-free sensors that could be obtained from a semiconductor wafer. As die and wafer sizes have increased (e.g. 300 mm) to meet the demands for the ever-expanding integrated circuits for computers and optical sensors for cameras and video recorders, so has the size of solid state fingerprint sensors. Additionally, the planar fabrication techniques allow additional functions to be incorporated either on the same substrate, or easily integrated into an interconnect substrate using semiconductor bonding techniques. Similarly, the small live-scan optical sensors now use electronic imagers and have memory on board to allow additional functionality to be implemented in the sensor. The added functionality means that the entire finger sensor, the associated electronics for processing the images, and the interface with the personal computer can be incorporated into a small, rugged assembly. In fact, most of these sensors now use the universal serial bus (USB) as the preferred means for connection. No more expensive frame grabbers and complicated hardware installation; just plug and play for fingerprint capture and matching.

2.8.2 Personal Access Protection

With the availability of affordable and small finger scanners and access protection software, fingerprint verification can be easily incorporated into most computers. Indeed, existing applications provide all the tools necessary to verify the persons accessing a computer or computer network from a central server. These sophisticated applications include, in many cases, the active evaluation of the positioning of the finger (as low-cost sensors may have an area as small as 0.5 × 0.5 in) to ensure that the finger is placed in the correct position for a match to be successful. This level of integration is a reflection of the increased awareness of the sensor manufactures to the operational issues that must be considered for implementations to be successful.

Personal access protection can, and probably should, extend beyond the initial logon to the computer or network. Most, if not all, computer users connect to the Internet to obtain information, send and receive email, transact business and make purchases. With all the concern over identity theft, there is an urgent need to incorporate some form of validation of personal identity, particularly for credit card purchases over the Internet. Although a number of biometrics could be used to verify the identity of a person across a network, fingerprints are particularly appropriate for remote transactions. With a fingerprint scanner attached to a personal computer, a person can register fingerprint data with an Internet sales organization or with a "third party" security service organization that hosts the data. This data could be used to verify that the person completing a transaction at a remote computer is the person authorized to use an offered credit card. Personal fingerprint scanners are ultimately reaching the $10 price point as manufacturing techniques are maturing and volume of production is increasing. This makes fingerprints a very affordable security technology for personal use.

2.8.3 Banking Security

The banking industry has been evaluating the use of fingerprints for many years for credit card commerce. Unlike the Internet environment, where a sales organization intrinsically has a substantial computer capability in place, most retail organizations have limited computer capability to support business transactions. Their computers (if they exist beyond the cash register) are used for maintenance of inventory data. The cost of fingerprint sensors has not been low enough until recently to make fingerprint identity verification a cost-effective part of their business operation. But with the cost of fingerprint sensors dropping, and with the software to enroll and compare fingerprints becoming reasonably robust, cost-effective commercial systems are increasingly being deployed for verifying customers at the point of sale (POS) terminals. The key requirement is that fingerprint data needs to be captured at the time of credit card issuance and stored on the card (typically requiring about 300 bytes). The computer at the point of sale is required only to verify that the print captured at the checkout counter matches the data encoded on the credit card. This does not eliminate the need, however, to verify that the credit card account is still authorized and within its spending limit.

Another important aspect of the banking business is the ubiquitous automatic teller machines (ATMs) and money access centers (MACs), now estimated to exceed 300,000 in number. There has been considerable evaluation of competing biometric technologies to verify the identity of persons using these machines. The two most commonly mentioned biometric technologies include fingerprint and iris scanning. Planar-constructed fingerprint scanners, capable of integration into small assemblies, seem ideally suited for incorporation into the mechanical enclosures of the ATMs. These sensors can be affordably replaced if they are damaged intentionally or through extended use (e.g. 3,000+ uses each day). The simplicity of the available miniaturized live-scan fingerprint scanner interface (USB connection), the robust capture techniques and proven matching accuracy make fingerprint verification a serious competitor for the ATM application.

2.8.4 Business-to-Business Transactions

With more and more businesses sharing information with manufacturers, suppliers etc. via the computer and the Internet, and with the potential for misuse of such information by unauthorized persons, there is an opportunity for fingerprint-based verification systems to provide increased security for business-to-business transactions. The fingerprints of persons authorized to conduct business transactions or access business data can either be registered at the web sites of the businesses involved or managed by a security service providing a broad range of identity validation needs for individuals. This function can be integrated with the verification of individuals for credit card purchases so that data need not be replicated for each separate verification application. The key elements are the small, low-cost fingerprint sensor and the personal computer, integrated to service the wide range of business applications that benefit from validation of a person's identity across a network.

References

[1] J. Berry, The history and development of fingerprinting, in H. C. Lee and R. E. Gaensslen (eds) *Advances in Fingerprinting Technology*. CRC Press, Florida, 1994, pp. 1–38.

[2] S. Cole, *Suspect Identities*. Harvard University Press, Cambridge, MA, 2001.

[3] J. Purkinji, A commentary on the physiological examination of the organs of vision and the cutaneous system, *Ph.D. Thesis*, Breslau University, 1823.

[4] H. T. Rhodes, *Alphonse Bertillon*. Abelard-Schuman, New York, 1956.

[5] C. E. Chapel, *Fingerprinting: A Manual of Identification*. Coward McCann, New York, 1941.

[6] W. Herschel, Skin furrows of the hand. *Nature*, **23**, 76, November 25, 1880.

[7] H. Faulds, On the skin furrows of the hand. *Nature*, **22**, 605, October 28, 1880.

[8] F. Galton, Personal identification and description. *Nature*, June 21 and 28, 1888, pp. 173–177, 201–202.

[9] F. Galton, *Fingerprints*, 1892.

[10] E. R.Henry, *Classification and Uses of Fingerprints*. Routledge, London, 1900.

[11] H. Cummins and R. Kennedy, Purkinji's observations (1823) on fingerprints and other skin features. *Am. J. Police Sci.*, **31**(3), 1940.

[12] J. D. Woodward, Biometrics: privacy's foe or privacy's friend? *Proc. IEEE*, **85**(9), 1480–1492, 1997.

[13] D. Mintie, Biometrics for state identification applications – operational experiences. *Proc. CTST'98*, Vol. 1, pp. 299–312. See also `http://www.dss.state.ct.us/digital.htm`.

[14] G. Koehler, Biometrics: a case study – using finger image access in an automated branch. *Proc. CTST'98*, Vol. 1, pp. 535–541.

[15] United States Department of Justice, Federal Bureau of Investigation, *The Science of Fingerprinting*. Government Printing Office, Washington, DC, 1988.

[16] M. Trauring, On the automatic comparison of finger ridge patterns, *Nature*, **197**, 938–940, 1963.

[17] J. Wegstein, Automated fingerprint identification. *Technical Note 538*, Center for Computer Sciences and Technology, National Bureau of Standards, Washington, DC, August 1970.

[18] Criminal Justice Information Services, Appendix F: IAFIS image quality specifications for scanners, in *Electronic Fingerprint Transmission Specification CJIS-RS-0010(v7)*, January 29, 1999. Available online at `http://www.fbi.gov/hq/cjisd/iafis/efts70/cover.htm`.

[19] National Institute of Standards and Technology, "American National Standard for Information Systems – Data format for the interchange of fingerprint, facial, & scar mark & tattoo (SMT) information. *NIST Special Publication 500-245, ANSI/NIST-ITL 1-2000*, Information Technology Laboratory, National Institute of Standards and Technology, Gaithersburg, MD, September 2000. Available online at `http://www.itl.nist.gov/iad/894.03/fing/fing.html`.

[20] L. Hong, Y. Wan and A. K. Jain, Fingerprint image enhancement: algorithms and performance evaluation. *IEEE Trans. Pattern Analysis and Machine Intelligence*, **20**(8), 777–789, August 1998.

[21] D. Maltoni, D. Maio, A. K. Jain and S. Prabhakar, *Handbook of Fingerprint Recognition*. Springer, New York, June 2003.
[22] D. Gabor, Theory of communication. *J. Inst. Electrical Eng.*, **93**(3) 429–457, 1946.
[23] A. M. Bazen and S. H. Gerez, Systematic methods for the computation of the directional fields and singular points of fingerprints. *IEEE Trans. Pattern Analysis and Machine Intelligence*, **24**(7), 905–919, 2002.
[24] D. Maio and D. Maltoni, Ridge-line density estimation in digital images. *Proc. 14th International Conference on Pattern Recognition (ICPR)*, Brisbane (Australia), August 1998, pp. 534–538.
[25] A. K. Jain, L. Hong, S. Pankanti and R. Bolle, An identity authentication system using fingerprints. *Proc. IEEE*, **85**(9), 1365–1388, 1997.
[26] A. Rosenfeld, *Digital Image Processing*, 2nd edn, Academic Press, 1997.
[27] D. Maio and D. Maltoni, Direct gray-scale minutiae detection in fingerprints. *IEEE Trans. Pattern Analysis Machine Intelligence*, **19**(1), 27–40, 1997.
[28] J. L. Wayman, Multi-finger penetration rate and ROC variability for automatic fingerprint identification systems, in N. Ratha and R. Bolle (eds), *Automatic Fingerprint Recognition Systems*. Springer-Verlag, 2003.
[29] M. Kawagoe and A. Tojo, Fingerprint pattern classification. *Pattern Recognition*, **17**(3), 295–303, 1984.
[30] K. Karu and A. K. Jain, Fingerprint classification. *Pattern Recognition*, **29**(3), 389–404, 1996.
[31] P. Perona, Orientation diffusions. *IEEE Trans. Image Processing*, 7(3), 457–467, March 1998.
[32] The Interpol AFIS Expert Working Group, Data format for the interchange of fingerprint, facial & SMT information. *ANSI/NIST ITL 1-2000*, Version 4, 2002. Available online at `http://www.interpol.int/Public/Forensic/fingerprints/RefDoc/default.asp`.
[33] National Research Council, *The Evaluation of Forensic DNA Evidence*. National Academy Press, Washington, DC, 1996.
[34] S. Pankanti, S. Prabhakar and A. K. Jain, On the individuality of fingerprints. *IEEE Transactions on PAMI*, **24**(8), 1010–1025, 2002. Available online at `http://biometrics.cse.msu.edu/publications.html`.
[35] *Daubert vs. Merrell Dow Pharmaceuticals*, 113 S. Ct. 2786, 1993.
[36] *US vs. Byron Mitchell*, Criminal Action No. 96-407, US District Court for the Eastern District of Pennsylvania, July 1999.
[37] J. L. Wayman, When bad science leads to good law, in J. L. Wayman (ed) *U.S. National Biometric Test Center Collected Works: 1997–2000*. San Jose State University, 2000.
[38] Applied Management and Planning Group, *Los Angeles County DPSS AFDC AFIRM Demonstration Project: Final Evaluation Report*, July 1997.
[39] D. Mintie (ed.), *Biometrics in Human Services User Group Newsletter*. Department of Social Services, State of Connecticut. Available online at `http://www.dss.state.ct.us/digital.htm`.
[40] American Civil Liberties Union, *Letter to President George W. Bush*, February 11, 2002. Available online at `http://archive.aclu.org/congress/l021102a.html`
[41] Electronic Privacy Information Center, Your papers, please: from the state driver's license to a national identification system. *Watching the Watchers – Policy Report #1*, February, 2002. Available online at `http://www.epic.org/privacy/id_cards/yourpapersplease.pdf`.

[42] National Research Council, *ID's Not That Easy*. National Academy Press, Washington, DC, 2002.

[43] L. O'Gorman, Practical systems for personal fingerprint authentication, *IEEE Computer*, February 2000, pp. 58–60.

[44] Federal Bureau of Investigation, Wavelet scalar quantization (WSQ) gray-scale fingerprint image compression specification. *FBI Report Number IAFIS-IC-00110v2*, Criminal Justice Information Services, February 16, 1993.

[45] D. Stoney, Measurement of fingerprint individuality, in H. C. Lee and R. E. Gaensslen (eds) *Advances in Fingerprint Technology*. CRC Press, Florida, 1994.

Iris Recognition 3

Richard Wildes

3.1 Introduction

Biometrics bear the promise of providing widely applicable approaches to personal verification and identification[1]. For such approaches to be widely applicable, they must be highly reliable even while avoiding invasiveness in evaluating a subject of interest. Reliability has to do with the ability of the approach to support a signature that is unique to an individual and that can be captured in an invariant fashion time and time again. Biometrics have the potential for high reliability because they are based on the measurement of an intrinsic physical property of an individual. For example, fingerprints provide signatures that appear to be unique to an individual and reasonably invariant with the passage of time, whereas faces, while fairly unique in appearance can vary significantly with the vicissitudes of time and place. Invasiveness has to do with the ability to capture the signature while placing as few constraints as possible on the subject of evaluation. In this regard, acquisition of a fingerprint signature is invasive as it requires that the subject makes physical contact with a sensor, whereas images of a subject's face that are sufficient for recognition can be acquired at a comfortable distance and, in certain scenarios, covertly.

Considerations of reliability and invasiveness suggest that the human iris is a particularly interesting structure on which to base a biometric approach for personal verification and identification. From the point of view of reliability, the spatial patterns that are visually apparent in the human iris are highly distinctive to an individual [1, 36]; see for example Figure 3.1. Further, the appearance of any one iris suffers little from day-to-day variation. From the point of view of invasiveness, the iris is an overt body that can be imaged at a comfortable distance from a subject with the use of extant machine vision technology [64]. Owing to these features of reliability and (non)invasiveness, iris recognition is a promising approach to biometric-based verification and identification of people. Indeed,

---

1 Throughout this discussion, the term "verification" will refer to recognition with respect to a specified database entry. The term "identification" will refer to recognition with respect to a larger set of alternative entries.

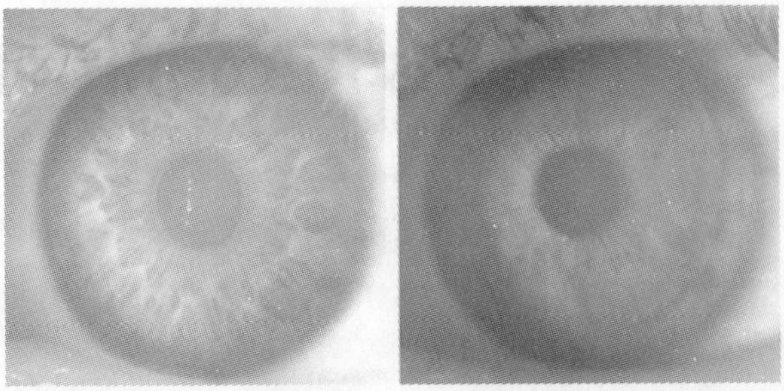

Figure 3.1 The distinctiveness of the human iris. The left and right panels show images of the left iris of two individuals. Even to casual inspection, the imaged patterns in the two irises are markedly different. R. Wildes, Iris recognition: an emerging biometric technology. *Proceedings of the IEEE*, **85**(9), 1348–1363, 1997 (© 1997 IEEE)

significant strides have been made toward bringing iris recognition out of the laboratory and into real-world deployment [56, 46].

Early use of the iris as the basis for a biometric-based approach to recognizing persons can be traced to efforts to distinguish inmates in the Parisian penal system by visually inspecting their irises, especially the patterning of color [4]. More recently, the concept of automated iris recognition was proposed [23]; however, it does not appear that this team ever developed and tested a working system. Early work toward actually realizing a system for automated iris recognition was carried out at Los Alamos National Laboratories in the USA [34]. Subsequently, a number of research groups developed and documented prototype iris recognition systems working with highly cooperative subjects at close distances [5, 14, 38, 40, 53, 57, 59, 65, 73].

These systems have shown promising performance on diverse databases of hundreds of iris images. More recently, efforts have been aimed at allowing for iris recognition at somewhat greater distances and with less active participation on the part of the subject [18, 26]. Again, these systems have shown interesting levels of performance.

More anecdotally, a notion akin to automated iris recognition came to popular attention in the James Bond film *Never Say Never Again*, as characters are depicted having images of their eye captured for the purpose of identification [25].

This chapter unfolds along the following lines. This section has served to introduce the notion of iris recognition as the basis for a biometric approach to verifying or identifying persons. The Section 3.2 reviews relevant facts about the anatomy and physiology of the iris. Section 3.3 provides an overview of approaches to sensing for the capture of iris images to drive recognition. Section 3.4 describes approaches to representing and matching iris signatures. Section 3.5 describes extant iris recognition systems, including the evaluation of their performance. Finally, section six provides a look at future directions. Throughout this discussion, the iris recognition systems and components developed by Daugman [12–16] and Wildes *et al.* [64–69] will serve as the major sources of illustration owing to the fact that they are the best documented approaches in the public domain literature.

3.2 Anatomical and Physiological Underpinnings

To appreciate the richness of the iris as a pattern for recognition, it is useful to consider its structure in a bit of detail. In gross terms, the iris is part of the uveal, or middle, coat of the eye. It is a thin diaphragm stretching across the anterior portion of the eye and supported by the lens; see Figure 3.2. This support gives it the shape of a truncated cone in three dimensions. At its base, the iris is attached to the eye's cilliary body. At the opposite end it opens into the pupil, typically slightly to the nasal side and below center. The cornea lies in front of the iris and provides a transparent, protective covering.

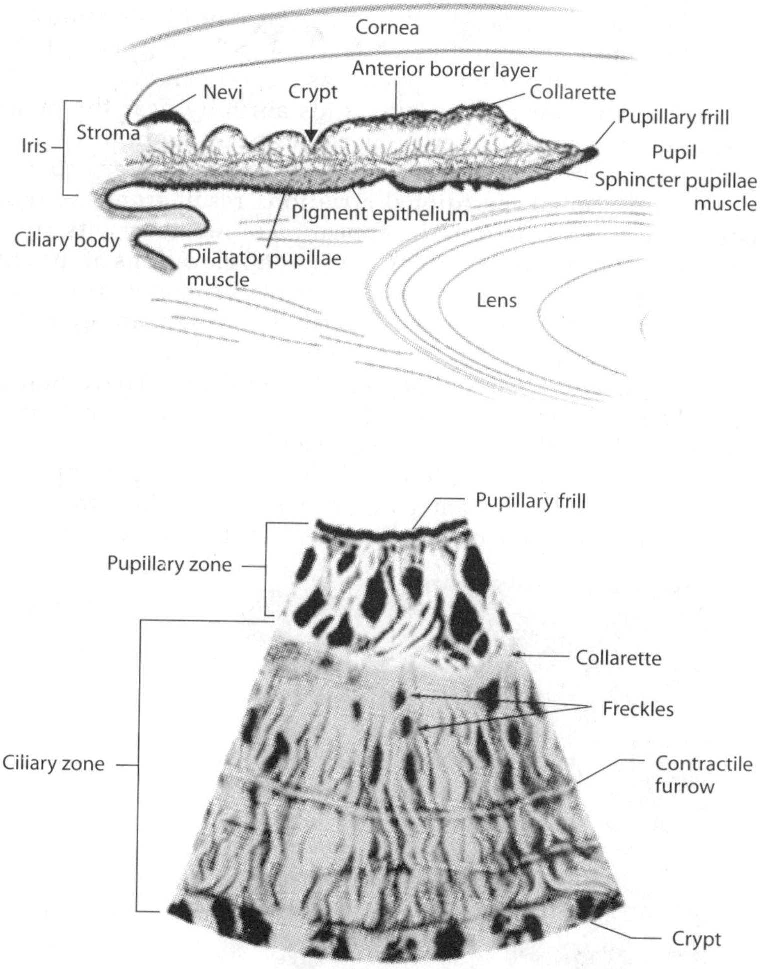

Figure 3.2 Anatomy of the human iris. The upper panel illustrates the structure of the iris seen in a transverse section. The lower panel illustrates the structure of the iris seen in a frontal sector. The visual appearance of the human iris derives from its anatomical structure. R. Wildes, Iris recognition: an emerging biometric technology. *Proceedings of the IEEE*, **85**(9), 1348–1363, 1997 (© 1997 IEEE)

At a finer grain of analysis, the iris is composed of several layers. The posterior surface is composed of heavily pigmented epithelial cells that make it impenetrable to light. Anterior to this layer two muscles are located that work in cooperation to control the size of the pupil. The stromal layer is next; it consists of collagenous connective tissue, arranged in arch-like processes. Corkscrew shaped blood vessels also are present in this layer, arranged along the radial direction. Finally, the anterior border layer completes the stratification.

The anterior border layer is distinguished from the stromal layer by its increased density, especially in terms of chromatophores, i.e. individual pigment cells.

The visual appearance of the iris is a direct result of its multilayered structure. The anterior surface of the iris is seen to be divided into a central pupillary zone and a surrounding cilliary zone. The border of these two areas is termed the collarette; it appears as a zigzag circumferential ridge resulting as the anterior border layer ends abruptly near the pupil. The cilliary zone contains many interlacing ridges resulting from stromal support. Contractile lines that are present in this region can vary with the state of the pupil. Additional meridional striations result from the radiating vasculature. Further variations in appearance owe to crypts (irregular atrophy of the anterior border layer), nevi (small elevations of the anterior border layer) and freckles (local collections of chromatophores). In comparison, the pupillary zone can be relatively flat. Often, however, it shows radiating spoke-like processes and a pigment frill where the posterior layer's heavily pigmented tissue shows at the pupil boundary. Significantly, an image taken of the iris with a standard video camera can capture many of the anatomical details just described; see Figure 3.3.

Iris color results from the differential absorption of light impinging on the pigmented cells in the anterior border layer. When there is little pigmentation in the anterior border layer, light reflects back from the

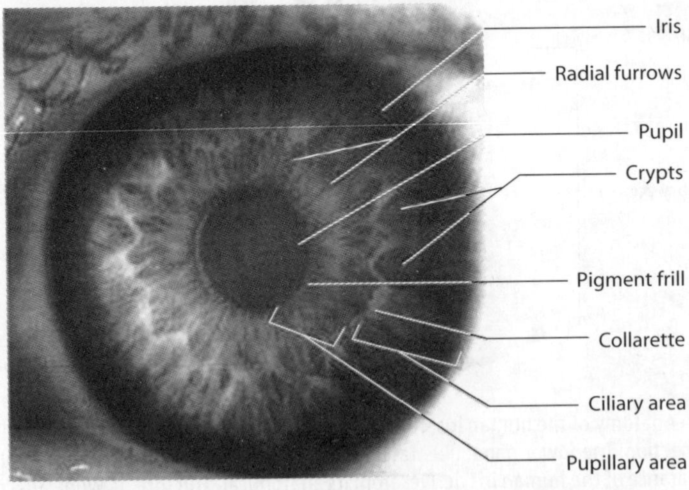

Figure 3.3 Anatomy of the iris visible in an optical image.

posterior epithelium and is scattered as it passes through the stroma to yield a blue appearance. Progressive levels of anterior pigmentation lead to darker colored irises.

Claims that the structure of the iris is unique to an individual and is stable with age come from two main sources. The first source of evidence comes from clinical observations. During the course of examining large numbers of eyes, ophthalmologists [23] and anatomists [1] have noted that the detailed spatial pattern of an iris, even the left and right irises of a single person, seems to be unique. Further, in cases with repeated observations, the patterns seem to vary little, at least past childhood. The second source of evidence comes from developmental biology [37,41]. In this literature one finds that while the general structure of the iris is genetically determined, the particulars of its minutiae are critically dependent on circumstances (e.g. the initial conditions in the embryonic precursor to the iris). Therefore they are highly unlikely to be replicated via the natural course of events. For example, the shape of the collarette depends on the particulars of how the anterior border layer recedes to allow for the mature pupil. Rarely, the developmental process goes awry, yielding only a rudimentary iris (aniridia) or a marked displacement (corectopia) or shape distortion (colobloma) of the pupil [37, 47]. Developmental evidence also bears on issues of stability with age. Certain parts of the iris (e.g. the vasculature) are largely in place at birth; whereas, others (e.g. the musculature) mature around two years of age [1, 37]. Of particular significance for the purposes of recognition is the fact that pigmentation patterning continues until adolescence [1, 49, 62]. Also, the average pupil size (for an individual) increases slightly until adolescence [1]. Following adolescence the healthy iris varies little for the rest of life, although slight depigmentation and shrinking of the average pupillary opening are standard with advanced age [1, 47]. Various diseases of the eye can drastically alter the appearance of the iris [45, 47]. Also, certain drug treatments for eye disease (e.g. prostoglandin-based treatment of glaucoma) may alter iris pigmentation. Further, it appears that intensive exposure to certain environmental contaminants (e.g. metals) can alter iris pigmentation [45, 47]. However, these conditions are rare. Claims that the iris changes with more general states of health (iridology) have been discredited [3, 70].

Another interesting aspect of the physical characteristics of the iris from a biometric point of view has to do with its moment to moment dynamics. Due to the complex interplay of the iris's muscles, the diameter of the pupil is in a constant state of small oscillation at a rate of approximately 0.5 Hz, a movement referred to as hippus [1, 17]. Potentially, this movement could be monitored to make sure that a live specimen is being evaluated. Further, since the iris reacts very quickly to changes in impinging illumination (e.g. on the order of hundreds of milliseconds for contraction), monitoring the reaction to a controlled illuminant could provide similar evidence. In contrast, upon morbidity the iris contracts and hardens, facts that may have ramifications for forensics.

On the whole, the anatomy and physiology of the iris suggest that it provides a rich source for biometric-based personal verification and identification. The iris tissue shows substantial visually apparent spatial detail. Moreover, the patterns of the minutiae are highly distinctive to an individual and, following childhood, typically stable with age. Nevertheless, it is important to note that large-scale studies that specifically address the distinctiveness and stability of the iris, especially for biometrics, have yet to be performed. Further details of iris structure can be found in the biomedical literature (e.g. [1, 17, 50]).

3.3 Sensing

Due to the differential nature of how light is reflected from the iris minutiae, optical sensing is well suited to acquisition of an iris image for recognition purposes. Still, acquisition of a high-quality iris image, while remaining non-invasive to human subjects, is one of the major challenges of automated iris recognition. Given that the iris is a relatively small (typically about 1 cm in diameter), dark object, and that people are very sensitive about their eyes, this matter requires careful engineering. Several points are of particular concern. First, it is desirable to acquire images of the iris with sufficient resolution and sharpness to support recognition. Second, it is important to have good contrast in the iris pattern while restricting illumination to be within limits of eye safety and comfort. Third, the iris must be well framed (i.e. centered) without unduly constraining the subject (i.e. preferably without requiring the subject to employ an eyepiece, chin rest or other contact positioning that would be invasive). Further, as an integral part of this process, artifacts in the acquired images (due to specular reflections, optical aberrations etc.) should be eliminated as much as possible.

In response to these challenges of optically imaging an iris for recognition, two kinds of approach have been developed. One type of approach makes use of passive sensing while requiring considerable active participation by the subject to self-position for centering and focus. Examples of this approach have been demonstrated at sensor to subject distances under 0.5 m. A second type of approach makes use of active sensing to acquire images of the iris with only modest participation on the part of the subject, e.g. the subject must simply stand still and look forward, while the system automatically adjusts its optical parameters to best accommodate the subject. Examples of this approach have been demonstrated to distances of 0.75 m. In the remainder of this section examples of each approach are presented.

Functional diagrams of two passive sensor rigs that have been developed for iris image acquisition are shown in Figure 3.4. Both of the depicted systems respond to the fundamental issue of spatial resolution using standard optics. For example, the apparatus reported by Daugman captures images with the iris diameter typically between 100 and 200 pixels from a distance of 46 to 15 cm using a standard lens and video camera with video rate capture. Similarly, the apparatus reported by Wildes *et al.* images the iris with approximately 256 pixels across the diameter from 20 cm using a standard lens and a silicon-intensified camera (to enable imaging with low illumination levels [28]) with video rate capture. Due to the need to keep the illumination level

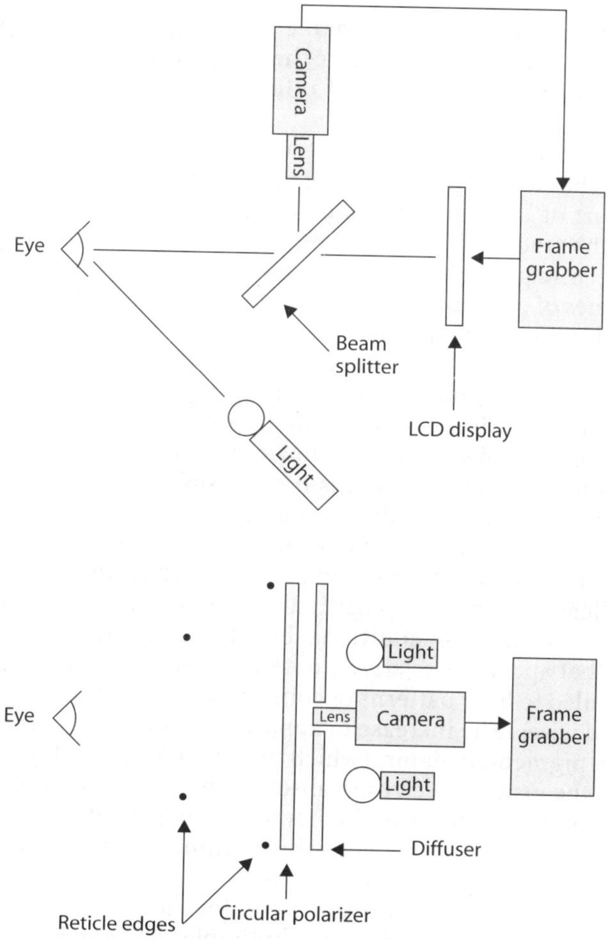

Figure 3.4 Passive sensing approaches to iris image acquisition. The upper diagram shows a schematic diagram of the Daugman [14] image acquisition rig. The lower diagram shows a schematic diagram of the Wildes *et al.* [65] image acquisition rig. R. Wildes, Iris recognition: an emerging biometric technology. *Proceedings of the IEEE*, **85**(9), 1348–1363, 1997 (© 1997 IEEE)

relatively low for subject comfort and safety, the optical aperture cannot be too small (e.g. f-stop 11). Therefore, both systems have fairly small depths of field, approximately 1 cm. Motion blur due to eye movements typically is not problematic, given the video rate capture and the assumption that the subject is cooperating in attempting to maintain a steady gaze. Empirically, the overall spatial resolution that results from these designs appears to be sufficient to support iris recognition. Unfortunately, neither Daugman nor Wildes *et al.* went so far as to document the combined camera/lens modulation transfer function for their particular optical assemblies.

Interestingly, both systems essentially eschew color information in their use of monochrome cameras with 8 bit gray level resolution. Presumably, color information could provide additional discriminatory power. Also, color could be of use for initial coarse indexing through large iris databases. For now, it is interesting to note that empirical studies to date suggest the adequacy of monochrome level information alone (see, for example, Section 3.5).

In order to cope with the inherent variability of ambient illumination, extant approaches to iris image sensing provide a controlled source of illumination as a part of their method. Such illumination of the iris must be concerned with the trade-off between revealing the detail in a potentially low-contrast pattern (e.g. due to relatively uniform dense pigmentation of dark irises) and issues of eye comfort and safety [44]. As originally documented, the Daugman and Wildes *et al.* sensor systems both made use of illumination with spectral energy concentrated in the visible range. More recently, illumination for iris recognition has tended to concentrate on near infrared sources, principally to decrease invasiveness and with an aim of realizing a source that is invisible to human subjects. To date, however, while the resulting light typically is not annoying to subjects, it is not invisible, as a dull red glow is perceived: the use of commercially available near-infrared sources that are adequate to illuminate the iris for recognition has not made it possible to achieve total invisibility to the human eye; the bandwidth of available sources overlaps sufficiently with the sensitivity of human photoreceptors to drive perception. Significantly, an additional benefit of iris imaging in the infrared is that irises that appear as relatively dark and patternless in the visible spectrum are revealed to have patterns of comparable richness to other more obviously textured irises. The increase in apparent detail is due to the fact that the principal iris pigment, melanin (which, when concentrated, yields dark colored irises in the visible spectrum), absorbs poorly in the infrared and hence allows the structural patterns of the iris to be imaged with greater contrast.

An interesting difference between the illumination solutions described in Daugman and Wildes *et al.* has to do with the former's use of a compact (unpolarized) source, while the latter employs a diffuse polarized source. The compact source yields a particularly simple solution. Further, by careful positioning of the light source below the operator, reflections of the point source by eye glasses can be avoided in the imaged iris. However, without placing undue restriction on the subject, it is difficult to reliably position the specular reflection at the eye's cornea outside the iris region. Therefore this design requires that the region of the image where the point source is seen (the lower quadrant of the iris as the system was originally instantiated) must be omitted during matching, since it is dominated by artifacts. The latter design results in an illumination rig that is more complex; however, certain advantages result. First, the use of matched circular polarizers at the light source and the camera essentially eliminates the specular reflection of the light source [2]. This allows for more of the iris detail to be available for

2 Light emerging from the circular polarizer will have a particular sense of rotation. When this light strikes a specularly reflecting surface (e.g. the cornea) the light that is reflected back is still polarized, but has reversed sense. This reversed sense light is not passed through the camera's filter and is thereby blocked from forming an

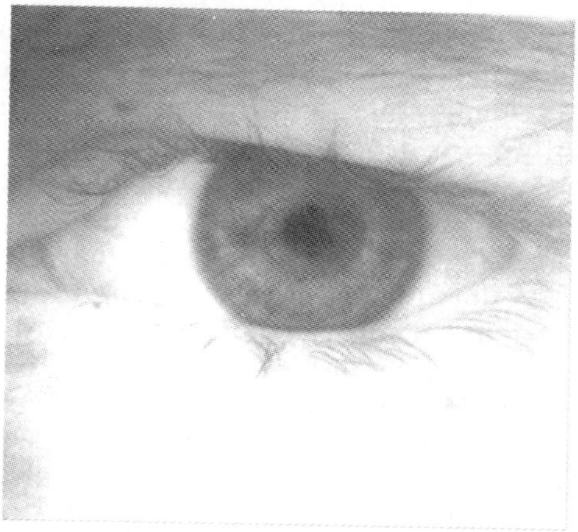

Figure 3.5 Captured iris image. An example iris image as captured by the Wildes *et al.* passive sensor is shown. Notice that this initial stage of sensing captures not only the iris but also surrounding portions of the eye region. Subsequent processing must more precisely localize the iris *per se*. R. Wildes, Iris recognition: an emerging biometric technology. *Proceedings of the IEEE*, **85**(9), 1348–1363, 1997 (© 1997 IEEE)

subsequent processing. Secondly, the coupling of a low light level camera (a silicon intensified camera [29]) with a diffuse illuminant allows for a level of illumination that is entirely unobjectionable to human subjects.

Positioning of the iris for image capture is concerned with framing all of the iris in the camera's field of view with good focus. Both the Daugman and Wildes *et al.* systems require the subject to self-position their eye region in front of the camera. Daugman's system provides the subject with live video feedback via a miniature LCD display, placed in line with the camera's optics via a beam splitter. This allows the subject to see what the camera is capturing and adjust their position accordingly. During this process the system is continually acquiring images. Once a series of images of sufficient quality is acquired, one image is automatically forwarded for subsequent processing. Image quality is assessed by looking for high-contrast edges marking the boundary between the iris and the sclera. In contrast, the Wildes *et al.* system provides a reticle to aid the subject in positioning.

As the subject maneuvers, the relative misalignment of cross-hairs provides continuous visual feedback regarding the accuracy of the current position. Once the subject has completed the alignment, image capture is activated via a button press. An example acquired iris image, as captured by the Wildes *et al.* approach is shown in Figure 3.5.

---

image. In contrast, the diffusely reflecting parts of the eye (e.g. the iris) scatter the impinging light. This light is passed through the camera's filter, and is subsequently available for image formation [33]. Interestingly, a similar solution using crossed polarizers (e.g. vertical at the illuminant and horizontal at the camera) is not appropriate for this application: the birefringence of the eye's cornea yields a low-frequency artifact in the acquired images [11].

Subjectively, both of the described approaches to positioning are fairly easy for a human subject to master. However, since the potential for truly non-invasive assessment is one of the intriguing aspects of iris recognition, it is worth underlining the degree of operator participation that is required in these systems. While physical contact is avoided, the level of required cooperativity may still prevent the systems from widespread application. It is to this limitation that an active sensing approach to iris image acquisition can respond, as documented next.

Research initiated at Sarnoff Corporation [26] and subsequently transferred to Sensar Incorporated [18] for refinement and commercialization has yielded the most non-invasive approach to iris image capture that has been documented to date. For capture, a subject merely needs to stand still and face forward with their head in an acquisition volume of 60° vertical by 45° horizontal and a distance of approximately 0.38 to 0.76 m, all measured from the front-center of the acquisition rig. Capture of an image that has proven suitable to drive iris recognition algorithms can then be achieved totally automatically, typically within 2–10 seconds.

The developed approach makes use of active vision techniques whereby a wide field of view binocular camera apparatus localizes the head and eye within the entire acquisition volume and then drives a narrow field of view camera to point, zoom and focus on the area immediately surrounding the iris. The image captured by the narrow field of view camera is then used for recognition purposes. Illumination is provided by a pair of LED near-infrared (narrow band, peak energy approximately 880 nm) illumination panels that cover the entire acquisition volume. The resulting lighting is not annoying to humans and is perceived as a dull red glow. A schematic diagram of the approach is shown in Figure 3.6.

The wide field of view binocular camera apparatus is charged with localizing the eye region of a subject in three dimensions with respect to the

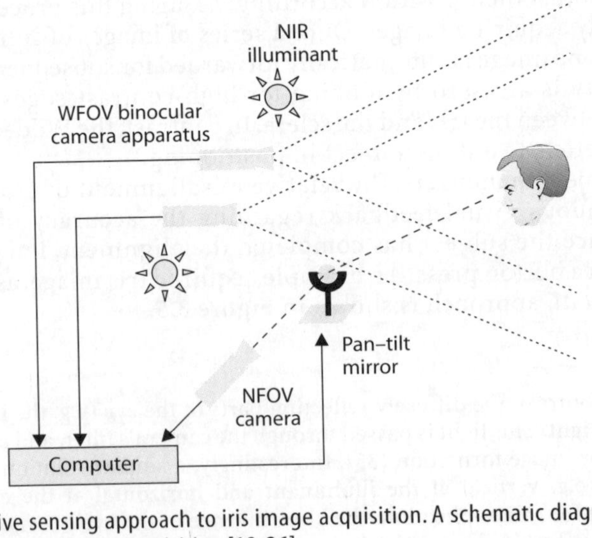

Figure 3.6 Active sensing approach to iris image acquisition. A schematic diagram of an active vision approach to iris image acquisition [18, 26].

image acquisition platform. To accomplish this task, standard techniques in binocular stereo from computer vision are employed [21, 60]. Two standard monochrome video cameras outfitted with wide field of view lenses are employed. The cameras are arranged with relative horizontal displacement, parallel optical axes, overlapping fields of view that cover the acquisition volume and geometric calibration that allows pixel measurements to be converted to scene measurements. Owing to the geometry of this situation, three-dimensional points in the viewed scene project to the images as two-dimensional features; the two-dimensional features associated with a scene point are imaged with a spatial displacement in the two images that depends on the distance to the cameras. Correspondingly, measured displacement between matched image features allows for recovery of the three-dimensional location of a projected scene point through a process of triangulation. For the particular system of interest, matching between points in the acquired image pair is accomplished via a correlation search algorithm that systematically shifts and compares small spatial windows across the images of concern. During an initial stage of processing, the overall head region is localized as that region in 3-space that is closest to the acquisition platform. Subsequently, a facial feature template matching algorithm operates within the head region to localize the eye region. The coordinates of the eye region are then used to drive the narrow field of view apparatus to capture a high-resolution iris image. Operations associated with the wide field of view apparatus are performed on a specialized image processing accelerator board. Code for the described algorithms running on this system allows for continuous generation of 3D coordinates for the eye at a rate of 2 Hz.

The narrow field of view apparatus consists of a pan/tilt mirror assembly, a fixed focal length lens with computer-controlled focus axis and a standard monochrome video camera. Eye position estimates delivered by the wide field of view apparatus are mapped to a pan/tilt/focus triple via a look-up-table generated as part of system calibration. After an initial motorized pan/tilt/focus adjustment, processing local to the narrow field of view apparatus serves to refine the capture quality through better centering, focusing and a rotational transformation that compensates for apparent torsional error arising from the pan/tilt steering mirror. All processing for this apparatus is performed without any special acceleration on a 166 MHz PC processor. Final capture time is data-dependent, but is typically between 2–10 seconds. Acquisition with this overall approach has allowed for capture of images of quality similar to that achieved with passive image acquisition platforms (e.g. 300 pixels on the iris diameter and modulation transfer function empirically capable of supporting iris recognition), albeit while requiring far less subject participation.

3.4 Iris signature representation and matching

Following image acquisition, the portion of the image that corresponds to the iris needs to be localized from its surroundings. The iris image data can then be brought under a representation to yield an iris signature for matching against similarly acquired, localized and represented irises. The next three subsections of this chapter discuss issues in localization, representation and matching.

3.4.1 Localization

Without placing undue constraints on the human subject, image acquisition of the iris cannot be expected to yield an image containing only the iris. Rather, image acquisition will capture the iris as part of a larger image that also contains data derived from the immediately surrounding eye region; see, for example Figure 3.5. Therefore, prior to performing iris pattern matching, it is important to localize that portion of the acquired image that corresponds to an iris. In particular, it is necessary to localize that portion of the image derived from inside the limbus (the border between the sclera and the iris) and outside the pupil. Further, if the eyelids are occluding part of the iris, then only that portion of the image below the upper eyelid and above the lower eyelid should be included.

Interestingly, the image contrast of these various iris boundaries can be quite variable. For example, owing to the relative spectral reflectance of the sclera and iris (in particular its melanin pigment) the limbic boundary is typically imaged with higher contrast in the visible than the infrared portion of the spectrum. For the case of the pupillary boundary, the image contrast between a heavily pigmented iris and its pupil can be quite small. Further, while the pupil typically is darker than the iris, the reverse relationship can hold in cases of cataract: the clouded lens leads to a significant amount of backscattered light. Like the other boundaries, eyelid contrast can be quite variable, depending on the relative pigmentation in the skin and the iris. The eyelid boundary also can be irregular due to the presence of eyelashes. Taken together, these observations suggest that iris localization must be sensitive to a wide range of edge contrasts, robust to irregular borders and capable of dealing with variable occlusion.

Reference to how the Daugman and Wildes *et al.* approaches perform iris localization further illustrates the issues. Both of these systems make use of first derivatives of image intensity to signal the location of edges that correspond to the borders of the iris. Here, the notion is that the magnitude of the derivative across an imaged border will show a local maximum due to the local change of image intensity. Also, both systems model the various boundaries that delimit the iris with simple geometric models. For example, they both model the limbus and pupil with circular contours. The Wildes *et al.* system also explicitly models the upper and lower eyelids with parabolic arcs. In initial implementation, the Daugman system simply excluded the upper and lower most portions of the image where eyelid occlusion was most likely to occur; subsequent refinements include explicit eyelid localization. In both systems, the expected configuration of model components is used to fine-tune the image intensity derivative information. In particular, for the limbic boundary the derivatives are filtered to be selective for vertical edges. This directional selectivity is motivated by the fact that, even in the face of occluding eyelids, the left and right portions of the limbus should be visible and oriented near the vertical (assuming that the head is in an upright position). Similarly, the derivatives are filtered to be selective for horizontal information when locating the eyelid borders. In contrast, since the entire (roughly circular) pupillary boundary is expected to be present in the image, the derivative information is used in a more isotropic fashion for localization of this structure. In practice, this fine tuning of the image information has proven to be critical for accurate localization. For example, without such tuning the fits can be driven astray by competing image structures (e.g. eyelids interfering with limbic localization).

The two approaches differ mostly in the way that they search their parameter spaces to fit the contour models to the image information. In order to understand how these searches proceed, let $I(x, y)$ represent the image intensity value at location (x, y) and let circular contours (for the limbic and pupillary boundaries) be parametrized by center location, (x_c, y_c), and radius, r. The Daugman approach fits the circular contours via gradient ascent on the parameters (x_c, y_c, r) so as to maximize

$$\left| \frac{\partial}{\partial r} G(r) * \oint_{r, x_c, y_c} \frac{I(x,y)}{2\pi r} \mathrm{d}s \right|$$

where $G(r) = (1/\sqrt{2\pi}\sigma)\exp[-(r - r_0)^2 / 2\sigma^2]$ is a radial Gaussian with center r_0 and standard deviation σ that smooths the image to select the spatial scale of edges under consideration, $*$ symbolizes convolution, ds is an element of circular arc and division by $2\pi r$ serves to normalize the integral. In order to incorporate directional tuning of the image derivative, the arc of integration, ds, is restricted to the left and right quadrants (i.e. near vertical edges), when fitting the limbic boundary. This arc is considered over a fuller range when fitting the pupillary boundary; however, the lower quadrant of the image is still omitted due to the artifact of the specular reflection of the illuminant in that region (see Section 3.3). Following localization of the circular boundaries, the eyelids are localized as being within the limbic boundary by fitting a spline contour parametrization in a fashion analogous to that used for the circular parametrization. In implementation, the contour-fitting procedure is discretized with finite differences serving for derivatives and summation used to instantiate integrals and convolutions. More generally, fitting contours to images via this type of optimization formulation is a standard machine vision technique, often referred to as active contour modeling; see, for example, [35, 54, 72].

The Wildes *et al.* approach performs its contour fitting in two steps. First, the image intensity information is converted into a binary edge-map. Second, the edge points vote to instantiate particular contour parameter values. The edge-map is recovered via gradient-based edge detection [52, 60]. This operation consists of thresholding the magnitude of the image intensity gradient, i.e. $|\nabla G(x,y) * I(x,y)|$, where $\nabla \equiv (\partial/\partial x, \partial/\partial y)$ while

$G(x,y)=(1/2\pi\sigma^2)\exp\{-[(x-x_0)^2+(y-y_0)^2]/2\sigma^2\}$ is a two-dimensional Gaussian with center (x_0, y_0) and standard deviation σ that smooths the image to select the spatial scale of edges under consideration. In order to incorporate directional tuning, the image intensity derivatives are weighted to favor certain ranges of orientation prior to taking the magnitude. For example, prior to contributing to the fit of the limbic boundary contour, the derivatives are weighted to be selective for vertical edges. The voting procedure is realized via Hough transforms [29, 30] on parametric definitions of the iris boundary contours. In particular, for the circular limbic or pupillary boundaries and a set of recovered edge points, (x_j, y_j), $j=1,\ldots,n$, a Hough transform is defined as

$$H(x_c, y_c, r) = \sum_{j=1}^{n} h(x_j, y_j, x_c, y_c, r)$$

where

$$h(x_j, y_j, x_c, y_c, r) = \begin{cases} 1 & \text{if } g(x_j, y_j x_c, y_c, r) = 0 \\ 0 & \text{otherwise} \end{cases}$$

with

$$g(x_j, y_j x_c, y_c, r) = (x_j - x_c)^2 + (y_j - y_c)^2 - r$$

For each edge point (x_j, y_j), $g(x_j, y_j, x_c, y_c, r) = 0$ for every parameter triple (x_c, y_c, r) that represents a circle through that point. Correspondingly, the parameter triple that maximizes H is common to the largest number of edge points and is a reasonable choice to represent the contour of interest. In implementation, the maximizing parameter set is computed by building $H(x_c, y_c, r)$ as an array that is indexed by discretized values for x_c, y_c and r. Once populated, the array is scanned for the triple that defines its largest value. Contours for the upper and lower eyelids are fit in a similar fashion using parametrized parabolic arcs in place of the circle parametrization $g(x_j, y_j, x_c, y_c, r)$. Just as the Daugman system relies on standard techniques for iris localization, edge detection followed by a Hough transform is a standard machine vision technique for fitting simple contour models to images [52, 60].

Both approaches to localizing the iris have proven to be successful in the targeted application. The histogram-based approach to model fitting should avoid problems with local minima that the active contour model's gradient descent procedure might experience. However, by operating more directly with the image derivatives, the active contour approach avoids the inevitable thresholding involved in generating a binary edge map. More generally, both approaches are likely to encounter difficulties if required to deal with images that contain broader regions of the surrounding face than the immediate eye region. For example, such images are likely to result

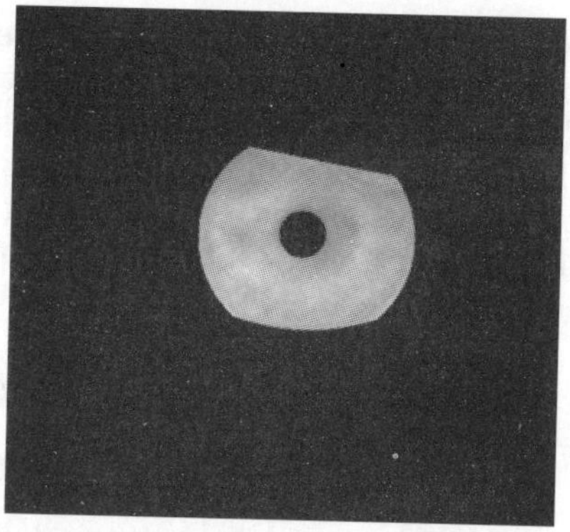

Figure 3.7 Illustrative results of iris localization. Given an acquired image, it is necessary to separate the iris from the surroundings. Taking as input the iris image shown in Figure 3.5, automated processing delineates that portion which corresponds to the iris. R. Wildes, Iris recognition: an emerging biometric technology. *Proceedings of the IEEE*, **85**(9), 1348–1363, 1997 (© 1997 IEEE)

from image acquisition rigs that require less subject participation than those currently in place. Here, the additional image "clutter" is likely to drive the current, relatively simple, model fitters to poor results. Recent extensions to the Wildes *et al.* approach for iris localization take steps along these directions [9]. Still, complete solutions to this type of situation most likely will entail a preliminary coarse eye localization procedure to seed iris localization proper, e.g. as provided by the active sensing approach descibed in Section 3.3 [18, 26]. In any case, following successful iris localization, the portion of the captured image that corresponds to the iris can be delimited. Figure 3.7 provides an example result of iris localization as performed by the Wildes *et al.* system.

3.4.2 Representation

The distinctive spatial characteristics of the human iris are manifest at a variety of scales. For example, distinguishing structures range from the overall shape of the iris to the distribution of tiny crypts and detailed texture. To capture this range of spatial detail, it is advantageous to make use of a multiscale representation. Both of the approaches to iris signature representation that are under discussion make use of bandpass image decompositions to avail themselves of multiscale information.

The Daugman approach makes use of a decomposition derived from application of a two-dimensional version of Gabor filters [24, 31] to the image data. Since the Daugman system converts to polar coordinates, (r, θ), during matching, it is convenient to give the filters in a corresponding form as

$$H(r,\theta) = \mathrm{e}^{-\mathrm{i}\omega(\theta-\theta_0)}\mathrm{e}^{-(r-r_0)^2/\alpha^2}\,\mathrm{e}^{-\mathrm{i}(\theta-\theta_0)^2/\beta^2}$$

where α and β co-vary in inverse proportion to ω to generate a set of quadrature pair frequency selective filters, with center locations specified by

(r_0, θ_0). These filters are particularly notable for their ability to achieve good joint localization in the spatial and frequency domains. Further, owing to their quadrature nature, these filters can capture information about local phase. Following the Gabor decomposition, the Daugman approach compresses its representation by retaining only the sign of the convolution profile. For a filter given with bandpass parameters α, β and ω and location (r_0, θ_0) a pair of bits $(h_{\Re}, h_{\Im})$ are generated according to

$$h_{\Re} = 1 \text{ if } \Re\left(\int_\rho \int_\psi e^{-i\omega(\theta_0-\psi)} e^{-(r_0-\rho)^2/\alpha^2} e^{-i(\theta_0-\psi)^2/\beta^2} I(\rho,\psi)\rho \, d\rho \, d\psi\right) \geq 0$$

$$h_{\Re} = 0 \text{ if } \Re\left(\int_\rho \int_\psi e^{-i\omega(\theta_0-\psi)} e^{-(r_0-\rho)^2/\alpha^2} e^{-i(\theta_0-\psi)^2/\beta^2} I(\rho,\psi)\rho \, d\rho \, d\psi\right) < 0$$

$$h_{\Im} = 1 \text{ if } \Im\left(\int_\rho \int_\psi e^{-i\omega(\theta_0-\psi)} e^{-(r_0-\rho)^2/\alpha^2} e^{-i(\theta_0-\psi)^2/\beta^2} I(\rho,\psi)\rho \, d\rho \, d\psi\right) \geq 0$$

$$h_{\Im} = 0 \text{ if } \Im\left(\int_\rho \int_\psi e^{-i\omega(\theta_0-\psi)} e^{-(r_0-\rho)^2/\alpha^2} e^{-i(\theta_0-\psi)^2/\beta^2} I(\rho,\psi)\rho \, d\rho \, d\psi\right) < 0$$

with $\Re(\cdot)$ and $\Im(\cdot)$ capturing the real and imaginary filter outputs, respectively. As originally realized, the parameters $r_0, \theta_0, \alpha, \beta$ and ω are sampled so as to yield a 256 byte signature that serves as the basis for subsequent processing. Subsequent developments augment this code with an equal number of masking bytes that serve to distinguish areas that arise from iris tissue as opposed to artifacts (e.g. specular reflections, eyelashes); how the mask is computed is not specified [16].

The Wildes *et al.* approach makes use of an isotropic bandpass decomposition derived from application of Laplacian of Gaussian (LoG) filters [27, 31] to the image data. The LoG filters can be specified via the form

$$-\frac{1}{\pi\sigma^4}\left(1-\frac{\rho^2}{2\sigma^2}\right)e^{-\rho^2/2\sigma^2}$$

with σ the standard deviation of the Gaussian and ρ the radial distance of a point from the filter's center. In practice, the filtered image is realized as a Laplacian pyramid [8, 31]. This representation is defined procedurally in terms of a cascade of small support Gaussian-like filters. In particular, let $\mathbf{w} = [1\ 4\ 6\ 4\ 1\}]/16$ be a one-dimensional mask and $W = \mathbf{w}^T\mathbf{w}$ be the two-dimensional mask that results from taking the outer product of $\mathbf{w}$ with itself. Given an image of interest, I, the construction of a Laplacian pyramid begins by convolving I with W so as to yield a set of low-pass filtered images g_k according to

$$g_k = (W * g_{k-1})_{\downarrow 2}$$

with $g_0 = I$ and $(\cdot)_{\downarrow 2}$ symbolizing downsampling by a factor of 2 in each image dimension. The kth level of the Laplacian pyramid l_k is formed as the difference between g_k and g_{k+1}, with g_{k+1} expanded before subtraction so that it matches the sampling rate of g_k. The expansion is accomplished by upsampling and interpolation:

$$l_k = g_k - 4W * (g_{k+1})_{\uparrow 2}$$

where $(\cdot)_{\uparrow 2}$ indicates upsampling by a factor of 2 via insertion of a row and column of zeros between each row and column of the original image. The generating kernel W is used as the interpolation filter and the factor of 4 is necessary because $\frac{3}{4}$ of the samples in the image are newly inserted zeros. The resulting Laplacian pyramid, constructed with four levels, serves as the iris signature for subsequent matching. The difference of Gaussians that the construction of this representation entails yields a good approximation to Laplacian of Gaussian filtering [43]. Additionally, it is of note for efficient storage and processing, as lower frequency bands are subsampled successively without loss of information beyond that introduced by the filtering. In implementation, Laplacian pyramid construction follows in a straightforward fashion from its procedural definition.

By retaining only the sign of the Gabor filter output, the representational approach that is used by Daugman yields a remarkably parsimonious representation of an iris. Indeed, a representation with a size of 256 bytes can be accommodated on the magnetic stripe affixed to the back of standard credit/debit cards [7]. In contrast, the Wildes *et al.* representation is derived directly from the filtered image for size on the order of the number of bytes in the iris region of the originally captured image. However, by retaining more of the available iris information the Wildes *et al.* approach might be capable of making finer-grained distinctions between different irises. Alternatively, by retaining more information in the representation, the Wildes *et al.* approach may show superior performance if less information is available in the captured iris image, e.g. due to reduced resolution imaging conditions. Since large-scale studies of iris recognition are currently lacking, it is too early to tell exactly how much information is necessary for adequate discrimination in the face of sizable samples from the human population. In any case, in deriving their representations from bandpass filtering operations, both approaches capitalize on the multiscale structure of the iris. For the sake of illustration, an example multiscale representation of an iris as recovered by the Wildes *et al.* approach is shown in Figure 3.8.

3.4.3 Matching

Iris matching can be understood as a three-stage process. The first stage is concerned with establishing a spatial correspondence between two iris signatures that are to be compared. Given correspondence, the second stage is concerned with quantifying the goodness of match between two iris

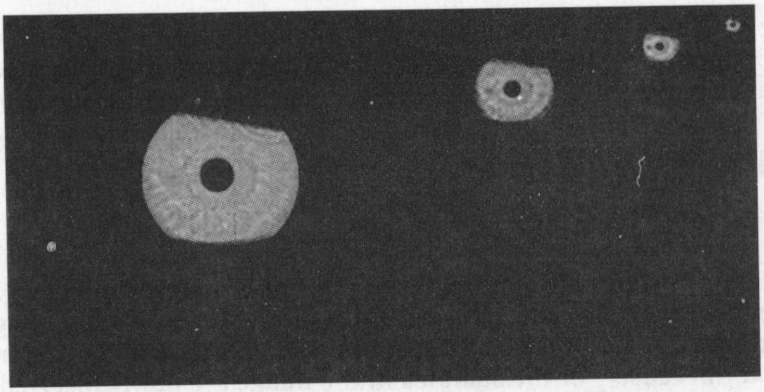

Figure 3.8 Multiscale signature representation for iris matching. Distinctive features of the iris are manifest across a range of spatial scales. Pattern matching is well served by a bandpass decomposition spanning high to low spatial frequency. A compact representation results from successive subsampling of lower frequency bands. The localized iris of Figure 3.7 is shown under such a multiscale representation. R. Wildes, Iris recognition: an emerging biometric technology. *Proceedings of the IEEE*, **85**(9), 1348–1363, 1997 (© 1997 IEEE)

signatures. The third stage is concerned with making a decision about whether or not two signatures derive from the same physical iris, based on the goodness of match. The remainder of this section describes these three stages in detail.

3.4.3.1 Correspondence

In order to make a detailed comparison between two iris signatures it is necessary to establish a precise correspondence between characteristic structures across the pair. Given the combination of required subject participation and the capabilities of sensor platforms currently in use, the key geometric degrees of freedom that must be compensated for in the underlying iris data are shift, scaling and rotation. Shift accounts for offsets of the eye in the plane parallel to the camera's sensor array. Scale accounts for offsets along the camera's optical axis. Rotation accounts for deviation in angular position about the optical axis. Another degree of freedom of potential interest is that of pupil dilation. The size of the pupil varies with the level of ambient illumination, subject arousal and various other influences [1, 17, 50]. As noted in Section 3.2, the details of an iris's pattern can vary with the state of pupil size.

Both the Daugman and Wildes *et al.* approaches compensate for shift, scaling and rotation in the underlying iris data. For both systems, iris localization is charged with isolating an iris in a larger acquired image and thereby essentially accomplishes alignment for image shift. Daugman's system uses radial scaling to compensate for overall size as well as a simple model of pupil variation based on linear stretching. The scaling serves to map Cartesian image coordinates (x, y) to polar image coordinates (r, θ) according to

$$x(r,\theta) = (1-r)x_{\mathrm{p}}(\theta) + rx_{\mathrm{l}}(\theta)$$
$$y(r,\theta) = (1-r)y_{\mathrm{p}}(\theta) + ry_{\mathrm{l}}(\theta)$$

where r lies on $[0, 1]$ and θ is cyclic over $[0, 2\pi]$, while $(x_{\mathrm{p}}(\theta), y_{\mathrm{p}}(\theta))$ and $(x_{\mathrm{l}}(\theta), y_{\mathrm{l}}(\theta))$ are the coordinates of the pupillary and limbic boundaries in the direction θ. Rotation is compensated for by brute force search: explicitly shifting an iris signature in θ by various amounts during matching.

The Wildes *et al.* approach uses an image registration technique to compensate for both scaling and rotation. This approach geometrically projects an image, $I_{\mathrm{a}}(x, y)$, into alignment with a comparison image, $I_{\mathrm{c}}(x, y)$, according to a mapping function $(u(x,y), v(x,y))$ such that, for all (x,y), the image intensity value at $(x,y) - (u(x,y), v(x,y))$ in I_{a} is close to that at (x,y) in I_{c}. More precisely, the mapping function (u, v) is taken to minimize

$$\int_x \int_y (I_{\mathrm{c}}(x,y) - I_{\mathrm{a}}(x-u, y-v))^2 \,\mathrm{d}x\mathrm{d}y$$

while being constrained to capture a similarity transformation of image coordinates (x,y) to (x', y'), i.e.

$$\begin{pmatrix} x' \\ y' \end{pmatrix} = \begin{pmatrix} x \\ y \end{pmatrix} - sR(\phi)\begin{pmatrix} x \\ y \end{pmatrix}$$

with s a scaling factor and $R(\phi)$ a matrix representing rotation by ϕ. In implementation, given a pair of iris images, I_{a} and I_{c}, the warping parameters, s and ϕ, are recovered via an iterative minimization procedure [2]. As originally implemented, this approach did not compensate for the effects of pupil dilation. Instead, the fact that a controlled (visible) illuminant was always in place during image capture was relied upon to bring pupils to a single size for an individual operator (ignoring effects of arousal etc.).

As with much of the processing that the two approaches under consideration perform, the methods for establishing correspondences between two irises seem to be adequate for controlled assessment scenarios. Once again, however, more sophisticated methods may prove to be necessary in more relaxed scenarios. For example, a simple linear stretching model of pupil dilation does not capture the complex physical nature of this process, e.g. the coiling of blood vessels and the arching of stromal fibers [48, 71]. Similarly, more complicated global geometric compensations will be necessary if full perspective distortions (e.g. foreshortening) become significant.

3.4.3.2 Match goodness

Given the fairly controlled image acquisitions that currently are enforced in iris recognition systems and the collateral success of extant correspondence mechanisms, an appropriate match metric can be based on direct pointwise comparisons between primitives in the corresponding signature representations. The Daugman approach quantifies this matter by computing the percentage of mismatched bits between a pair of iris representations, i.e. the normalized Hamming distance [32]. Letting A and B be two iris signatures to be compared, this quantity can be calculated as

$$\frac{1}{2048}\sum_{j=1}^{j=2048} A_j \oplus B_j$$

with subscript j indexing bit position and $\oplus$ denoting the exclusive-OR operator. (The exclusive-OR is a Boolean operator that equals 1 if and only if its operands differ.) The result of this computation is then used as the goodness of match, with smaller values indicating better matches. The exclusive-OR of corresponding bits in the acquired and database iris representations can be calculated with negligible computational expense. This allows the system to compare an acquired representation with interesting numbers of database entries (e.g. a raw comparison rate of approximately 10^5 per second using a 300 MHz processor.) As implemented, this comparison rate is exploited to yield a brute force solution not just to verification, but also to identification, i.e. sequential examination of each record in moderate size databases. While this search ability is impressive, identification in the presence of significantly larger databases might require a cleverer indexing strategy.

The Wildes *et al.* system employs a somewhat more elaborate procedure to quantify the goodness of match. The approach is based on normalized correlation between two signatures (i.e. pyramid representations) of interest. In discrete form, normalized correlation can be defined in the following fashion. Let $p_1[i,j]$ and $p_2[i,j]$ be two image arrays of size $n \times m$. Further, let $\mu_1 = (1/nm)\Sigma_{i=1}^{n}\Sigma_{j=1}^{m} p_1[i,j]$ and $\sigma_1 = [(1/nm)\Sigma_{i=1}^{n}\Sigma_{j=1}^{m}(p_1[i,j]-\mu_1)^2]^{1/2}$ be the mean and standard deviation for the intensities of p_1, respectively. Also, let μ_2 and σ_2 be similarly defined with reference to p_2. Then the normalized correlation between p_1 and p_2 can be defined as

$$\frac{\sum_{i=1}^{n}\sum_{j=1}^{m}(p_1[i,j]-\mu_1)(p_2[i,j]-\mu_2)}{nm\sigma_1\sigma_2}.$$

Normalized correlation captures the same type of information as standard correlation (i.e. integrated similarity of corresponding points in the regions); however, it also accounts for local variations in image intensity that corrupt standard correlation [60]. This robustness comes about as the mean intensities are subtracted in the numerator of the correlation ratio, while the standard deviations appear in the denominator. In implementation, the correlations are performed discretely over small blocks of pixels (8×8) in each spatial frequency band of the Laplacian pyramid representations. A goodness of match is subsequently derived for each band by combining the block correlation values via the median statistic. Blocking combined with the median operation allows for local adjustments of matching and a degree of outlier detection and thereby provides robustness against mismatches due to noise, misalignment and occlusion (e.g. a stray eyelash). As implemented, this method has been applied to the verification task only.

3.4.3.3 Decision

The final subtask of matching is to evaluate the goodness of match values to make a final judgement as to whether two signatures under consideration do (authentic) or do not (impostor) derive from the same physical iris. In the Daugman approach, this amounts to choosing a separation point in the space of (normalized) Hamming distances between iris signatures: Distances smaller than the separation point will be taken as indicative of authentics; those larger will be taken as indicative of impostors[3]. An appeal to statistical decision theory [39, 58] is made in an attempt to provide a principled approach to selecting the separation point. There, given appropriate distributions for the two events to be distinguished (i.e. authentic vs. impostor), the optimal decision strategy is defined by taking the separation as the point at which the two distributions cross-over. This decision strategy is optimal in the sense that it leads to equal probability of false accept and false reject errors. (Of course, even with a theoretically "optimal" decision point in hand, one is free to choose either a more conservative or more liberal criterion according to the needs of a given installation.) In order to calculate the cross-over point, sample populations of impostors and authentics were each fit with parametrically defined distributions. This was necessary since no data, i.e. Hamming distances, were observed in the cross-over region. Binomial distributions [19] were used for the empirical fits. A binomial distribution is given as

$$p(k) = \binom{n}{k} P^k (1-P)^{n-k}$$

where

$$\binom{n}{k} = \frac{n!}{(n-k)!k!}$$

is the number of k combinations of n distinguishable items. This formula gives the probability of k successes in n independent Bernoulli trials. A Bernoulli trial, in turn, is defined to generate an experimental value of a discrete random variable v according to the distribution

---

3 As documented, both the Daugman and Wildes *et al.* approaches remain agnostic about how to deal with cases that lie at their separation points, where the goodness of match is supposed to be equally supportive of deciding authentic or impostor. In empirical evaluations, it appears that neither system has been confronted with this situation (see Section 3.5).

$$p_v(v_0) = \begin{cases} 1-P & v_0 = 0 \\ P & v_0 = 1 \\ 0 & \text{otherwise} \end{cases}$$

with an outcome of $v = 1$ taken as a success and an outcome of $v = 0$ taken as a failure. The use of a binomial distribution was justified for the case of impostor matches based on the distinctiveness of different irises. That is, the matching of bits between a pair of representations from different irises was taken to be a series of Bernoulli trials. However, not all of the bit matches were taken as independent due to the presence of inherent correlations in iris structure as will as correlations introduced during processing. Significantly, no such justification was given for the modeling of the authentics.

In the Wildes *et al.* approach, the decision making process must combine the four goodness of match measurements that are calculated by the previous stage of processing (i.e. one for each pass band in the Laplacian pyramid representation that comprises a signature) into a single accept/reject judgement. Here, recourse is made to standard techniques from pattern classification. In particular, the notion that is appealed to is to combine the values in a fashion so that the variance within a class of iris data is minimized, while the variance between different classes of iris data is maximized. The linear function that provides such a solution is well known and is given by Fisher's Linear Discriminant [20, 22]. This function can be defined in the following fashion. Let there be n d-dimensional samples $\mathbf{q}$, n_{a} of which are from a set $\mathcal{A}$ and n_i of which are from a set $\mathcal{I}$. For example, in the current application each sample corresponds to a set of multiscale goodness of match measurements, while the classes to be distinguished are the authentics and impostors. Fisher's linear discriminant defines a weight vector ω such that the ratio of between class variance to within class variance is maximized for the transformed samples $\omega^{\mathrm{T}}\mathbf{q}$. To formalize this notion, let $\mu_{\mathrm{a}} = (\Sigma_{\mathbf{q}\in\mathcal{A}}\mathbf{q})/n_i$ be the d-dimensional mean for $\mathbf{q} \in \mathcal{A}$ and similarly for μ_i. A measure of variance within a class of data can be given in terms of a scatter matrix with the form

$$S_{\mathrm{a}} = \sum_{\mathbf{q}\in\mathcal{A}} (q-\mu_{\mathrm{a}})(q-\mu_{\mathrm{a}})^{\mathrm{T}}$$

for $\mathcal{A}$ and with S_i similarly defined for $\mathcal{I}$. The total within class scatter is given as $S_{\mathrm{w}} = S_{\mathrm{a}} + S_i$. A corresponding measure of variance between classes can be defined in terms of the scatter matrix

$$S_{\mathrm{b}} = (\mu_{\mathrm{a}} - \mu_i)(\mu_{\mathrm{a}} - \mu_i)^{\mathrm{T}}$$

With the preceding definitions in hand, the expression

$$\frac{\omega^{\mathrm{T}} S_{\mathrm{b}} \omega}{\omega^{\mathrm{T}} S_{\mathrm{w}} \omega}$$

describes the ratio of between to within class variance of the transformed samples $\omega\mathbf{q}$. Finally, the use of a little calculus and linear algebra leads to the conclusion that the ω which maximizes this ratio is given as

$$\omega = S_{\mathrm{w}}^{-1}(\mu_{\mathrm{a}} - \mu_i)$$

Interestingly, S_{b} does not appear in this formula for ω since it simply scales the overall result, without otherwise changing the separation. To apply this discriminant function to classification, a separation point must be defined in its range: values above this point will be taken as derived from class $\mathcal{A}$; values below this point will be taken as derived from class $\mathcal{I}$. In the current application, the separation point is taken as the midpoint between the transformed means of the samples from $\mathcal{A}$ and $\mathcal{I}$, i.e. $\frac{1}{2}\omega^{\mathrm{T}}(\mu_{\mathrm{a}} + \mu_i)$. If the probabilities of the measurements given either class are normally distributed and have equal variance, (i.e. $p(\mathbf{q}|\mathcal{A}) = (1/\sqrt{2\pi\sigma})\exp[-(|\mathbf{q} - \mu_{\mathrm{a}}|^2)/2\sigma^2]$ with σ^2 the variance [19], and similarly for $\mathcal{I}$), then this choice of separation point can be shown to be optimal (i.e. equal probability of false accept and false reject errors). However, it is heuristic for the case of iris match measurements where these assumptions are not known to hold. In implementation, the discriminant was defined empirically based on a set of iris training data.

While both of the decision methods have performed well to date, the underlying data modeling assumptions need to be rigorously evaluated against a larger corpus of data. Both of the methods rely on the assumptions that the impostor and authentic populations can each be modeled with single distributions. A basic tenet of iris recognition is that different irises are highly distinct. Therefore, it is reasonable to view the distribution of impostors as varying about a central tendency dictated by some notion of independence, e.g. a 50% chance of individual bits matching in the Daugman approach or low correlation values for the Wildes *et al.* approach. Indeed, empirically this seems to be the case for both approaches. However, there is no such theoretical underpinning for modeling the authentics with a single distribution. In fact, one might argue that authentics would be best modeled by a mixture of distributions [63], perhaps even one distribution for repeat occurrences of each iris. From an empirical point of view, it is of concern that the current decision strategies are derived from rather small samples of the population (i.e. of the order of 10^2 or 10^3). This matter is exacerbated by the fact that little data has been reported in the cross-over regions for the decisions, exactly the points of most concern. To properly resolve these issues it will be necessary to consider a larger sample of iris data than the current systems have employed.

3.4.3.4 A caveat

Both of the reviewed approaches to matching are based on methods that are closely tied to the recorded image intensities. More abstract representations may be necessary to deal with greater variation in the appearance of any one iris, e.g. as might result from more relaxed image acquisition. One way to deal with greater variation would be to extract and match sets of features that are expected to be more robust to photometric and geometric distortions in the acquired images. In particular, features that bear a closer and more explicit relationship to physical structures of the iris might exhibit the desired behavior. For example, preliminary results indicate that multiscale blob matching could be valuable in this regard [40, 66]. This approach relies on the correspondence between the dark and light blob structures that typically are apparent in iris images and iris structures such as crypts, freckles, nevi and striations. If current methods in iris pattern matching begin to break down in future applications, then such symbolic approaches will deserve consideration. However, it is worth noting that the added robustness that these approaches might yield will most likely come with increased computational expense.

3.5 Systems and performance

Following on the foregoing discussion, the main functional components of extant iris recognition systems consist of image acquisition and signature representation/matching; see Figure 3.9. Both the Daugman and Wildes *et al.* approaches have been instantiated in working systems and have been awarded US patents [15, 68, 69]. Initial laboratory versions of both systems have been realized with commercially available hardware components (optics, illumination, computer workstation to support image processing) and custom image processing software. Similarly, the described approach to active image acquistion has been instantiated in a working system and awarded a US patent [10]. An initial laboratory version of this system was realized with commercially available hardware components, bolstered with a special-purpose image processing accelerator board and custom image processing software. This system was further refined and packaged so that it could be subjected to field trials [46]. Finally, the Daugman iris recognition approach has also been introduced as a commercial product, first through IriScan and subsequently through Iridian [56]. This system embodies largely the same approach as that of the laboratory system, albeit with further optimization and use of special-purpose hardware for a more compact product.

Two reports of laboratory-based experiments with the Daugman system are available. In the first experiment [14], 592 irises were represented as derived from 323 persons. An average of approximately 3 images were taken of each iris. (The time lags involved in repeat captures of a single iris were not reported.) The irises involved spanned the range of common iris colors: blue, hazel, green and brown. This preparation allowed for evaluation of authentics and impostors across a representative range of iris pigmentations and with some passage of time. In the face of this data set, the system exhibited no false accepts and no false rejects. In an attempt to analyze the data from this experiment, binomial distributions were fit to both the observed

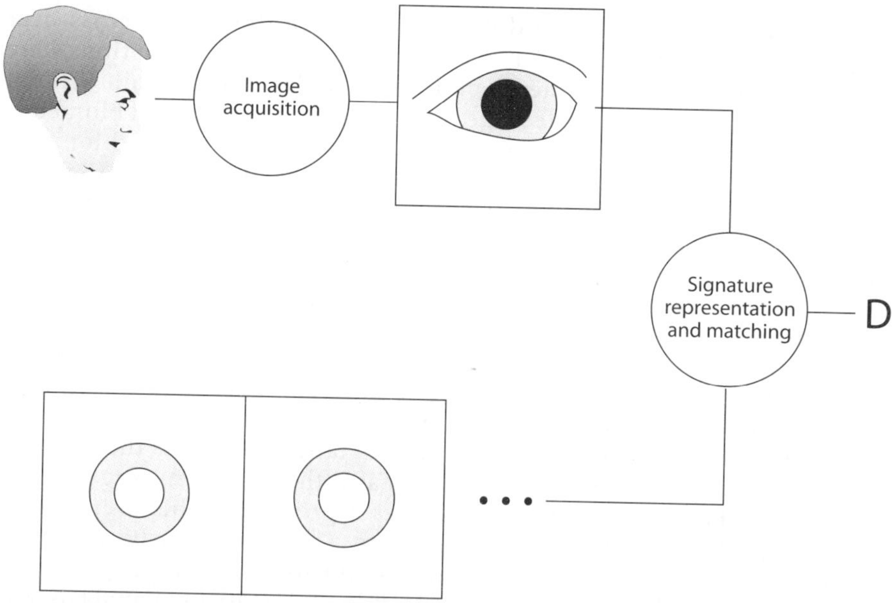

Figure 3.9 Schematic diagram of iris recognition system. Given a subject to be evaluated (left of upper row) relative to a database of iris records (left of lower row), recognition proceeds in two major steps. The first step is image acquisition, which yields an image of the subject's eye region. The second step is signature representation and matching, which produces a decision, D. For verification, the decision is a yes/no response relative to a particular pre-specified database entry; for identification, the decision is a record (possibly null) that has been indexed relative to a larger set of entries.

authentic and impostor scores, i.e. as previously described during the discussion of matching. The fits were used to calculate the cross-over error rate for false accepts and false rejects as 1 in 131,000. Further, d', a measure of the separability of two distributions used in statistical decision theory [39, 58], calculated as the absolute difference of the means, divided by a conjoint measure of the standard deviations, was found to be 8.4.

In the second experiment with the Daugman system [16], 2,150 iris images were compared, including 10 images of the same iris for 70 subjects. In this case the images were acquired with a variety of different image acquisition platforms. Other details of the experiment were unspecified (e.g. nature of image acquisition platforms, time lag between repeat acquisitions). Here it was found that d' was decreased to 7.3. It also was noted that for a subset of iris images acquired under "ideal" conditions (same camera with fixed zoom, same illumination, same subject to sensor distance), it was possible to increase d' to 14.1. Interpretation of the reported statistics requires caution. As noted during the discussion of matching, justification for fitting the observed data with binomial distributions for calculating cross-over error rates is problematic. From a theoretical point of view, it is not clear why a binomial distribution is appropriate for the case of authentics. From an empirical point of view, the fits are based on small samples of the populations and data is lacking in the critical cross-over region. Similarly, the calculation of d' assumed that the distributions were well characterized by their means and standard deviations, again without appropriate justification. Indeed, general theoretical analyses of biometric data suggests that they are not well characterized in this fashion, e.g. authentic distributions are typically multimodal [63]. With regard to visual inspection of the particular distributions at hand, it appears that at the very least they exhibit a pronounced skew. Nevertheless, it is worth noting that for all cases it was possible to select empirically a single decision point that allowed perfect separation of the authentic and impostor distributions.

The Wildes *et al.* laboratory system also has been the subject of empirical evaluation [65]. In this study, a total of 60 different irises were represented as derived from 40 persons. For each iris 10 images were captured: 5 at an initial session and 5 approximately 1 month latter. Of note is the fact that this sample included identical twins. Again, the common range of iris colors (blue, hazel, green and brown) was represented. This preparation allowed for the same types of comparison as the previously described experiments. There were no observed false positives or false negatives in the evaluation of this corpus of data. In this case, statistical analysis was eschewed owing to the small sample size. However, at a qualitative level, the data for authentics and impostors were well separated. In subjective reports, subjects found the system to be unobjectionable.

The laboratory version of the active vision approach to iris image acquisition has been evaluated in terms of its ability to support iris recognition by using it as the image acquisition "front end" to the Daugman approach to signature representation and matching. The details of these studies are less well documented; nevertheless, they are interesting to review as corroborating evidence of the efficacy of iris recognition, especially as the images are acquired while making fewer demands on the subject. In one study [26], iris images were acquired from 618 subjects on two occasions; once for enrollment and once for verification (time lag unspecified). The verification was 98.9% successful, with all failures coming about as false rejections. In all cases, evaluation was complete within 10 seconds. The failures were reported as being due to specular reflections from the eye/eyewear or, in one case, the subject being outside the capture volume of the apparatus. The field test prototype of this approach has also been the subject of tests, both in the lab and in the field [46]. Laboratory tests (with unspecified number of subjects and other details) led to no false accepts and a false reject rate of approximately 0.5%. Field trials wherein the system was used by the Nationwide Building Society in Swindon, UK, ran for six months with over 1,000 participants. Unfortunately, no quantitative data on recognition accuracy was reported; however, user acceptance was reported to be above 90%.

All of the tests described so far were conducted by iris recognition system developers. Two additional tests of iris recognition have been conducted by independent evaluation teams. Both of these tests have considered commercial iris recognition systems marketed by IriScan/Iridian. In the first of these tests, a preproduction system was evaluated [6]. In this study the system was installed in a public space at Sandia National Laboratories, USA. Subjects consisted of volunteers from the Sandia community. The study was conducted in two phases. In the first phase, 199 irises were represented as derived from 122 people. Following enrollment the subjects made a total of 878 attempts to use the system in identification mode over a period of 8 days. Of these attempts, 89 false rejects were recorded; however, for 47 of these cases the subject made a retry and all but 16 of these were accepted. All of these errors were traced to either reflections from eyewear that obscured the iris or user difficulty (e.g. difficulty in self-positioning). No false accepts were recorded. In the second phase, 96 of the people involved in the first phase attempted an identification relative to a database with 403 entries, none of which corresponded to the subjects in question. Once again, no false accepts were recorded. In subjective evaluation, subjects found the system generally unobjectionable; however, some reports of discomfort with the illuminant were reported.

In the second independent test, a commercially available hardware system was evaluated using algorithms specially modified to support testing [42]. This test was conducted in the UK at the National Physics Laboratory and also included evaluations of commercial systems for face, fingerprint, hand, vein and voice recognition. The general test scenario was that of verification in a normal office environment (albeit with controlled, near-constant ambient illumination), with cooperative non-habituated users. The evaluation used 200 volunteers from the test site, extended over a period of three months and was conducted in accordance with accepted testing standards in the biometrics community [61]. The "typical" separation between enrollment and verification was one to two months. Various statistics were compiled for all systems, including failure to enrol rate, failure to acquire rates, false match vs. false non-match rates and user throughput (defined in terms of time differences logged between consecutive transactions). In terms of failure to enrol, iris recognition achieved a 0.5% rate (third worst among systems evaluated) as it failed to enrol a blind eye. No failures to acquire were logged for iris recognition. Because the images were selected for storage based on a pre-set matching score threshold, it was impossible to plot an ROC for false match vs. false non-match rates. However, using the provided threshold no false matches were observed in approximately two million cross-comparisons. At this same threshold approximately 2% false non-matches were observed. To put these numbers somewhat in perspective, when an evaluated fingerprint system had its decision threshold set to achieve just one false match (< 0.001%), its false non-match rate was approximately 7%: apart from iris recognition this was the lowest false non-match rate for a single false match rate across all tested systems. Finally, the median throughput for iris recognition was 10 seconds, a number comparable to that of other systems.

Overall, the two iris recognition systems that are being used for illustration have performed remarkably well under preliminary testing. Empirical tests of other approaches to automated iris recognition also are generally positive [5, 38, 53, 59, 73]. However, given that experiments were conducted on samples of the order of 10^2, or in one case 10^3, (i.e. number of irises in the experiments) from a population on the order of 10^{10} (i.e. total number of human irises), one must be cautious in the extrapolation of these results. Nevertheless, the results speak in favor of iris recognition as a promising biometric technology.

3.6 Future directions

Future directions for iris recognition can be thought of in terms of two broad categories of endeavors: technology development and scientific inquiry into biometric-based human identification. In terms of technology development, two major directions for future research are present. One of these directions comes from consideration of what can be accomplished if one is willing to accept iris image capture under relatively constrained situations, e.g. that required by the described passive acquisition systems. Under such restrictions, further developments could be focused on yielding ever more compact systems that can be easily incorporated into consumer products where access control is desired (e.g. automobiles, personal computers, various handheld devices). While requiring careful attention to engineering detail (e.g. in miniaturization of optics, algorithmic optimization), there should be no outstanding obstacles along this path. Preliminary results along these lines already have been reported [46].

The second major direction for technology development arises as one attempts to push the operational envelop of iris recognition to include more unconstrained acquisition scenarios. Can iris recognition be performed at greater subject to sensor distances while remaining unobtrusive? How much subject motion can be tolerated during image capture? Can performance be made more robust to uncontrolled ambient illumination? Is it possible for iris recognition to be accomplished covertly, i.e. with the subject totally unaware that they are under observation? The development of systems that can respond to these queries will entail consideration of computer vision and image processing techniques for optical and illumination design, image stabilization, target detection, tracking, image enhancement and control. To some extent, extant technology can be exploited to marshal initial attacks along relevant paths. It is likely, however, that additional basic research in computer vision and image processing will be required to fully respond to the challenges at hand. As a step along these directions, Figure 3.10 shows an iris image that was captured at 10 m subject to sensor distance. Here, a commercially available video camera with 1 m focal length lens was coupled with a semi-collimated near infrared illumination source to yield an image with resolution and contrast that has proven successful to drive iris recognition at closer distances.

A complementary direction for future research comes about if one thinks in terms of the science of biometrics-based human identification. At the most basic level, little is known about the intrinsic information content of the human iris in support of human identification. While the general

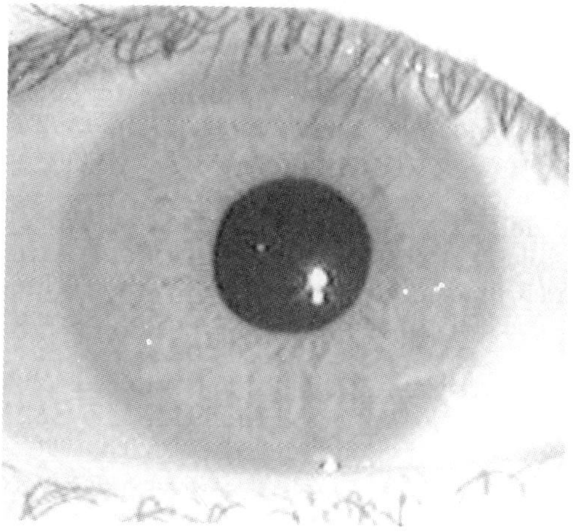

Figure 3.10 Toward iris recognition at a distance. An interesting direction for future research in iris recognition is to relax constraints observed by extant systems. As a step in this direction, an iris image captured at 10 m subject to sensor distance is shown.

anatomical basis of iris features are known (as described in Section 3.2), studies still need to be conducted to reveal the discriminatory information content of these structures apropos human identification. In particular, at the level of individual algorithmic approaches to iris signature representation and matching, studies need to be performed that reveal exactly what information is required for recognition. For example, one could attempt to construct something akin to a modulation transfer function (MTF) for a given approach that specifies the minimal image requirements (e.g. in terms of spatial frequency content) that are required to support recognition at various levels of performance. A significant number of human irises will need to be sampled to produce such characterizations.

At a more operational level of performance analysis, studies of iris recognition systems need to be performed wherein details of acquisition are systematically manipulated, documented and reported. Parameters of interest include, geometric and photometric aspects of the experimental stage (e.g. MTF of the optical platform, level of illumination at the iris, subject to sensor distance, subject attitude relative to sensor, arrangement of ambient illuminants), length of time monitored and temporal lag between template construction and recognition attempt. Similarly, details of captured irises and relevant personal accessories need to be properly documented in these same studies (e.g. eye color, eyewear). Along these lines, it is important that recognition results derived from executing iris recognition algorithms on this data be reported in a meaningful fashion, i.e. through observation of accepted standards for reporting recognition rates in the biometrics community, such as Receiver Operator Characteristics (ROCs) [63] and Rank Order Analyses [51]. More generally, if iris recognition is to make solid scientific advances, then future tests of iris recognition systems must conform to accepted practices in the evaluation of biometric devices [55, 61, 63].

References

[1] F. H. Adler, *Physiology of the Eye*. C. V. Mosby, Saint Louis, Missouri, 1965.

[2] J. R. Bergen, P. Anandan, K. Hanna and R. Hingorani, Hierarchical model-based motion estimation. *European Conference on Computer Vision 1991*, pp. 5–10.

[3] L. Berggren, Iridology. A critical review *Acta Ophthalmoligica*, **63**, 1–8, 1985.

[4] A. Bertillon, La couleur do l'iris. *Revue Scientifique*, **63**, 65–73, 1885.

[5] W. W. Boles and B. Boashash, A human identication technique using images of the iris and wavelet transform. *IEEE Trans. Signal Processing*, **46**(4), 1185–1188, 1998.

[6] F. Bouchier, J. S. Ahrens and G. Wells, Laboratory evaluation of the IriScan prototype biometric identifier. *Technical Report SAND96-1033*, Sandia National Laboratories, Albuquerque, New Mexico, 1996.

[7] R. Bright, *Smartcards: Principles, Practice, Applications*. Ellis Horwood, Chichester, 1988.

[8] P. J. Burt and E. Adelson, The Laplacian pyramid as a compact image code. *IEEE Trans. Commun.*, **31**(4), 532–540, 1983.

[9] T. A. Camus and R. P. Wildes, Reliable and fast eye finding in close-up images. *Proceedings of the IAPR International Conference on Pattern Recognition*, 2002, in press.

[10] T. A. Chmielewski, G. A. Vonho and M. Negin, Compact image steering device. *US Patent 5,717,512*, 1998.

[11] W. T. Cope, M. L. Wolbarsht and B. S. Yamanishi, The corneal polarization cross. *J. Opt. Soc. Am. A, Opt. Image. Sci.*, **68**(6), 1139–1140, 1978.

[12] J. G. Daugman, Biometric signature security system. *Technical Report*, Harvard University, Cambridge, Massachusetts, 1990.

[13] J. G. Daugman, High confidence personal identification by rapid video analysis of iris texture. *Proceedings of the IEEE International Carnahan Conference on Security Technology*, 1992, pp. 1–11.

[14] J. G. Daugman, High confidence visual recognition of persons by a test of statistical independence. *IEEE Trans. Pattern Analysis and Machine Intelligence*, **15**(11), 1148–1161, 1993.

[15]. J. G. Daugman, Biometric personal identification system based on iris analysis. *US Patent 5,291,560*, 1994.

[16] J. G. Daugman, Statistical richness of visual phase information: update on recognizing persons by iris patterns. *Int. J. Computer Vision*, **45**(1), 25–38, 2001.

[17] H. Davson, *The Physiology of the Eye*, 2nd edn. Little, Brown and Company, 1963.

[18] M. Della Vecchia, T. Chmielewski, T. Camus, M. Salganico and M. Negin, Methodology and apparatus for using the human iris as a robust biometric. *Proceedings of the SPIE Conference on Ophthalmic Technologies*, 1998.

[19] A. W. Drake, *Fundamentals of Applied Probability Theory*. McGraw-Hill, New York, 1986.

[20] R. O. Duda, P. E. Hart and D. G. Stork, *Pattern Classification*, 2nd edn. Wiley, New York, 2001.

[21] O. Faugeras, *Three-Dimensional Computer Vision*. MIT Press, Cambridge, MA, 1993.

[22] R. A. Fisher, The use of multiple measurements in taxonomic problems. *Annals of Eugenics*, 7(2), 179–188, 1936.

[23] L. Flom and A. Sar, *Iris recognition system*. US Patent 4,641,349, 1987.

[24] D. Gabor, Theory of communication. *J. Inst. Electrical Engineers*, **93**, 429–459, 1946.
[25] S. B. Grimes and I. Kershner, *Never Say Never Again*. Warner Brothers, 1983.
[26] K. Hanna, R. Mandelbaum, L. Wixson, D. Mishra and V. Paragano, A system for non-intrusive human iris acquisition. *Proceedings of the IAPR Workshop on Machine Vision Applications*, 1996, pp. 259–263.
[27] B. K. P. Horn, *Robot Vision*. Massachusetts Institute of Technology Press, Cambridge, Massachusetts, 1986.
[28] P. Horowitz and W. Hill, *The Art of Electronics*, 2nd edn. Cambridge University Press, New York, 1988.
[29] P. V. C. Hough, Method and means for recognizing complex patterns. *US Patent 3,069,654*, 1962.
[30] J. Illingworth and J. Kittler, A survey of the Hough transform. *Computer Vision Graphics and Image Processing*, **44**, 87–116, 1988.
[31] B. Jahne and H. Hausbecker (eds), *Computer Vision and Applications: A Guide for Students and Practitioners*. Academic Press, London, 2000.
[32] N. S. Jayant and P. Noll, *Digital Coding of Waveforms*. Prentice Hall, Engelwood Cliffs, NJ, 1984.
[33] F. A. Jenkins and H. E. White, *Fundamentals of Optics*. Macmillan, New York, 1976.
[34] R. G. Johnson, Can iris patterns be used to identify people? *Chemical and Laser Sciences Division LAPR-12331-PR*, Los Alamos National Laboratory, Los Alamos, California, 1991.
[35] M. Kass, A. Witkin and D. Terzopoulos, Snakes: active contour models. *Int. J. Computer Vision*, **1**(4), 321–331, 1987.
[36] A. L. Kroeber, *Anthropology*. Harcourt, Brace, Jovanovich, New York, 1948.
[37] P. C. Kronfeld, The gross anatomy and embryology of the eye. In H. Davson (ed.) *The Eye*, Vol 1., pp. 1–66. Academic Press, London, 1968.
[38] L. Ma, Y. Wang and T. Tan, Iris recognition using circular symmetric filters. *Proceedings of the IAPR International Conference on Pattern Recognition*, 2002, in press.
[39] N. A. Macmillan and C. D. Creelman, *Detection Theory: A User's Guide*. Cambridge University Press, Cambridge, 1991.
[40] A Malickas, Personal communication, 1994
[41] I. Mann, *The Development of the Human Eye*. Grune and Stratton, 1950.
[42] T. Mansfield, G. Kelly, D. Chandler and J. Kane, Biometric product testing final report. Centre for Mathematics and Scientfiic Computing, *CESG/BWG X92A/4009309*, National Physical Laboratory, Middlesex, 2001.
[43] D. Marr, *Vision*. W. H. Freeman & Company, New York, 1982.
[44] R. Matthes and D. Sliney (eds), *Measurements of Optical Radiation Hazards*. International Commission on Non-Ionizing Radiation Protection, New York, 1998.
[45] D. Miller, *Ophthalmology*. Houghton Mifflin, Boston, Massachusetts, 1979.
[46] M. Negin, T. Chmielewski, M. Salganico, T. Camus, U. von Seelen and G. Zhang, An iris biometric system for public and personal use. *IEEE Computer*, February, 70–75, 2000.
[47] F. W. Newell, *Ophthalmology Principles and Practice*, 7th edn. C. V. Mosby, Saint Louis, Missouri, 1991.
[48] D. A. Newsome and I. E. Loewenfeld, Iris mechanics II: Influence of pupil size on the details of iris structure. *Am. J. Ophthalmol.*, **71**(2), 553–573, 1971.
[49] G. Olivier, *Practical Anthropology*, C. C. Thomas, Springfield, Illinois, 1969.

[50] C. W. Oyster, *The Human Eye Structure and Function*. Sinauer Associates, New York, 1999.
[51] P. J. Phillips, H. J. Moon, S. A. Risvi and P. J. Rauss, The FERET evaluation methodology for face recognition algorithms. *IEEE Trans. Pattern Analysis and Machine Intelligence*, **10**(22), 137–143, 2000.
[52] W. K. Pratt, *Digitial Image Processing*, 2nd edn. Wiley, New York, 1991.
[53] R. Sanchez-Reillo, C. Sanchez-Avila and J. Martin-Pereda, Minimal template size for iris recognition. *Proceedings of the IEEE International Conference on BMES/EMBS*, p. 792, 1999.
[54] J. A. Sethian, *Level Set Methods and Fast Marching Methods*. Cambridge University Press, Cambridge, 1999.
[55] W. Shen, M. Suretta and R. Khanna, Evaluation of automated biometrics-based identification and verification systems. Proc. IEEE, **85**(9), 1464–1479, 1997.
[56] J. E. Siedlarz, Iris: more detailed than a fingerprint. *IEEE Spectrum*, February, 27, 1994.
[57] J. Stotz, Personal communication, 1994.
[58] W. P. Tanner and J. A. Swets, A decision-making theory of visual detection. *Psychological Review*, **61**, 401–409, 1954.
[59] C. Tisse, L. Martin, L. Torres and M. Robert, Person identification technique using human iris recognition. *Proceedings of the International Conference Vision Interface*, 2002, pp. 294–299.
[60] E. Trucco and A. Verri, *Introductory Techniques for 3-D Computer Vision*. Prentice Hall, Upper Saddle River, NJ, 1998.
[61] UK Government Biometrics Working Group, Best practices in testing and reporting of biometric devices. Available online at `http://www.cesg.gov.uk/site/ast/biometrics/media/BestPractice.pdf`.
[62] H. P. Wasserman, *Ethnic Pigmentation*. Elsevier Publishing, New York, 1974.
[63] J. L. Wayman, Technical testing and evaluation of biometric identification devices. In A. Jain, R. Bolle and S. Pankanti (eds), *Biometrics Personal Identification in Networked Society*, pp. 345–368. Kluwer Academic Publishers, Norwell, MA, 1999.
[64] R. P. Wildes, Iris recognition: an emerging biometric technology. *Proc. IEEE*, **85**(9), 1348–1363, 1997.
[65] R. P. Wildes, J. C. Asmuth, G. L. Green, S. C. Hsu, R. J. Kolczynski, J. R. Matey and S. E. McBride, A machine vision system for iris recognition. *Machine Vision and Applications*, **9**, 1–8, 1996.
[66] R. P. Wildes, J. C. Asmuth, G. L. Green, S. C. Hsu, R. J. Kolczynski, J. R. Matey and S. E. McBride, Iris recognition for security access control: Final report. *Technical Report*, National Information Display Laboratory, Princeton, New Jersey, 1992.
[67] R. P. Wildes, J. C. Asmuth, G. L. Green, S. C. Hsu, R. J. Kolczynski, J. R. Matey and S. E. McBride, A system for automated iris recognition. *Proceedings of the IEEE Workshop on Applications of Computer Vision*, 1994, pp. 121–128.
[68] R. P. Wildes, J. C. Asmuth, S. C. Hsu, R. J. Kolczynski, J. R. Matey and S. E. McBride, Automated non-invasive iris recognition system and method. *US Patent 5,572,596*, 1996.
[69] R. P. Wildes, J. C. Asmuth, S. C. Hsu, R. J. Kolczynski, J. R. Matey and S. E. McBride, Automated non-invasive iris recognition system and method II. *US Patent 5,751,836*, 1998.
[70] R. S. Worrall, Iridology: diagnosis or delusion. *The Skeptical Inquirer* Spring, 23–35, 1983.

[71] H. J. Wyatt, A minimum-wear-and-tear meshwork for the iris. *Vision Research*, **40**, 2167–2176, 2000.
[72] A. L. Yuille, P. Hallinan and D. Cohen, Feature extraction from faces using deformable templates, *Int. J. Computer Vision*, **8**(2), 99–112, 1992.
[73] Y. Zhu, T. Tan and Y. Wang, Biometric personal identification based on iris patterns. *Proceedings of the IAPR International Conference on Pattern Recognition*, 2000, pp. 805–808, 2000.

4 Face Recognition

Chengjun Liu and Harry Wechsler

4.1 Introduction

Face recognition falls into the broadly defined area of biometrics, which is concerned with the verification and recognition of a person's identity by means of unique appearance or behavioral characteristics. Appearance characteristics include hand, fingerprint, eye, iris, retina and face, while behavioral characteristics include signature, voice, keystroke and grip. Automated fingerprint recognition, speaker and speech recognition, and iris and retina recognition are all examples of "active" biometric tasks. Face recognition, however, is usually "passive", as it does not require people's cooperation to look into an iris scanner, to place their hands on a fingerprint reader, or to speak to a close-by microphone. The unobtrusive nature of face recognition makes it more suitable for wide range surveillance and security applications. In particular, an automated face recognition system is capable of capturing face images from a distance using a video camera, and the face recognition algorithms can process the data captured: detect, track and finally recognize people sought, such as terrorists or drug traffickers.

Face recognition involves computer recognition of personal identity based on geometric or statistical features derived from face images [11, 12, 18, 69, 77]. Even though humans can detect and identify faces in a scene with little or no effort, building an automated system that accomplishes such objectives is very challenging. The challenges are even more profound when one considers the large variations in the visual stimulus due to illumination conditions, viewing directions or poses, facial expression, aging, and disguises such as facial hair, glasses or cosmetics. The enormity of the problem has involved hundreds of scientists in interdisciplinary research, but the ultimate solution remains elusive [57, 58, 64, 79]. Face recognition research provides cutting edge technologies in commercial, law enforcement and military applications. An automated vision system that performs the functions of face detection, verification and recognition will find countless unobtrusive applications, such as airport security and access control, building (i.e. embassies) surveillance and monitoring, human–computer intelligent interaction and perceptual interfaces, and smart environments at home, in the office, and in cars [12, 18, 57, 59, 69, 79].

4.2 Background

An automated face recognition system includes several related face processing tasks, such as detection of a pattern as a face, face tracking in a video sequence, face verification, and face recognition. Face detection generally learns the statistical models of the face and non-face images, and then applies a two-class classification rule to discriminate between face and non-face patterns. Face tracking predicts the motion of faces in a sequence of images based on their previous trajectories and estimates the current and future positions of those faces. While face verification is mainly concerned with authenticating a claimed identity posed by a person, such as "Is she the person who she claims to be?", face recognition focuses on recognizing the identity of a person from a database of known individuals.

Figure 4.1 shows a block diagram of the overall face recognition system. When an input image is presented to the face recognition system, the system first performs face detection and facial landmark detection, such as the detection of the centers of the eyes. The system then implements the normalization and cropping procedures, which perform the following three tasks: (1) spatial normalization, which aligns the centers of the eyes to predefined locations and fixes the number of pixels between the eyes (interocular distance) via rotation and scaling transformations; (2) facial region extraction, which crops the facial region that contains only the face, so that the performance of face recognition is not affected by the factors not related to the face itself, such as hair styles; and (3) intensity normalization, which converts the facial region to a vector by concatenating its rows (or columns), and then normalizes the pixels in the vector to zero mean and unit variance. Finally, the system extracts features with high discriminating power for face recognition.

Performance evaluation is an important factor for a face recognition system. The strength and weakness of an automated face recognition system are evaluated using standard databases and objective performance statistics. The face recognition vendor tests [63] are designed to evaluate state-of-the-art vendor face recognition systems, and the detailed performance of those competing vendor systems can be found in the reports [9] and [63]. The FRVT 2002 [63], for example, reported that (1) under normal indoor illumination, the current state-of-the-art vendor face recognition systems reach 90% verification rate at a false accept rate of 1%; (2) in outdoor illumination, the best vendor system can only get 50% verification rate at a 1% false accept rate; and (3) the three-dimensional morphable models technique [10] is capable of improving non-frontal face recognition. These results suggest that illumination and pose are still challenging

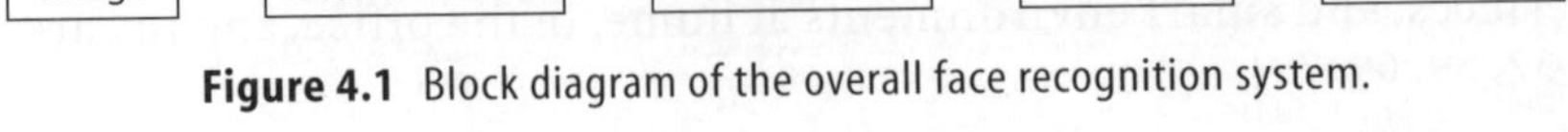

Figure 4.1 Block diagram of the overall face recognition system.

research areas for face recognition, and a 3D model-based approach provides a promising method for dealing with pose variations.

4.3 Face Detection

Face detection is the first stage of an automated face recognition system, since a face has to be located in the overall image before it is recognized. Earlier efforts had been focused on correlation or template matching, matched filtering, sub-space methods, deformable templates etc. [60, 82]. For comprehensive surveys of these early methods, see [12], [69] and [73]. Recent approaches emphasize data-driven learning-based techniques, such as statistical modeling methods [41, 53, 70, 71, 74], neural network-based learning methods [67,68,74], statistical learning theory and Support Vector Machine (SVM) based methods [31, 32, 54], Markov random field based methods [17, 66], and color-based face detection [33].

Statistical methods usually start with the estimation of the distributions of the face and non-face patterns, and then apply a pattern classifier or a face detector to search over a range of scales and locations for possible human faces. Neural network-based methods, however, learn to discriminate the implicit distributions of the face and non-face patterns by means of training samples and the network structure, without involving an explicit estimation procedure.

Moghaddam and Pentland [53] applied unsupervised learning to estimate the density in a high-dimensional eigenspace and derived a maximum likelihood method for single face detection. Rather than using Principal Component Analysis (PCA) for dimensionality reduction, they implemented the eigenspace decomposition as an integral part of estimating the conditional Probability Density Function (pdf) in the original high-dimensional image space. Face detection is then carried out by computing multiscale saliency maps based on the maximum likelihood formulation.

Sung and Poggio [74] presented an example-based learning method by means of modeling the distributions of face and non-face patterns. To cope with the variability of face images, they empirically chose six Gaussian clusters to model the distributions for face and non-face patterns, respectively. The density functions of the distributions are then fed to a multiple layer perceptron for face detection.

Scheiderman and Kanade [70] proposed a face detector based on the estimation of the posterior probability function, which captures the joint statistics of local appearance and position as well as the statistics of local appearance in the visual world. To detect side views of a face, profile images were added to the training set to incorporate such statistics [71].

Liu [41] recently presented a Bayesian Discriminating Features (BDF) method for multiple frontal face detection. The BDF method, which is trained on images from only one database yet works on test images from diverse sources, displays robust generalization performance. First, the method derives a discriminating feature vector by combining the input image, its 1-D Harr wavelet representation, and its amplitude projections. Then statistical modeling estimates the conditional probability density functions, or pdfs, of the face and non-face classes, respectively. While the face class is usually modeled as a multivariate normal distribution, the non-face class is much more difficult to model due to the fact that it includes "the rest of the world". The estimation of such a broad category is, in practice, intractable. However, the BDF method derives a subset of the non-faces that lie closest to the face class, and then models this particular subset as a multivariate normal distribution. Finally, the Bayes classifier applies the estimated conditional pdfs to detect multiple frontal faces in an image.

Rowley *et al.* [67] developed a neural network-based upright, frontal face detection system, which applies a retinally connected neural network to examine small windows of an image and decide whether each window contains a face. The face detector, which was trained using a large number of face and non-face examples, contains a set of neural network-based filters and an arbitrator which merges detections from individual filters and eliminates overlapping detections. In order to detect faces at any degree of rotation in the image plane, the system was extended to incorporate a separate router network, which determines the orientation of the face pattern. The pattern is then derotated back to the upright position, which can be processed by the early developed system [68].

Hsu *et al.* [33] presented a color-based face detection method under variable illumination and complex background. First, the method applies a lighting compensation technique and a nonlinear color transformation to detect skin regions in a color image. Then it generates face candidates based on the spatial arrangement of the skin patches. Finally, the method constructs eye, mouth and boundary maps to verify those candidates. Experiments show that the method is capable of detecting faces over a wide range of facial variations in color, position, scale, orientation, pose and expression [33].

4.4 Face Recognition: Representation and Classification

Robust face recognition schemes require both low-dimensional feature representation for data compression purposes and enhanced discrimination abilities for subsequent image classification. The representation methods usually start with a dimensionality reduction procedure, since the high dimensionality of the original space makes the statistical estimation very difficult, if not impossible, due to the fact that the high-dimensional space is mostly empty. The discrimination methods often try to achieve high separability between different patterns. Table 4.1 shows some popular representation and classification techniques and some methods that apply these techniques for face recognition.

4.4.1 Some Representation Techniques and Their Applications to Face Recognition

Principal Component Analysis is commonly used for deriving low-dimensional representations of input images. Specifically, PCA derives an orthogonal projection basis that directly leads to dimensionality reduction and possibly to feature selection [38]. Applying PCA technique to face recognition, Turk and Pentland [76] developed the well-known "Eigenface" method, where the eigenfaces correspond to the eigenvectors associated with the largest eigenvalues of the face covariance matrix. The eigenfaces thus define a feature space, or "face space", which drastically reduces the dimensionality of the original space, and face recognition is then carried out in the reduced space.

PCA, an optimal criterion for dimensionality reduction, however, does not necessarily provide for good discrimination, since no discrimination criteria are considered by the PCA procedure. To improve the discrimination power of PCA, one can integrate PCA, the optimal representation criterion, with the Bayes classifier, the optimal classification criterion [44]. This method, named Probabilistic Reasoning Models (PRM) [44], applies first PCA for dimensionality reduction, and then uses the within-class scatter to estimate the covariance matrix for each class in order to derive the conditional probability density functions. Finally, the PRM method applies the Maximum *A Posteriori* (MAP) rule for classification. The MAP decision rule optimizes the class separability in the sense of Bayes error and improves upon the PCA-based methods, which apply a criterion not related to the Bayes error.

Shape and texture ('shape-free' image) coding usually applies a two-stage process once the face has been located [8, 14, 16, 24, 40, 45]. Coding starts by annotating the face using important internal and face boundary points. Once these control points are located, they are aligned using translation, scaling and rotation transformations as necessary. The average of these aligned control points defines the mean shape. The next stage then

Table 4.1 Some representation and classification techniques and their applications to face recognition.

	Techniques	Face recognition methods
Representation methods	PCA	Eigenfaces [76], PRM [44]
	Shape and texture	Method [24], EFC [45]
	Gabor wavelets	Method [39], GFC [46], IGF [47]
Recognition methods	Bayes/MAP	Method [52], PRM [44]
	FLD/LDA	Fisherfaces [6], methods [75], [25], EFM [44]
	ICA	Method [22], EICA [42]
	Graph matching	Elastic bunch graph [80]

triangulates the annotated faces and warps each face to the mean shape. The first stage yields the shape, while the second stage yields the texture.

Beymer [8] introduced a vectorized image representation consisting of shape and texture. Vetter and Poggio [78] used such a vectorized face representation for image synthesis from a single example view. Craw *et al.* [15] and Lanitis *et al.* [40] developed Mahalanobis distance classifiers for face recognition using the shape and texture representation. The Mahalanobis distance is measured with respect to a common covariance matrix for all classes in order to treat variations along all axes as equally significant by giving more weight to components corresponding to smaller eigenvalues [15]. Note that the weighting procedure does not differentiate the between-class scatter from the within-class scatter and it suppresses the former while reducing the latter. To address this issue and to better distinguish the different roles of the two scatters, Edwards *et al.* [24] presented yet another Mahalanobis distance classifier by using the pooled within-class covariance matrix. Liu and Wechsler [45] developed an Enhanced Fisher Classifier (EFC), which applies the enhanced Fisher model on the integrated shape and texture features. Shape encodes the feature geometry of a face while texture provides a normalized shape-free image. The dimensionalities of the shape and the texture spaces are first reduced using principal component analysis, constrained by the Enhanced Fisher Model (EFM) for enhanced generalization. The corresponding reduced shape and texture features are then combined through a normalization procedure to form the integrated features that are processed by the EFM for face recognition.

The Gabor wavelets, whose kernels are similar to the 2D receptive field profiles of the mammalian cortical simple cells, exhibit desirable characteristics of spatial locality and orientation selectivity [26]. The biological relevance and computational properties of Gabor wavelets for image analysis have been described in [19, 20, 50 and 36]. Lades *et al.* [39] applied the Gabor wavelets for face recognition using dynamic link architecture (DLA). This starts by computing the Gabor jets, and then performs a flexible template comparison between the resulting image decompositions using graph matching. Based on the 2D Gabor wavelet representation and labeled elastic graph matching, Lyons *et al.* [48, 49] proposed an algorithm for two-class categorization of gender, race and facial expression. The algorithm includes two steps: registration of a grid with the face using either labeled elastic graph matching [39, 80] or manual annotation of 34 points on every face image [49]; and categorization based on the features extracted at grid points using Linear Discriminant Analysis (LDA). Donato *et al.* [22] recently compared a method based on Gabor representation with other techniques and found that the former gave better performance.

Liu and Wechsler [46] presented a Gabor–Fisher Classifier (GFC) method for face recognition. The GFC method, which is robust to illumination and facial expression variability, applies the enhanced Fisher linear discriminant model or EFM [44] to an augmented Gabor feature vector derived from the Gabor wavelet transformation of face images. To encompass all the features produced by the different Gabor kernels one concatenates the resulting Gabor wavelet features to derive an augmented Gabor feature vector. The dimensionality of the Gabor vector space is then reduced under the eigenvalue selectivity constraint of the EFM method to derive a low-dimensional feature representation with enhanced discrimination power. Liu and Wechsler [46] recently developed an Independent Gabor Features (IGF) method for face recognition. The IGF method derives the independent Gabor features, whose independence property facilitates the application of the PRM method [44] for classification.

4.4.2 Some Classification Techniques and Their Applications to Face Recognition

The Bayes classifier yields the minimum error when the underlying probability density functions are known. This error, called the Bayes error, is the optimal measure for feature effectiveness when classification is of concern, since it is a measure of class separability [27]. The MAP Bayes decision rule thus optimizes the class separability in the sense of the Bayes error and should yield the optimal classification performance [27].

Moghaddam *et al.* [52] proposed a probabilistic similarity measure for face image matching based on a Bayesian analysis of image deformations. The probability density functions for the intra-object and extra-object classes are estimated from training data and used to compute a similarity measure. The PRM method, introduced by Liu and Wechsler [44], improves face recognition performance by integrating PCA (the optimal representation criterion) and the Bayes classifier (the optimal classification criterion).

Fisher Linear Discriminant (FLD), or the LDA, is a commonly used criterion in pattern recognition and recently in face recognition [6, 25, 44, 75]. Intuitively, FLD derives a projection basis that separates the different class means as far as possible and compresses the same classes as compactly as possible. Based on FLD, a host of face recognition methods have been developed to improve the classification accuracy and the generalization performance [6, 25, 44, 45, 46, 75]. The Fisherfaces method [6], similar to the methods presented by Swets and Weng [75] and by Etemad and Chellappa [25], first applies PCA to derive a low-dimensional space, where FLD is implemented to derive features for face recognition. These FLD-based methods, however, are superior to the Eigenface approach for face recognition only when the training images are representative of the range of face (class) image variations; otherwise, the performance difference between the Eigenface and Fisherface approaches is not significant [75].

The FLD procedure, when implemented in a high-dimensional PCA space, often leads to overfitting [44]. Overfitting is more likely to occur for the small training sample size scenario, which is the typical situation for face recognition [61]. One possible remedy for this drawback is to generate additional data artificially and thus increase the sample size [25]. Another solution is to analyze the reasons for overfitting and propose new models with improved generalization abilities [44]. The EFM method, developed by Liu and Wechsler [44], addresses (concerning PCA) the range of principal components used and how it affects performance, and (regarding FLD) the reasons for overfitting and how to avoid it. The EFM method improves the generalization performance of the FLD-based scheme by balancing the spectral energy criterion for sufficient representation and the eigenvalue spectral requirement for good generalization. This requirement suggests that the selected PCA eigenvalues account for most of the spectral energy of the raw data when the (trailing) eigenvalues of the within-class scatter matrix in the reduced PCA subspace are not too small [44].

Independent Component Analysis (ICA), a powerful technique for blind source separation, [13, 35, 37], is also applied to possible feature selection and face recognition by Donato *et al.* [22]. According to Barlow [3–5], an important characteristic of sensory processing in the brain is 'redundancy reduction', or reducing dependency and deriving independent features. Such independent features might be learned under the criterion of sparseness [5] or independence [4]. Field [26] described a compact coding scheme in terms of sparse distributed coding, whose neurobiological implications were examined in [56]. The resulting sparse image code possesses a high degree of statistical independence among its outputs [55]. Bell and Sejnowski [7] developed an unsupervised learning method that is based on information maximization to separate statistically independent components in the inputs.

Donato *et al.* [22] applied a neural network approximation to demonstrate the possible application of Independent Component Analysis (ICA) to face recognition. Liu and Wechsler [42] described an Enhanced ICA (EICA) method and its application to face recognition. The EICA method derives the independent components of face images by using a statistical algorithm rather than a neural network approximation. EICA, whose enhanced generalization performance is achieved using a sensitivity analysis, operates in a reduced PCA space, whose dimensionality is determined using an eigenvalue spectrum analysis. The motivation for this aspect of EICA is that during whitening, the eigenvalues of the covariance matrix appear in the denominator and that the small trailing eigenvalues mostly encode noise. As a consequence the whitening component, if used in an uncompressed image space, would fit for misleading variations and thus generalize poorly to new data. Liu and Wechsler [42] have also assessed the performance of the EICA alone or when combined with other discriminant criteria such as the Bayesian framework or the FLD criterion. Discriminant analysis shows that the ICA criterion, when carried out in the properly compressed and whitened space, performs better than the Eigenface and Fisherface methods for face recognition, but its performance deteriorates when augmented by additional criteria such as the MAP rule of the Bayes classifier or the FLD. The reason for the last finding is that the Mahalanobis distance embedded in the MAP classifier duplicates to some extent the whitening component, while using FLD is counter to the independence criterion intrinsic to EICA.

Graph matching has the potential for achieving face recognition invariant to affine transformations or localized facial expression changes. However, the graph nodes have to be manually defined to associate the corresponding nodes in the different graphs [80]. Lades *et al.* [39] presented a dynamic link architecture (DLA) for face recognition. Wiskott *et al.* [80] further expanded on DLA when they developed an elastic bunch graph matching method to label and recognize human faces. Faces are represented by labeled graphs, based on a Gabor wavelet representation of face images. The graphs of new faces are extracted by an elastic graph matching process and can be compared using a simple similarity function [80].

4.5 Kernel-Based Methods and 3D Model-based Methods for Face Recognition

Kernel-based methods, such as kernel PCA, kernel FLD, and SVM, overcome the limitations of the linear approaches by nonlinearly mapping the input space to a high-dimensional feature space. Theoretical justifications of applying kernel-based methods to face recognition stem from Cover's theorem on the separability of patterns, which states that nonlinearly separable patterns in an input space are linearly separable with high probability if the input space is transformed nonlinearly to a high-dimensional feature space [30]. Computationally, kernel methods take advantage of the Mercer equivalence condition and are feasible because the dot products in the high-dimensional feature space are replaced by a kernel function in the input space, while computation is related to the number of training examples rather than the dimension of the feature space.

Scholkopf *et al.* [72] showed that kernel PCA outperforms PCA using an adequate nonlinear representation of input data. Mika *et al.* [51] presented a kernel FLD method whose linear classification in the feature space corresponds to a powerful nonlinear classification in the input space. Phillips [62] proposed an SVM-based face recognition algorithm for both face verification and recognition, and demonstrated its superior performance over a PCA-based method. Yang [81] presented face recognition results by applying the kernel FLD method to two data sets: the AT&T data set containing 400 images of 40 subjects and the Yale data set containing 165 images of 11 subjects. Experimental results show that the kernel FLD method achieves lower error rates in face recognition than ICA, Eigenface or Fisherface methods [81].

3D methods, or 3D model-based methods [2, 10, 34, 83], provide potential solutions to pose invariant face recognition [63]. 3D face models are usually derived from laser-scanned 3D heads (range data) or reconstructed using shape from shading [2, 10, 34, 83]. Hsu and Jain [34] proposed a method of modeling 3D human faces based on a triangular mesh model and individual facial measurements encoding both shape and texture information. The method provides a potential solution to face recognition with variations in illumination, pose and facial expression. Zhao and Chellappa [84] presented a method which applies a 3D model to synthesize a prototype image from a given image acquired under different lighting and viewing conditions.

4.6 Learning the Face Space

The idea of learning the face space has been motivated by natural scene encoding, which takes advantage of intrinsic image statistics while seeking to derive a natural ("universal") basis [28, 55]. The derived basis functions have been found to closely approximate the receptive fields of simple cells in the mammalian primary visual cortex. The receptive fields resemble various derivative-of-Gaussian (DoG) functions: spatially localized, oriented and bandpass [55]. Barlow [4] argues that such receptive fields might arise from unsupervised learning, subject to redundancy reduction or minimum entropy coding. Olshausen and Field [55] derive localized oriented receptive fields based on a criterion of sparseness, while Bell and Sejnowski [7] use an independence criterion to derive qualitatively similar results.

The rationale behind a natural basis is to allow for the efficient derivation of suitable image representations corresponding to the intrinsic structure of sensory signals. The intrinsic structures are essential for processes such as image retrieval and object recognition. Once the natural basis has been derived, no additional training is necessary and both the training and the test images on future tasks are represented in terms of the already available natural basis. The natural basis, however, also has its drawbacks, i.e. it might be too general to properly encode for a specific task. Regarding face recognition, the class of objects to be represented is quite specific, i.e. human face images, possibly indexed by gender, ethnicity and age; one should thus seek and learn the face space basis rather than a "universal" and all-encompassing natural basis. This observation also fits with knowledge that the "bias/variance dilemma may be circumvented if we are willing to purposely introduce bias, which then makes it possible to eliminate the variance or reduce it significantly" [29]. Learning low-dimensional representations of visual objects with extensive use of prior knowledge has also been recently suggested by Edelman and Intrator [23] who claim that "perceptual tasks such as similarity judgment tend to be performed on a low dimensional representation of the sensory data. Low dimensionality is especially important for learning, as the number of examples required for attaining a given level of performance grows exponentially with the dimensionality of the underlying representation space".

4.6.1 Evolutionary Pursuit

The problem we address now is that of learning the face space(s) from large and diverse populations using evolution as the driving force. The dimensions ("BASIS") for the face space, to be evolved using Genetic Algorithms (GAs), are such that their "fitness" is enhanced and driven by the classification/discrimination ("cognitive") and representational ("perceptual") factors referred to earlier, and the interplay between the complexity, the cost ("dimension") and the (categorical) density of the face space on one hand, and the trade-offs between faithful face reconstruction ("representation") and the expected classification accuracy ("guaranteed risk") for the face classifier, on the other hand. The quality of the face space is also driven by the diversity encountered while learning the face space. Characteristic of both co-evolution and active learning methods, challenging training samples would be boosted and thus given extra weight while assessing the fitness of the face space.

The fundamental problem, that of finding the proper mix of cognitive ("classification") and perceptual ("preprocessing") processes, and in the process deriving the optimal projection basis for face encoding, can be addressed using Evolutionary Pursuit (EP) [43]. In analogy to (exploratory) pursuit methods from statistics, EP seeks to learn an optimal face space for the dual purpose of data compression and pattern classification. The challenges that EP has successfully met on limited population types are characteristic of sparse functional approximation and statistical learning theory. Specifically, EP increases the generalization ability of the face recognizer as a result of handling the trade-off between minimizing the empirical risk encountered during training ("performance accuracy"), and narrowing the predicted risk ("confidence interval") for reducing the guaranteed risk during future testing on unseen probe images. The prediction risk, corresponding to the penalty factor from regularization methods, measures the generalization ability of the object classifier, and it is driven by the regularization index corresponding to class separation. EP implements strategies characteristic of genetic algorithms for searching the space of possible solutions in order to determine the optimal projection basis. EP starts by projecting the original images into a lower-dimensional and whitened PCA space. Directed but random rotations of the basis vectors in this space are then searched by GAs where evolution is driven by a fitness function defined in terms of performance accuracy ("empirical risk") and class separation ("confidence interval").

Evolutionary computation represents an emerging methodology motivated by natural selection. Evolution takes place by maintaining one or more populations of individuals, each of them a candidate solution, and competing for limited resources in terms of placing offsprings in future generations. The competition is implemented via selection mechanisms that choose from the dynamically changing populations due to the birth and death of individuals. The selection mechanisms evaluate the fitness value 10 of individuals based on some fitness criteria (fitness functions), while the population evolves via genetic operators that reflect the concept of inheritance (offsprings resemble their parents). When the fitness functions lack an analytical form suitable for gradient descent or the computation involved is prohibitively expensive, as is the case when the solution space is too large to search exhaustively, one alternative is to use (directed) stochastic search methods for nonlinear optimization and variable selection. The unique exploration (variations farther away from an existing population) and exploitation (minor variations of the more fit parents) ability of evolutionary computation guided by fitness values has made it possible to analyze very complex search spaces.

Learning the face space requires EP to search through a large number of possible subsets of rotated axes in a properly whitened PCA space. The rotation angles (represented by strings of bits) and the axis indicators (indicating whether the axes are chosen or not) constitute the form of the search space whose size (2 to the power of the length of the whole string) is too large to search exhaustively. The number and choice of (nonorthogonal) axes in the subsets and the angles of rotations are evolved using genetic algorithms. GAs work by maintaining a constant-sized population of candidate solutions known as individuals ("chromosomes"). The power of genetic algorithms lies in their ability to exploit, in a highly efficient manner, information about a large number of individuals. The search underlying GAs are such that breadth and depth – exploration and exploitation – are balanced according to the observed performance of the individuals evolved so far. By allocating more reproductive occurrences to above average individual solutions, the overall effect is to increase the population's average fitness.

Evolution is driven by a fitness function formulated as follows:

$$\varsigma(F) = \varsigma_a(F) + \lambda\varsigma_g(F)$$

where F encompasses the parameters (such as the number of axes and the angles of rotations defining each chromosome solution) subject to learning, the first term $\varsigma_a(F)$ records performance accuracy, i.e. the empirical risk, the second term $\varsigma_g(F)$ is the generalization index, i.e. the predicted risk, and λ is a positive constant that indicates the importance of the second term relative to the first one. Accuracy indicates the extent to which learning has been successful so far, while the generalization index gives an indication of the expected fitness on future trials. By combining those two terms together with a proper weight factor λ, GA can evolve balanced results with good recognition performance and generalization abilities. The fitness function has a similar form to the cost functional used by regularization theory [65] and to the cost function used by sparse coding [55]. The cost functional of the former method exploits a regularization parameter to control the compromise between a term of the solution's closeness to the data and a term indicating the degree of regularization ('quality') of the solution, while the cost function of the latter method uses a positive constant to achieve a balance between a term of information preserving and a term assessing the sparseness of the derived code.

4.6.2 Face Recognition Using Evolutionary Pursuit

We consider now the application of the EP method to learning the face space for face recognition [43]. The experimental data consists of a subset of 1,107 images from the FERET database, with three frontal images for each of 369 subjects. For the first 200 subjects, the third image is acquired at low illumination, while for the remaining 169 subjects the three face images are acquired during different photo sessions. The last acquired image for each subject is called the "probe". Two images of each subject are used for training, with the remaining "probe" image used for testing. In other words, the training set includes 738 images while the test set has 369 images. The images are cropped to a size of 64 × 96 pixels and the eye coordinates are manually located. The image background is uniform and the face regions are not masked. Masking, as it has been usually implemented, deletes areas of the image outside the face outline, retaining only the face proper. The effect of such deletions on recognition performance is discussed in [45]. Shape-free face recognition methods avoid this problem by using the shape of the outline encoded by a number of control points for subsequent alignment and normalization [14].

Starting from a 30-dimensional PCA space, the EP method derives 26 vectors as the optimal basis for the learned face space. Note that while for PCA the basis vectors have a natural order (the descending magnitudes of the eigenvalues associated with each vector), this is not the case with the projection basis derived by EP due to the rotations involved during the evolutionary process. The natural order characteristic of the principal components reflects the representational aspect of PCA and its relationship to spectral decomposition. The very first principal components encode global image characteristics, in analogy to low-frequency components. EP, on the other hand, is a procedure geared primarily towards recognition and generalization. It is also worth pointing out that while PCA derives orthogonal basis vectors, EP's basis vectors are usually not orthogonal. Orthogonality is a constraint for optimal reduced-space signal representation, but not a requirement for pattern recognition. Actually, non-orthogonality has been known to have great functional significance in biological sensory systems [21].

The EP face space approach using 26 basis vectors yields 92% recognition performance at this database size when testing on "sequestered" face images unavailable during training. This compares favorably against Eigenface and Fisherface methods [43]. To assess the statistical significance of the experimental results, we implemented McNemar's test [1] to determine whether or not there is strong statistical evidence to indicate that the EP method improves recognition performance over Eigenface and Fisherface methods. We found that the EP method improves face recognition performance at a statistically significant level.

4.7 Conclusion

This chapter has surveyed recent research in face detection and recognition, discussed performance of the current face recognition systems, and presented promising research directions. In particular, face detection methods reviewed include statistical, neural network-based and color-based approaches. Face recognition methods surveyed include PCA-based approaches, shape and texture-based approaches, Gabor wavelet-based approaches, approaches applying the Bayes classifier or MAP, FLD or LDA, ICA, and graph matching. Some kernel-based methods and 3D model-based methods for face recognition are discussed. These methods provide new research directions for potential solutions to facial recognition under conditions of pose and illumination variation, which recent vendor tests show are challenging issues for face recognition. Finally, a method of learning the face space using evolutionary pursuit is also presented.

References

[1] A. Agresti (ed.), *Categorical Data Analysis*. John Wiley & Sons, 1990.

[2] J. J. Atick, P. A. Griffin and A. N. Redlich, Statistical approach to shape from shading: reconstruction of three-dimensional face surfaces from single two-dimensional images. *Neural Computation*, **8**, 1321–1340, 1996.

[3] H. B. Barlow, Possible principles underlying the transformation of sensory messages. In W. Rosenblith (ed.), *Sensory Communication*, pp. 217–234. MIT Press, 1961.

[4] H. B. Barlow, Unsupervised learning. *Neural Computation*, **1**, 295–311, 1989.

[5] H. B. Barlow, T. P. Kaushal and G. J. Mitchison, Finding minimum entropy codes. *Neural Computation*, **1**(3), 412–423, 1989.

[6] P. N. Belhumeur, J. P. Hespanha and D. J. Kriegman, Eigenfaces vs. Fisherfaces: recognition using class specific linear projection. *IEEE Trans. Pattern Analysis and Machine Intelligence*, **19**(7), 711–720, 1997.

[7] A. J. Bell and T. J. Sejnowski, An information-maximization approach to blind separation and blind deconvolution. *Neural Computation*, 7, 1129–1159, 1995.

[8] D. Beymer, Vectorizing face images by interleaving shape and texture computations. *A.I. memo No. 1537*, Artificial Intelligence Laboratory, MIT, 1995.

[9] D. M. Blackburn, J. M. Bone and P. J. Phillips, FRVT 2000 evaluation report. *Technical report*, February, 2001. Available online at `http://www.frvt.org/`.

[10] V. Blanz and T. Vetter, A morphable model for the synthesis of 3D faces. *Proc. SIGGRAPH 99*, 1999.

[11] R. Brunelli and T. Poggio, Face recognition: features vs. templates. *IEEE Trans. Pattern Analysis and Machine Intelligence*, **15**(10), 1042–1053, 1993.

[12] R. Chellappa, C. L. Wilson and S. Sirohey, Human and machine recognition of faces: a survey. *Proc. IEEE*, **83**(5), 705–740, 1995.

[13] P. Comon, Independent component analysis, a new concept? *Signal Processing*, **36**, 287–314, 1994.

[14] T. F. Cootes, G. J. Edwards and C. J. Taylor, Active appearance models. *Proc. Fifth European Conference on Computer Vision*, University of Freiburg, Germany, June 2–6, 1998.

[15] I. Craw, N. Costen, T. Kato and S. Akamatsu, How should we represent faces for automatic recognition? *IEEE Trans. Pattern Analysis and Machine Intelligence*, **21**(8), 725–736, 1999.

[16] I. Craw and D. Tock. The computer understanding of faces. In V. Bruce and M. Burton (eds), *Processing Images of Faces*. Ablex Publishing Corporation, 1992.

[17] S. C. Dass and A. K. Jain, Markov face models. *Eighth IEEE International Conference on Computer Vision*, pp. 680–687, 2001.

[18] J. Daugman, Face and gesture recognition: overview. *IEEE Trans. Pattern Analysis and Machine Intelligence*, **19**(7), 675–676, 1997.

[19] J. G. Daugman, Two-dimensional spectral analysis of cortical receptive field profiles. *Vision Research*, **20**, 847–856, 1980.

[20] J. G. Daugman, Uncertainty relation for resolution in space, spatial frequency, and orientation optimized by two-dimensional cortical filters. *J. Opt. Soc. Am.*, **2**(7), 1160–1169, 1985.

[21] J. G. Daugman, An information-theoretic view of analog representation in striate cortex. In E. Schwartz (ed.), *Computational Neuroscience*, pp. 403–424. MIT Press, 1990.

[22] G. Donato, M. S. Bartlett, J. C. Hager, P. Ekman and T. J. Sejnowski, Classifying facial actions. *IEEE Trans. Pattern Analysis and Machine Intelligence*, **21**(10), 974–989, 1999.

[23] S. Edelman and N. Intrator, Learning as extraction of low-dimensional representations. In D. Medin, R. Goldstone and P. Schyns (eds), *Mechanisms of Perceptual Learning*. Academic Press, 1990.

[24] G. J. Edwards, T. F. Cootes and C. J. Taylor, Face recognition using active appearance models. *Proc. Fifth European Conference on Computer Vision*, University of Freiburg, Germany, June 2–6, 1998.

[25] K. Etemad and R. Chellappa, Discriminant analysis for recognition of human face images. *J. Opt. Soc. Am. A*, **14**, 1724–1733, 1997.

[26] D. Field, What is the goal of sensory coding. *Neural Computation*, **6**, 559–601, 1994.

[27] K. Fukunaga, *Introduction to Statistical Pattern Recognition*, 2nd edn. Academic Press, 1991.

[28] P. Hancock, R. Baddeley and L. Smith, The principal components of natural images. *Network*, **3**, 61–70, 1992.

[29] S. Haykin, *Neural Networks – A Comprehensive Foundation*. Macmillan College Publishing Company, Inc., 1994.

[30] S. Haykin, *Neural Networks – A Comprehensive Foundation*, 2nd edn. Prentice Hall, 1999.

[31] B. Heisele, T. Poggio and M. Pontil, Face detection in still gray images. *A.I. memo AIM-1687*, Artificial Intelligence Laboratory, MIT, 2000.

[32] P. Ho, Rotation invariant real-time face detection and recognition system. *A.I. memo AIM-2001-010*, Artificial Intelligence Laboratory, MIT, 2001.

[33] R.-L. Hsu, M. Abdel-Mottaleb and A. K. Jain, Face detection in color images. *IEEE Trans. Pattern Analysis and Machine Intelligence*, **24**(5), 696–706, 2002.

[34] R. L. Hsu and A. K. Jain, Face modeling for recognition. *International Conference on Image Processing*, 2001.

[35] A. Hyvarinen and E. Oja, A fast fixed-point algorithm for independent component analysis. *Neural Computation*, **9**, 1483–1492, 1997.

[36] J. Jones and L. Palmer, An evaluation of the two-dimensional Gabor filter model of simple receptive fields in cat striate cortex. *J. Neurophysiology*, 1233–1258, 1987.

[37] J. Karhunen, E. Oja, L. Wang, R. Vigario and J. Joutsensalo, A class of neural networks for independent component analysis. *IEEE Trans. Neural Networks*, **8**(3), 486–504, 1997.

[38] M. Kirby and L. Sirovich, Application of the Karhunen–Loeve procedure for the characterization of human faces. *IEEE Trans. Pattern Analysis and Machine Intelligence*, **12**(1), 103–108, 1990.

[39] M. Lades, J. C. Vorbruggen, J. Buhmann, J. Lange, C. von der Malsburg, R. P. Wurtz and W. Konen, Distortion invariant object recognition in the dynamic link architecture. *IEEE Trans. Computers*, **42**, 300–311, 1993.

[40] A. Lanitis, C. J. Taylor and T. F. Cootes, Automatic interpretation and coding of face images using flexible models. *IEEE Trans. Pattern Analysis and Machine Intelligence*, **19**(7), 743–756, 1997.

[41] C. Liu, A Bayesian discriminating features method for face detection. *IEEE Trans. Pattern Analysis and Machine Intelligence*, **25**(6), 725–740, 2003.

[42] C. Liu and H. Wechsler, Comparative assessment of independent component analysis (ICA) for face recognition. In *Proc. Second International Conference on Audio- and Video-based Biometric Person Authentication*, Washington DC, March 22–24, 1999.

[43] C. Liu and H. Wechsler, Evolutionary pursuit and its application to face recognition. *IEEE Trans. Pattern Analysis and Machine Intelligence*, **22**(6), 570–582, 2000.

[44] C. Liu and H. Wechsler, Robust coding schemes for indexing and retrieval from large face databases. *IEEE Trans. on Image Processing*, **9**(1), 132–137, 2000.

[45] C. Liu and H.Wechsler, A shape and texture based enhanced Fisher classifier for face recognition. *IEEE Trans. on Image Processing*, **10**(4), 598–608, 2001.

[46] C. Liu and H. Wechsler, Gabor feature based classification using the enhanced Fisher linear discriminant model for face recognition. *IEEE Trans. on Image Processing*, **11**(4), 467–476, 2002.

[47] C. Liu and H. Wechsler, Independent component analysis of Gabor features for face recognition. *IEEE Trans. on Neural Networks*, **14**(4), 2003.

[48] M. J. Lyons, J. Budynek, A. Plante and S. Akamatsu, Classifying facial attributes using a 2D Gabor wavelet representation and discriminant analysis. *Proc. Fourth IEEE International Conference on Automatic Face and Gesture Recognition*, 2000.

[49] M. J. Lyons, J. Budynek and S. Akamatsu, Automatic classification of single facial images. *IEEE Trans. Pattern Analysis and Machine Intelligence*, **21**(12), 1357–1362, 1999.

[50] S. Marcelja, Mathematical description of the responses of simple cortical cells. *J. Opt. Soc. Am.*, **70**, 1297–1300, 1980.

[51] S. Mika, G. Ratsch, J. Weston, B. Scholkopf and K. R. Muller, Fisher discriminant analysis with kernels. In Y. H. Hu, J. Larsen, E. Wilson and S. Douglas (eds), *Neural Networks for Signal Processing IX*, pp. 41–48. IEEE, 1999.

[52] B. Moghaddam, C. Nastar and A. Pentland, A Bayesian similarity measure for direct image matching. *Proc. 13th International Conference on Pattern Recognition*, Vienna, Austria, August, 1996.

[53] B. Moghaddam and A. Pentland, Probabilistic visual learning for object representation. *IEEE Trans. Pattern Analysis and Machine Intelligence*, **19**(7), 696–710, 1997.

[54] A. Mohan, C. Papageorgiou and T. Poggio, Example-based object detection in images by components. *IEEE Trans. Pattern Analysis and Machine Intelligence*, **23**(4), 349–361, 2001.

[55] B. A. Olshausen and D. J. Field, Emergence of simple-cell receptive field properties by learning a sparse code for natural images. *Nature*, **381**(13), 607–609, 1996.

[56] B. A. Olshausen and D. J. Field, Sparse coding with an overcomplete basis set: a strategy employed by v1? *Vision Research*, **37**, 3311–3325, 1997.

[57] S. Pankanti, R. M. Bolle and A. Jain, Guest editors' introduction: Biometrics – the future of identification. *Computer*, **33**(2), 46–49, 2000.

[58] A. Pentland, Looking at people: sensing for ubiquitous and wearable computing. *IEEE Trans. Pattern Analysis and Machine Intelligence*, **22**(1), 107–119, 2000.

[59] A. Pentland and T. Choudhury, Face recognition for smart environments. *Computer*, **33**(2), 50–55, 2000.

[60] A. Pentland, B. Moghaddam and T. Starner, View-based and modular eigenspaces for face recognition. *Proc. Computer Vision and Pattern Recognition*, 1994.

[61] P. J. Phillips, Matching pursuit filters applied to face identification. *IEEE Trans. Image Processing*, 7(8), 1150–1164, 1998.

[62] P. J. Phillips, Support vector machines applied to face recognition. In *Neural Information Processing Systems*, 1998.

[63] P. J. Phillips, P. Grother, R. J. Micheals, D. M. Blackburn, E. Tabassi and J. M. Bone, FRVT 2002: Evaluation report. *Technical report*, March, 2003. Available online at `http://www.frvt.org/`.

[64] P. J. Phillips, A. Martin, C. l. Wilson and M. Przybocki, An introduction to evaluating biometric systems. *Computer*, **33**(2), 56–63, 2000.

[65] T. Poggio, V. Torre and C. Koch, Computational vision and regularization theory. *Nature*, **317**, 314–319, 1985.

[66] R. J. Qian and T. S. Huang, Object detection using hierarchical MRF and map estimation. *Proc. Computer Vision and Pattern Recognition*, 1997.

[67] H. A. Rowley, S. Baluja and T. Kanade, Neural network-based face detection. *IEEE Trans. Pattern Analysis and Machine Intelligence*, **20**(1), 23–38, 1998.

[68] H. A. Rowley, S. Baluja and T. Kanade, Rotation invariant neural network-based face detection. *Proc. IEEE Computer Society Conference on Computer Vision and Pattern Recognition*, Santa Barbara, California, June 23–25, 1998.

[69] A. Samal and P. A. Iyengar, Automatic recognition and analysis of human faces and facial expression: a survey. *Pattern Recognition*, **25**(1), 65–77, 1992.

[70] H. Scheiderman and T. Kanade, Probabilistic modeling of local appearance and spatial relationships for object recognition. *Proc. IEEE Computer Society Conference on Computer Vision and Pattern Recognition*, Santa Barbara, California, June 23–25, 1998.

[71] H. Scheiderman and T. Kanade, A statistical method for 3D object detection applied to faces and cars. *Proc. IEEE Computer Society Conference on Computer Vision and Pattern Recognition*, 2000.

[72] B. Scholkopf, A. Smola and K. Muller, Nonlinear component analysis as a kernel eigenvalue problem. *Neural Computation*, **10**, 1299–1319, 1998.

[73] K. K. Sung, Learning and example selection for object and pattern detection. *PhD Thesis*, AI Lab, MIT, 1996.

[74] K. K. Sung and T. Poggio, Example-based learning for view-based human face detection. *IEEE Trans. Pattern Analysis and Machine Intelligence*, **20**(1), 39–51, 1998.

[75] D. L. Swets and J. Weng, Using discriminant eigenfeatures for image retrieval. *IEEE Trans. Pattern Analysis and Machine Intelligence*, **18**(8), 831–836, 1996.

[76] M. Turk and A. Pentland, Eigenfaces for recognition. *J. Cognitive Neuroscience*, **13**(1), 71–86, 1991.

[77] D. Valentin, H. Abdi, A. J. O'Toole and G. W. Cottrell, Connectionist models of face processing: a survey. *Pattern Recognition*, **27**(9), 1209–1230, 1994.

[78] T. Vetter and T. Poggio, Linear object classes and image synthesis from a single example image. *IEEE Trans. Pattern Analysis and Machine Intelligence*, **19**(7), 733–742, 1997.

[79] H. Wechsler, P. J. Phillips, V. Bruce, F. F. Soulie and T. S. Huang (eds), *Face Recognition: From Theory to Applications*. Springer-Verlag, 1998.

[80] L. Wiskott, J. M. Fellous, N. Kruger and C. von der Malsburg, Face recognition by elastic bunch graph matching. *IEEE Trans. Pattern Analysis and Machine Intelligence*, **19**(7):775–779, 1997.

[81] M. H. Yang, Kernel eigenfaces vs. kernel Fisherfaces: face recognition using kernel methods. *Proc. Fifth International Conference on Automatic Face and Gesture Recognition*, Washington DC, May, 2002.

[82] A. L. Yuille, Deformable templates for face recognition. *J. Cognitive Neuroscience*, **3**(1), 59–70, 1991.

[83] W. Zhao and R. Chellappa, 3D model enhanced face recognition. *International Conference on Image Processing*, 2000.

[84] W. Zhao and R. Chellappa, Symmetric shape-from-shading using self-ratio image. *Int. J. Computer Vision*, **45**(1), 55–75, 2001.

5 Elements of Speaker Verification

Herbert Gish

5.1 Introduction

The activities of automatic speaker verification and identification have a long history going back to the early 1960s [1–3] in which we see the emergence of various pattern recognition techniques and statistical methods and exploration of features and feature selection methods. However, theories were not well supported by experimental evidence, and the selected features were insufficient, perhaps because of limited computing resources. With advent of more powerful computer resources and larger and better annotated corpora such as the Switchboard Corpus [4] there has been steady progress over the years, using more sophisticated statistical models and training methods. In this chapter we review the current work on speaker verification (SV) technology, describing the basic processes and factors that affect the performance of the SV process.

5.1.1 The Speaker Verification Problem

The problem of speaker verification is that of corroborating the identity of speakers from their voices. Our basic assumption is that the voice production apparatus singularly characterizes the acoustic waves and emanations that we interpret as words and sounds with the identity of the speaker. In order to establish an identity from these acoustic events we need a model that characterizes a speaker's voice. Once we have created a model from the available speech samples we are then be able to verify that an utterance from an enrolled or target speaker (i.e. a speaker for which we have training data) by evaluating speech that is claimed to be from the speaker with the speaker model that we have created. If the collected speech data fits the model then we will affirm the speech as having come from our enrollee or, if not, reject the claim.

While the above paragraph gives the essence of speaker verification we must now face the steps necessary to accomplish this task. The first step is the representation of speech itself. The goal of this step is to represent the speech in such a way that we have features that characterize the speech process in an economical way suitable for the subsequent modeling process. Our desire for economy is to keep the dimensionality of the representation sufficiently low so as to reduce burdens on modeling process. At the same time we want to retain sufficient information in the representation such as to keep the impairment of the identification process to a minimum.

The representation that we choose provides us with the features from which we will construct the model for the speaker. Basically, the choice for the representation of speech defines a feature space and the training data fills up the feature space with feature vectors. When we are given a speech utterance that is asserted to be from a particular enrolled speaker the question becomes that of determining whether the collection of features generated by the utterance in question "matches" the training data.

In classical pattern recognition problems we are typically given a single feature vector and asked whether or not it is from the training set. The SV problem differs from the classical pattern recognition problem in that we are comparing the test collection to the training collection and not just determining the membership of a single observation.

While it is possible to formulate the speaker verification problem as just comparing test data to training data, and some speaker verification systems have been built in this fashion, it has been found to be much more effective to formulate the speaker verification problem as one in which we ask whether:

- the test data matches the training data from the target speaker, or
- the test data more closely matches the data from other speakers.

This comparative approach makes an important difference in performance and is justified on both theoretical grounds as well as intuitive grounds. This process of making a decision on the basis of a comparison is a means of model normalization, and more will be said about this later.

The intuition for the comparative approach is that if we are scoring data that has distortions and biases these will be mitigated by comparing the target speaker's model to the models of other speakers. The theoretical justification is a consequence of Bayes' Theorem. That is, if we want to determine whether our observation x was generated by target speaker T we need to compute $P(T|x)$, the probability of speaker T having produced the utterance that generated the observed features x. From Bayes' Theorem we can write

$$\mathrm{p}P(T \mid x) = \frac{p_T(x)P(T)}{p(x)}$$

where $P_T(x)$ is the probability of observing features x given using the model for speaker T, $P(T)$ is the prior probability that speaker T has spoken, and (very important) $p(x)$ is the probability of observing features x irrespective of who was speaking.

In a nutshell, a good deal of what transpires in speaker verification work deals with the features x that are generated for an utterance: selection of the models $P_T(x)$ and $p(x)$. In the Bayes formulation, $p(x)$ can actually contain the model for the target speaker T. If we consider $p(x)$ as an additive mixture of all speakers we can write

$$p(x) = \alpha p_T(x) + (1-\alpha) p_A(x)$$

where $P_A(x)$ represents all speakers other than the target speaker and α is the weighting on the target speaker model, and α is greater than 0 and less than 1.

We now can write

$$P(T \mid x) = \frac{P(T)}{\alpha + [(1-\alpha) p_A(x)] / p_T(x)}$$

which explicitly shows the dependence of the probability of the target speaker as a function of the ratio of the likelihoods for the two classes, i.e. the class of features for the target speaker T and class A of all speakers other than the target speaker, T.

In Figure 5.1 we illustrate our view of the underlying structure of the verification problem. We see that the target speaker fills up part of feature space, and other speakers (often called the cohort speakers or the universal background "speaker") will overlap the target speaker. The different regions of feature space correspond to different phonetic units and we may find that two speakers overlap in one part of space while they do not overlap very much in another.

If we again let $P_T(x)$ denote the probability density function (pdf) for the target speaker, i.e. the speaker that claims an identity, and let $P_A(x)$ denote the pdf for an alternate speaker or speakers (or a composite of speakers), then this alternate set when constructed from a collection of models from other individuals is termed a cohort set [5], and when it is created from a composite of speakers it is often referred to as a universal background model (UBM) [6, 7]. We form a score based on the pdfs for each of the

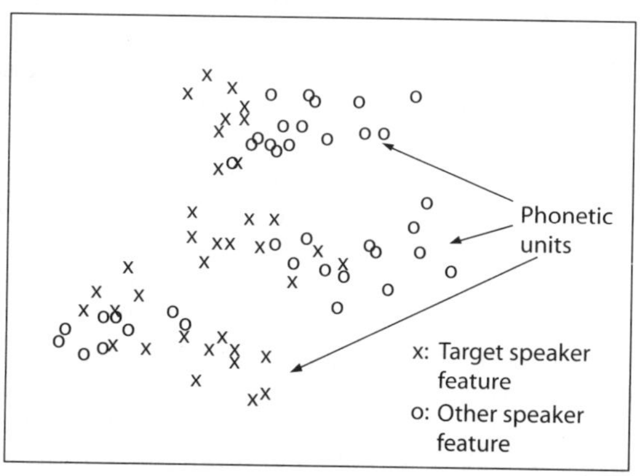

Figure 5.1 An illustration of feature space for speaker verification.

speakers as a first step in the classification process. The score is simply the log likelihood for each of the pdfs for the test data, i.e.

$$s_T = \sum_{i=1}^{N} \log p_T(x_i) \text{ for the target speaker and}$$

$$s_A = \sum_{i=1}^{N} \log p_A(x_i) \text{ for the alternate}$$

where N is the number of feature vectors, x_i, observed. A decision is made that the claimed identity is true if the score difference (the log likelihood ratio)

$$s_{T,A} = s_T - s_A \geq thr_{T-A}$$

otherwise the claim is rejected. Here thr_{T-A} is the threshold value for comparing the target T against the alternative A.

The setting of the threshold determines the size of the errors that the verification system will make. There are two types of error to be concerned with: (1) the error of missing a target, P_{M}, when the target speaker is rejected incorrectly and (2) the error of false acceptance or false alarm, P_{FA}, when an impostor speaker is accepted as the target speaker.

The adjustment of the threshold enables the trade-off between the two types of error. Increasing the threshold will reduce the frequency of false acceptances by demanding an increase in the difference in the scores of the target and alternative models. The price now paid for such a change in threshold is that it is now more difficult for acceptance to occur when the target is actually present, resulting in an increase in the false dismissals. This trade-off between the two errors is characterized by a ROC (Receiver Operating Characteristic) curve which is a plot of the probability of correct acceptance, which is just $1 - P_{\mathrm{M}}$ versus P_{FA}. Quite often a variant of the ROC is currently employed: the DET curve (the detection error trade-off curve [8]), since it allows for a more direct comparison of errors by plotting P_{M} versus P_{FA} directly. This is done on a logarithmic scale for easier reading of the smaller error region. In Figure 5.2 a sample DET curve has been plotted.

The use of the likelihood ratio between the pdf of the target speaker and that of the alternate choice is fundamental to statistical decision making [9]. How it is actually implemented will depend on a variety of considerations and can be quite important to system performance. For example, in speaker verification there are choices for this alternate model, usually consisting of a cohort set or, if enough data is available, a UBM. In either case it is believed that the best performance is achieved when this alternate model is drawn from a population of speakers that share characteristics with the target speaker, such as gender and channel. Keeping the alternate model narrowly focused does, however, allow impostors that are

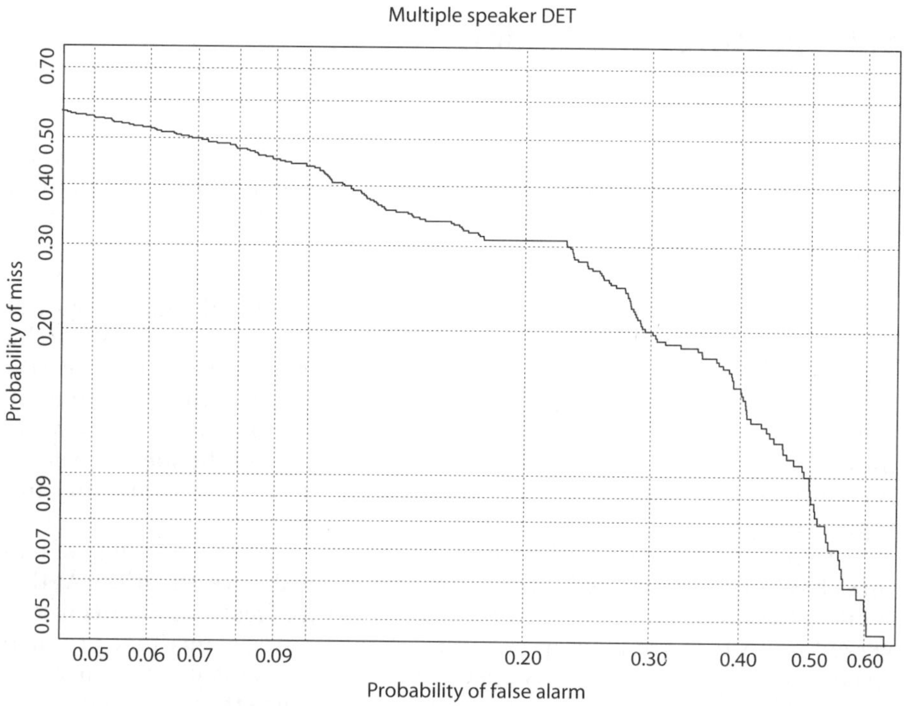

Figure 5.2 Example of a DET curve based on scores from multiple speakers.

not incorporated in the alternate model to be an outlier for the alternate model as well as the target model. An outlier is evaluated in the poorly estimated tails of the distributions and can easily give rise to false acceptances. This is an important practical consideration and can be mitigated by disallowing the acceptance of a target speaker when low values of likelihoods are observed.

Our characterization of speaker verification in terms of pdfs is perhaps the most important approach to the speaker verification problem, but there are a variety of other approaches, some of which will be noted below.

In the following we will consider the issues of modeling and coping with variability that are the main issues for designers of SV[1] systems. We will also focus, for the most part, on the problem of text-independent verification. "Text-independent" means that there is no *a priori* knowledge of the text spoken by the speaker.

1 In references to the literature we will not distinguish those papers that are dealing with the identification problem, i.e. the determination of who the speaker is, from the SV problem, since the underlying technology will apply to both tasks.

5.2 Features and Models

5.2.1 Speech Features

It is in the selection of the features that we encounter the distinctive characteristics of the speech process in defining this speaker verification pattern recognition problem. At the heart of the problem is to distinguish one person's speech production mechanism from another person's. Given that speech is considered, at least as an initial approximation, to be a time-varying linear filter (our vocal tract) operating on signal sources (glottal pulses or noise generators), a reasonable feature set would characterize the signal produced by the filter acting on the signal source. Since the spectral domain has achieved great success in characterizing filtering operations as well as in other aspects of speech analysis, it is natural that it forms the basis of the features used. The time-varying character of the speech process is captured by performing the spectral analysis at short time intervals and repeating the analysis periodically.

Within the framework of time-varying spectral methods, a variety of features have been considered for the problem of speaker verification. At this point in the evolution of speaker verification, a set of features has been converged upon that are employed by the majority of systems that perform speaker verification as well as speech recognition. This convergence has occurred because of a persistent edge in performance that goes to systems that utilize these features. These are the Mel-frequency Cepstral Coefficients, usually referred to as MFCCs. These speech features seem particularly well suited to density function models such as mixture models or Hidden Markov Models (HMMs).

The Mel-frequency represents a warping of the frequency band that is linear from 0 to 1 kHz and logarithmic at the higher frequencies, and the analysis is performed over this warped axis (in principle). This warping function has its origins in the human perception of tones. The speech signal is constrained to a bandwidth of about 150–3500 Hz for telephone speech and about 50 Hz–8 kHz for broadband communications. The cepstral features are obtained by analyzing the Mel-scaled speech every 0.01 seconds over a time interval of about 20 milliseconds (nominally) in duration. This yields the Mel-scale spectrum and the Fourier coefficients of the logarithm of this function are produced. Depending on the system there are about 12–20 such coefficients produced every 0.01 seconds. These coefficients, MFCCs, are used to generate cepstral derivatives by differencing techniques, resulting in about 24 to 40 features and sometimes second derivatives. One or more normalized energy features also employed to round out the basic feature set.

The cepstra are in effect the lower-order Fourier coefficients of the log magnitude spectrum. By staying with the low-order coefficients the variation in the spectrum, induced by the pulsed nature of the vocal tract excitation, has minimal effect.

5.2.2 Speaker Models

Speaker models enable us to generate the scores from which we will make decisions. As in any pattern recognition problem the choices are numerous, and include the likes of neural networks, Support Vector Machines, Gaussian Mixture Models (GMMs) and Hidden Markov Models (HMMs), as well as combinations of the various approaches. While some models are capable of exploiting dependency in the sequence of feature vectors from the target speakers, the current state-of-the-art modeling approaches, based on Gaussian Mixture Models, treat the sequence of feature vectors as independent random vectors. While we shall initially consider models that do not exploit time dependence, we will consider those that do in Section 5.3 when we briefly consider text dependence.

The choice of model to use for a particular application will depend on the specific circumstances in any particular situation, although the choice, even to the expert, may not be clear. The major factors that will influence any choice are the amount of data available for the speaker models and the nature of the verification problem (is it text-independent or text-dependent?) and the level of performance that is desired.

5.2.2.1 The Gaussian Mixture Model

At the heart of today's state-of-the-art speaker verification systems is the Gaussian Mixture Model (GMM). This is a probability density function (pdf) that itself consists of a sum of multivariate Gaussian density functions. Being a mixture of Gaussians, it has the ability to place mass where collections of data training data exist. Additionally, its parameters can be readily estimated by Maximum Likelihood (ML) training procedures including the EM (Estimate-Maximize) algorithm [10], a well-known iterative algorithm for finding the ML estimate. In practice, each speaker has a GMM that is trained for them individually and the likelihoods generated from the GMM form the basis for generating the speaker scores from which a decision is made with regard to a speaker's identity. In addition to ML training methods it can also be trained by discriminative and adaptive methods that we describe below.

The probability density function is given for speaker S by the sum (or "mixture") of M Gaussian pdfs:

$$p_S(x) = \sum_{i=1}^{M} w_{i,S}\, p_{i,S}(x)$$

where

$$p_S(x) = \frac{1}{(2\pi)^{D/2} |\Sigma_{i,S}|^{D/2}} \exp\left\{-\frac{1}{2}(x-\mu_{i,S})^t \Sigma_{i,S}^{-1}(x-\mu_{i,S})\right\}$$

is the multivariate Gaussian density function, D is the dimension of the feature vector x, $\mu_{i,S}$ is the mean of each Gaussian pdf (indexed by distribution and speaker) and the $w_{i,S}$ are positive weights that sum to unity. Typically the covariance matrix, Σ, for each of the terms in the mixture is taken as diagonal to reduce computational complexity and storage requirements.

5.2.2.2 Considering the Models

While there are a variety of ways of viewing model choices, an important and perhaps insightful way for speaker verification applications is through the approach taken to training them. The different approaches to training are connected to the amount of training data available and provide a good initial approach to dealing with the issues of model selection. A taxonomy of approaches to training includes class conditional training, adaptive training and discriminative training. This taxonomy should be considered as a rough guide, since it is possible for the models to fall into more than one category, and the categories can also show some similarities. Furthermore, there is a connection between the training procedure and model normalization, i.e. the use of the alternate model.

5.2.2.3 Class Conditional Modeling – Maximum Likelihood Training

When employing probability density functions for speaker models an important approach is that of class conditional density estimation. This is nothing more than training the models for the speakers individually from the data collected for the particular target speaker. In the case of GMMs as well as other families of probabilities, the parameters of the models are typically estimated by the Maximum Likelihood (ML) criterion. If we let $p_S(x;\theta)$ denote the pdf of observed features, x, for speaker S, then the ML estimate of the model parameters, θ, is specified by

$$\hat{\theta} = \arg\max_{\theta} p_S(x;\theta)$$

Quite often there is no direct method for determining the ML estimate, and often iterative methods are employed. An important iterative training method for GMMs is the previously noted EM algorithm [10], which is employed quite frequently in speaker verification as well as speech recognition problems.

While the model for the target speaker has been estimated in a class conditional manner, that is, without knowledge of the alternative choice or choices for the speaker, the target speaker model is still employed with the use of an alternative model (e.g. cohort speakers or UBM), when a decision is made about the speaker's claimed identity.

The alternative model will consist of a collection of speaker models that have been trained in much the same way that the model has been trained for the target speaker. It can be a large set used for all target speakers, but can also be a set of speakers that are selected from the larger set and work especially well with the target speaker. The likelihood generated from this alternative model is usually the average of the likelihoods of all the speakers in the set. There are many variants in their use, such as averaging just the scores from the top-scoring cohort speakers.

Another alternative, which we will discuss below, is the Universal Background Model (UBM), which is a composite model constructed from the utterances of many speakers. The speakers that are employed for this model will typically be selected to match the target speaker population in one or more ways, such as gender and channel. Reynolds *et al.* discuss some of the issues in building a UBM in [11]. We have already noted in Section 5.1.1 the potential difficulties of matching the target too closely.

5.2.2.4 Class Conditional Models – Discriminative Training

We have discussed above the training of class conditional models with the traditional Maximum Likelihood approach. The process is one of finding the model parameters, whether done by an EM algorithm or other algorithm, to maximize the likelihood of the model. If the models were correct and there was a sufficient amount of training data available we could, with no loss in performance, stay with training of models in the class conditional manner.

When we go to discriminative training procedures we are acknowledging the usually correct assumption of model and data inadequacy. This is not to say that discriminative training can solve all our problems with speaker verification, but that it is a useful alternative. The discriminative training of models is concerned with the important and relevant question of "How should I select the parameters of my model or models such that we maximize the performance on our goal of speaker separability?". In order to accomplish this we need to move away from the ML criterion and employ as a training criterion the actual loss function that will be employed in the evaluation of an SV system or some other criterion that is highly correlated with our performance metric.

The performance of an SV system is typically measured as some function of an ROC or DET curve. For example, this can be the detection probability at a certain false acceptance rate or the equal error rate operation point or the area under part of a ROC curve. If the model parameters can be directly trained to the specific measurement criterion that is employed then it can be expected that performance improvements can be obtained even when working with the same class of models such as GMMs or HMMs that can be used for class conditional training.

Some of the early work using discriminative training methods in speech applications was with speech recognition systems [12, 13]. Speaker verification systems have more recently explored the use of discriminative training methods, e.g. [14–17]. Rosenberg *et al.* [14] describe a speaker verification model in which the criterion of total number of errors was employed, i.e. the number of false acceptances and false dismissals. The performance gain of class conditional methods seemed substantial when employing HMMs, which we will briefly discuss below. Heck and Konig

[17] employ discriminative techniques to optimize a speaker verification system at a particular operating point.

5.2.2.5 Class Conditional Models via Adaptation

The process of creating speaker models via adaptation starts with a generic model for a speaker and uses the data collected from the target speaker to tailor the generic model to the speaker. If we consider the generic model as a prior probability distribution we can view the process of model adaptation in the Bayesian framework of parameter estimation. See Reynolds *et al.* [11] for a discussion of a system based on this approach.

The generic model, often referred to as a universal background model (UBM) is typically modeled with a GMM and trained with utterances from hundreds of speakers. To the extent that the collection of speakers is kept as similar as possible to the target speaker, an economy is achieved in modeling the UBM. For example, creating a model consisting of just males or females is basic. The use of GMMs with of the order of 2000 mixture components is not uncommon.

The GMM is a particularly flexible model for the purposes of speaker adaptation. The process of adapting a model to a target speaker is to calculate, for each feature vector, the likelihood for each of the terms of the UBM mixture model. This likelihood is then converted into the probability that the feature vector came from each of the terms. That is, $\Pr(i|x)$, the probability that the feature vector x is from the ith Gaussian pdf in the mixture is computed. This probability is then employed in the computation of new means and variances for each of the Gaussians in the mixture. These new means and variances are interpolated with the original values to give new parameters for the Gaussian pdfs in the mixture as well as new weights. If for some terms of the original mixture there are no non-zero probabilities for any of the features, that term will remain unchanged in the new model.

One of the advantages achieved by this modeling approach is the efficient use of training data. The process of adaptation is incorporating into the UBM the specific speaker differences. Encoding of differences can be much more data-efficient than training a complete model. Another aspect of the adaptation process is that when we look at it as a process of estimating differences between the target speaker and the UBM we are in some sense also creating a discriminatively trained model, albeit without an explicit discriminative criterion.

The use of adaptation methods in conjunction with a UBM is a way to make effective use of large amounts of available speaker data. If such data is not available or if the data that is available is not a good match to the conditions and channel of the target speaker then this approach becomes less useful.

5.2.2.6 Other Class Conditional Models

It is possible to employ this class conditional approach with other than pdf models for the speaker. For example, in nearest neighbor classification the collection of feature vectors themselves becomes the model for the speaker. In this case the classification process would take the feature data for a test speaker, and based on the distances of the features to the two classes, determine a score for the data. An approach based on these ideas was developed by Higgins *et al.* [18] and gave good performance.

5.2.2.7 Inherently Discriminative Approaches

Class conditional models that have discriminative training imposed on them via the imposition of a discriminative training criterion, such as we have discussed above, form one class of discriminatively trained models. The other class is that for which discrimination is inherent in the approach. Such approaches include neural networks, Support Vector Machines (SVM), and classification trees to name just a few (see [19] for a description of the previously mentioned techniques). What these approaches have in common is that, from the outset, the training process involves the separation of two or more classes from each other (e.g. the target speaker from the non-target speaker). This is done either implicitly or explicitly. The result is effectively to create a decision boundary between the classes of interest. In the case of SVM, a linear function in feature space is employed to separate the classes. Although this approach employs a linear function of features, the features can be nonlinear functions of the cepstra that can result in very nonlinear boundaries with respect to cepstra. Classification trees employ recursive feature selection methods for determining boundaries to separate classes. Neural networks do not directly go after a decision boundaries but rather create a model for *P*(speaker | observations) which is inherently discriminative, since this method of training the model needs observations from both speaker and non-speaker and the optimization criteria that are employed are usually strongly correlated with maximizing correct classifications.

These discriminative techniques can all be used as the basis for a speaker verification system. For example, Schmidt and Gish [20] applied SVM techniques to the SV problem and Farrell *et al.* [21] employed a system that combined classification trees with neural networks to perform verification. There are many other examples.

5.2.2.8 Model Selection

We have described above a wide variety of models that have been employed in various speaker identification applications. The selection of the appropriate model for a particular application is still an art, and a variety of approaches should be considered. Currently GMMs have shown themselves to be adaptable to a wide variety of situations and should be one of the contenders in almost any application. However, any of the discriminative approaches, such as neural networks, offer the possibility of being more effective in situations with limited training data, since they use all the training data to find the decision boundary between classes rather than modeling classes separately and then combining the separate models.

5.3 Additional Methods for Managing Variability

Statistical models, given enough training data, should be able to handle all the variability that we will encounter under test conditions. However, we are usually somewhat data-starved and we must resort to other techniques in order to improve SV performance. Below we will discuss some of the approaches.

5.3.1 Channel Normalization and Modeling

We consider the channel to be the entire communication link over which the speaker's voice travels to the listener. The fact that this channel can be different at the time of verification than at the time of speaker enrollment can result in major degradations in verification performance if not dealt with. Of course, the usual statistical modeling approaches cannot account for this source of variability because they have not heard the speaker on this particular channel during the enrollment process.

The modeling alternative is to perform some type of channel normalization. The process of normalization is that of mitigating the effect of the channel based upon its inherent characteristics. If we assume that the channel can be modeled by a linear, time-invariant filter, then we can readily predict the effect of the channel on our cepstral speech features. If we let $c_{s,n}$ denote a cepstral vector from only the speech at time n and let $c_{r,n}$ denote the corresponding received cepstral vector, then we can write

$$c_{r,n} = c_{s,n} + c_{ch,n}$$

where $c_{ch,n}$ is the cepstral contribution from the channel. The channel contribution is additive because the cepstral coefficients are the Fourier coefficients of a log spectrum. We also note that the contribution from the channel, $c_{ch,n}$, will not actually depend on the index, n, because of our assumption of time invariance of the channel.

Thus we see that time differences in the received cepstra will not depend on the channel component. In particular, if we removed the mean value of all the cepstra in an utterance we will have performed the standard channel normalization called Cepstral Mean Subtraction (CMS). While the use of this method of normalization can be quite effective, it does destroy some speaker information. Since speakers have their own average cepstral value, i.e. $c_{s,n}$ has a speaker-dependent mean value, CMS will remove this as well. However, when channel variation is an issue the loss in speaker information will be inconsequential to the improvements obtained by normalization.

While the assumptions of channel linearity and time-invariance may not be strictly true, the merits of such a normalization can be seen by its centering of data about the origin in feature space to partially compensate for the effects of channels that can cause shifts in the data.

5.3.1.1 Other Methods for Dealing with the Channel

When we observe a speaker over an unknown linear time-invariant channel we have seen that in cepstral feature space the speaker has been shifted to new place. We have also seen that we can compensate for this shift by altering the features through the process of cepstral mean subtraction so that this shift is eliminated, albeit at the cost of losing some speaker information.

Thus, by this process, we have created new features that are less sensitive to channel variability. The question some researchers have pursued is whether there are features that are highly robust to channel variation and at the same time carry significant amounts of speaker-dependent information.

One of the more important classes of feature candidates has been formant-related features [22,23]. A formant is a location of a peak in the short-term spectrum, i.e. a resonant frequency of the vocal tract. Variations in the glottal pulse timings [23] and fundamental frequency [24] have also been examined. While all these features do carry speaker information and do have a degree of channel invariance, difficulties in extracting these features and modeling them have limited their utility. Some recent work by Murthy *et al.* [25] using a front end that produced spectral slope information showed significant performance improvements.

Another approach to the channel variability problem is to treat the channel as a random disturbance and integrate its variability directly into the scoring process for the speaker. The challenge with this approach is to estimate a probabilistic model for the channel and perform the integration. Gish *et al.* [26] modeled the time-invariant channel as a Gaussian random vector and were able to compute the effect of the channel randomness on the speaker model, obtaining useful gains in performance.

A recent and fairly ambitious approach for dealing with the channel has been developed by Teunen *et al.* [27]. In their approach they have constructed a channel detector, employing it to detect the type of channel over which the speech is received. If the current channel differs from that used when modeling the speaker, then MAP adaptation methods are employed to create a model for the target speaker under the current channel type. This approach seems to make some headway on the mismatched-channel problem, but it currently does not appear to be more effective than a handset normalization technique that is discussed below.

5.3.1.2 Score Normalization

In much of the recent speaker verification literature much work has been devoted to the subject of score normalization. The aim of this work is to counter variability in scoring for the purpose of allowing scores from different target speakers to be evaluated as a single collection of scores. In such a situation of score combination, if target speaker B always has better scores with his model than any impostor, i.e. perfect verification performance, and has his scores combined with those of speaker C, who also has perfect verification performance, the evaluation of their scores as a single set could be rather poor unless some normalization is performed. This can happen because without normalization the impostor scores for speaker C can be, for example, greater than the non-impostor scores for speaker B.

The approaches to dealing with this problem involve simple transformations that convert the scores to zero mean and unit variance. That is, if $s_{T,A}(k)$ is the kth impostor score generated in training for the target speaker, and we let $\mu(s_{T,A}(k))$ denote the mean and $\sigma(s_{T,A}(k))$ denote the standard deviation of all the impostor scores against this target speaker, scores for this target can be normalized by

$$s_{T,A,\text{norm}}(k) = \frac{s_{T,A}(k) - \mu(s_{T,A}(k))}{\sigma(s_{T,A}(k))}$$

When these transformation parameters are generated using impostor speakers from a set withheld (or "sequestered") during the training process, the norm is called the Z-norm. When the impostor scores are generated during the testing phase from impostor speakers within the actual test data, the norm is referred to as the T-norm. Although it might also be possible to do a similar type of normalization using "genuine" scores for each target speaker, this is typically are not done due to lack of a sufficient number of target speaker scores.

Normalizations can also be generated when transmission is over a telephone channel where there are two different hand-set types (carbon button and electret). This is called the H-norm. In this case one needs to be able to detect the type of hand-set in use in the training and the test phases to be able to apply the appropriate normalizations.

In all the above cases the shift and scaling transformation employed cannot change the rank ordering of scores for a given speaker or channel condition. They will enable a verification system to operate with a detection threshold that is not dependent on the speaker or the channel. For an extensive discussion of normalization techniques, see Auckenthaler *et al.* [28].

5.3.2 Constraining the Text

Thus far we have been considering text-independent verification – the case where there is no constraint on the text spoken. In those situations, such as access control, where it is possible to have the speaker utter a known phrase, performance of an SV system can be greatly improved. When the text is known in advance there is the advantage of being able to have models that are targeted to the way the speaker says all parts of the utterance as well as having the sequential structure. These two factors in systems can bring about major gains over text-independent approaches.

One of the approaches originally employed in this text-dependent situation was a non-statistical, template-based method called dynamic time-warping (DTW). An example of such a template-based system is presented by Furui in [29]. In this type of approach there was an attempt to match the string of features produced by the speaker against a stored template. The matching allowed for what was assumed to be normal variability in speaker rhythm. A dynamic programming algorithm was employed to determine the best alignment between the feature strings in the received speech and the stored template. Such approaches are still in use today because of their ability to produce useful results with very limited amounts of training data and modest amounts of computation. When sufficient training data are available the method of choice is that of Hidden Markov Models (HMMs) [30] which form the basis of state-of-the-art speech recognition systems and are much more efficient than DTW methods in characterizing the variability in the speech process.

The HMM consists of a sequence of states with a GMM at each state. Words consist of a sequence of phonetic units and a three-state HMM is typically used to model a phonetic unit. Sometimes the HMM will be used to model an entire word. At the risk of oversimplification, we can consider an HMM to be a sequence of GMMs each tuned to a different part of each word. Because of the state transition probabilities in the HMM model, the states included for evaluation can be variable. Although with the use of constrained text the actual acoustic models have increased in complexity, almost everything else has remained the same. For example we are still concerned with cohort models, discriminative training and channel normalization. The previously cited paper by Rosenberg *et al.* [14] considers an HMM-based verification system that is also discriminatively trained. Below we will describe a verification systems that are based on HMM speech recognition technology (see Section 5.5).

5.4 Measuring Performance

We have already noted that we can measure the performance of a speaker verification system by its probability of a false dismissal versus the probability of false acceptance at a given threshold setting. As we change the system's detection threshold we end up with a collection of values that give the DET (Detection Error Trade-off) curve. While this is a fairly straightforward concept, how one should proceed from the collection of scores for individuals to the system DET curve is not uniquely defined.

For any individual, since there are typically only a few target speaker scores, a DET curve will be highly discontinuous. We show in Figure 5.3 a DET curve for a single target speaker for which we have a small number of target test samples and significantly more impostor scores. One approach that is used by the National Institute of Standards and Technology in their annual evaluations [31] is to treat all the individual scores from individual speakers as if they were from one speaker and create the composite DET curve. This requires that a great emphasis be placed on normalizing scores, as we have discussed above, in order to avoid severe degradations in measured performance due to the incommensurate scores from different target speakers.

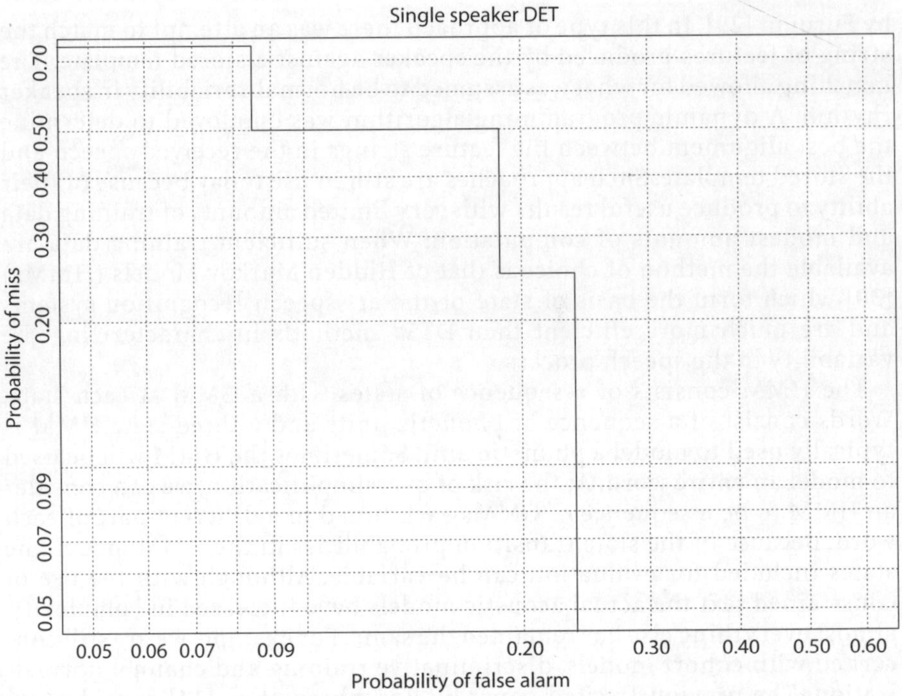

Figure 5.3 Example of a DET curve for a single speaker.

If one's system has speaker- and/or channel-dependent thresholds then such normalizations are not necessary and composite DET curves produced by employing such normalizations can actually be misleading indicators of system performance. In the case of systems with speaker-dependent thresholds (and no channel detection, for example) a more appropriate way of combining the scores would be by combining the individual DET curves, e.g. for a given P_{FA} average all the P_Ms. While this procedure does not require score normalizations, adjustments may be needed for variations in the number of scores provided by individual speakers. The point is, however, that the method of performance measurement should be in tune with the way the system will be employed. Also note that the two examples presented do not exhaust the possible methods for developing DET curves for collections of speakers from the various scores of individual speakers.

While the DET or ROC curves provide a great deal of information about SV system performance, it is often desirable to characterize the performance of a system by a single number in order to facilitate system comparisons. The number often employed is the equal error rate (EER), the point on the DET or ROC where the probability of a miss is equal to the probability of false acceptance. The EER has found wide acceptance because it is easy to understand, even though it may not be a desirable operating point for a practical system. Another single measure of system performance called the detection cost function (DCF) involves assigning costs to the two different types of error. Ultimately a user needs to determine the performance factors of most importance and develop the measures appropriate for the application.

5.4.1 How Well do These Systems Perform?

Clearly, there are so many factors that affect the performance of a system that any numbers presented must be highly qualified. The factors affecting performance include the amounts of training data and the duration of the evaluation data, the variability and types of channels employed, the protocols employed in certain systems such as using additional prompts for the target speaker, the constraints on the text, and others. Notwithstanding these caveats, it is probably useful to provide the reader with some numbers.

We can nominally think of text-independent verification systems with modest amounts of training data, test data and channel variation as having an EER in the 5% to 20% range. I include such a number in order for the reader to have some sense of performance range, as vague as it is. A system to be described below [39], which employed large amounts of training and test data and applied several sources of information in verifying speakers, measured less than a 1% EER. For text-dependent systems we can have EERs that can be an order of magnitude smaller that the numbers quoted for text-dependent systems, again depending on many factors.

5.5 Alternative Approaches

In the approaches described below, we stay with the statistical likelihood view of the speaker verification problem. The speech recognition approaches we describe bring feature time-dependence, as well as new knowledge sources, to bear on the verification problem. The systems that employ these new sources of information can bring about major improvements in verification performance if sufficient data is available for training and testing. The second approach we consider is more narrowly focused and shows that by judicious model selection, likelihood ratio approaches can create robust models that give excellent performance and require significantly less computation than GMMs.

5.5.1 Speech Recognition Approaches

The approaches that we have considered thus far, except when we were considering the constraints of text dependency, have treated the cepstra extracted from the speech as independent events in time, and have not exploited any of the higher level information in the speech process (e.g. phonetic categories), relying primarily on the general characteristics of an individual's vocal production mechanism for the purpose of identification. While there have been some attempts at using various categories of phonetic units in the verification process the discriminative gains obtained from their employment was mitigated by the inability to reliably extract them from speech over communication channels. However, with improvements in both speech and phonetic recognizers these approaches have become important alternatives to the memoryless approaches relying on low-level information.

An early advocate of the use of speech recognizers for speaker identification applications was Dragon Systems [32, 33]. Their approach focused on the recognizer's ability to provide higher-level acoustic knowledge (context-independent phonetic units) about speakers. To train the models they adapted their speaker-independent recognizer to target speakers by employing standard Baum–Welch HMM training procedures. The evaluation process entailed performing speech recognition on a test utterance and comparing the scores on the target speaker-trained phonetic unit with the scores obtained by the speaker-independent recognizer. This comparison provided normalized scores upon which the decision was made.

While the Dragon approach fared reasonably well in most of its direct comparisons to other techniques, it fell short of the performance obtained by state-of-the-art GMM approaches. In these comparisons, the scenarios were those of rather limited available training and test data – a difficult place to operate with a knowledge-intensive approach.

Related to the above was the use of a speech recognizer by Kimball *et al.* [34] for a text-dependent application, although the motivation for the choice was the limited amount of training data available for each speaker. In their application, only one training sample was available for each speaker. This training sample was used to adapt a speaker-independent recognizer by means of Maximum Likelihood Linear Regression (MLLR) methods. The adaptation parameters effectively became the speaker model. The score produced by the adapted recognizer on the test utterance became the target speaker's score that was normalized by the scores from cohort speakers. The system performed well with the interesting property that overall performance was unaffected when tests were performed with handsets different from those used in training.

5.5.2 Words (and Phonetic Units) Count

The two approaches described above both employed speech recognition systems and used the speech recognizer to produce acoustic scores for the target speaker. In the paper by Gauvain *et al.* [35], a speaker identification system based on a phone recognizer was described. The training of the acoustic models for target speakers was accomplished by adaptation and since the model was a fully connected HMM, i.e. transitions permitted between all phones at any time, the model was, to a degree, capturing the target speaker's use of language in learning the transitions between phones. This implicit learning of the speaker's language or idiolect can be quite important in improving performance.

While the approach presented by Gauvain incorporated the language patterns of the speaker implicitly, there has been more recent work incorporating the language patterns of the target speaker explicitly. Doddington [36] demonstrated that frequently occurring bigrams, i.e. word pairs, contain a great deal of speaker-specific information. His experiments were done on the Switchboard conversational speech corpus employing the transcriptions of the conversations. The bi-grams that were found to contain the speaker dependent information were typically content-free word pairs, such as "it were", "you bet" and "so forth". Extracting speaker information from the speaker's language, as one might expect, required much more training than has typically been employed in evaluations. The work of Andrews *et al.* [37] demonstrated that word-based speaker information could be captured by phonetic recognition systems. They showed that significant speaker information was contained in triphone (sequences of three phones) occurrence frequencies. They also demonstrated that useful results could be obtained even if the phonetic recognizer was not in the language of the target speaker.

The natural extension of the proceeding work was to combine language information with acoustic information. This extension was done by Andrews *et al.* [38] and showed that the two sources of information complement one another. The exploitation of large amounts of training and test data for speaker identification purposes was further advanced in a project that was part of the Johns Hopkins 2002 Speech and Language Workshop [39]. At this workshop, the use of acoustic, lexical and prosodic information was explored. The different components were effectively combined to achieve very good performance for speaker verification on conversational speech.

5.5.3 Models Exploring the Shape of Feature Space

Statistical modeling through the use of the likelihood ratio plays a key role in SV. Employing likelihood ratios, however, does not necessarily imply the use of GMMs. Under the assumption that a speaker's model is characterized by a single Gaussian pdf, combined with the assumption that the mean of the cepstral data is a source of noise due to channel variation, one is led to a likelihood ratio test for speaker models that score the covariance structure of the models being compared. Thus this model ignores the fine structure of the speech, working with the shape that a speaker's features form in feature space. In [40], Gish introduced the notion of comparing the shapes of feature spaces through the scoring of the relative eigenvalues of covariance matrices. Zilca *et al.* [41] consider measures of shape and T-norm scaling of scores in conjunction with channel detection to produce a near state-of-the-art system working in the cellular phone environment. This system requires significantly less computation than GMMs.

5.6 Summary

The problem of speaker verification is challenging due to the inherent variability in both the speech of individuals and the transmission channels.

Statistical modeling and pattern recognition techniques have been able to tame some of this variability. Different sources of speaker information are now being used effectively in combination to improve verification performance. While good progress has been made in recent years, the important problem of verifying the identity of a speaker over a channel different from that used to obtain the speaker's training data still offers challenges for the future.

References

[1] S. Pruzansky, Pattern-matching procedure for automatic talker recognition. *J. Acoust. Soc. Am.*, **35**, March, 354–358, 1963.

[2] J. E. Keith Smith, A decision-theoretic speaker recognizer. *64th Meeting of the Acoustical Society of America*, November 1962.

[3] S. Pruzansky, Talker recognition procedure based on analysis of variance. *J. Acoust. Soc. Am.*, **36**(11), 2041–2047, 1964.

[4] *Switchboard: A User's Manual*. Linguistic Data Consortium. Available online at `http://www.ldc.upenn.edu/Catalog/CatalogEntry.jsp?catalogId=LDC97S62`.

[5] A. L. Higgins, L. Bahler and J. Porter, Speaker verification using randomized phrase prompting. *Digital Signal Processing*, **1**, 89–106, 1991.

[6] M. J. Carey and E. S. Parris, Speaker verification using connected words. *Proc. Inst. Acoustics*, **14**(6), 95–100, 1992.

[7] D. A. Reynolds, Comparison of background normalization methods for text-independent speaker verification, *Proc. European Conference on Speech Communication and Technology*, September 1997, pp. 963–966.

[8] A. Martin, G. Doddington, T. Kamm, M. Ordowski and M. Przybocki, The DET curve in assessment of detection task performance. *Proc. Eurospeech-97*, Rhodes, Vol. 4, pp. 1895–1898.

[9] D. R. Cox and D. V. Hinkley, *Theoretical Statistics*. Chapman & Hall, 1974.

[10] A. P. Dempster, N. M. Laird and D. B. Rubin, Maximum likelihood from incomplete data via the EM algorithm. *J. Roy. Statistical Soc. B*, **39**, 1–38, 1977.

[11] D. R. Reynolds, T. F. Quatieri and R. Dunn, Speaker verification using adapted mixture models. *Digital Signal Processing*, **10**, 19–41, 2000.

[12] A. Ljolje, Y. Ephraim and L. R. Rabiner, Estimation of hidden Markov model parameters by minimizing empirical error rate. *IEEE International Conference on Acoustics, Speech, and Signal Processing*, Albuquerque, April 1990, pp. 709–712.

[13] L. R. Bahl, P. F. Brown, P. V. de Souza and R. L. Mercer, Maximum mutual information estimation of Hidden Markov Model parameters for speech recognition. *IEEE International Conference on Acoustics, Speech, and Signal Processing*, Tokyo, 1986, pp. 49–52.

[14] A. E. Rosenberg, O. Siohan, and S. Parthasarathy. Speaker verification using minimum verification error training. *IEEE International Conference on Acoustics, Speech, and Signal Processing*, Seattle, Washington, May 1998.

[15] C.-S. Liu, C.-H. Lee, W. Chou, A. E. Rosenberg and B.-H. Juang, A study on minimum error discriminative training for speaker recognition. *J. Acoust. Soc. Am.*, **97**(1), 637–648, 1995.

[16] O. Siohan, A. E. Rosenberg and S. Parthasarathy, Speaker identification using minimum classification error training. *IEEE International Conference on Acoustics, Speech, and Signal Processing*, Seattle, Washington, May 1998.

[17] L. Heck and Y. Konig, Discriminative training of minimum cost speaker verification systems. *Proc. RLA2C-ESCA Speaker Recognition and its Commercial and Forensic Applications*, Avignon, France, 1998, pp. 93–96.

[18] A. L. Higgins, L. G. Bahler and J. E. Porter, Voice identification using nearest-neighbour distance measure. *IEEE International Conference on Acoustics, Speech, and Signal Processing*, Vol. 2, Minneapolis, 1993, pp. 375–378.

[19] R. O. Duda and P. E. Hart and D. Stork, *Pattern Classification*. John Wiley, New York, 2001.

[20] M. Schmidt and H. Gish, Speaker identification via support vector classifiers. *IEEE International Conference on Acoustics, Speech, and Signal Processing*, Atlanta, 1996, pp. 1:105–108.

[21] K. R. Farrell, R. J. Mammone and K. T. Assaleh, Speaker recognition using neural networks and conventional classifiers. *IEEE Trans. Speech and Audio Processing*, **2**(1:II), 194–205, 1994.

[22] S. K. Das and W. S. Mohn, A scheme for speech processing in automatic speaker verification. *IEEE Trans. Audio Electroacoustics*, **AU-19**, 32–43, 1971.

[23] C. R. Jankowski Jr., T. F. Quatieri and D. A. Reynolds, Measuring fine structure in speech: application to speaker identification. *ICASSP'95*, pp. 325–328.

[24] M. J. Hunt, J. W. Yates and J. S. Bridle, *International Conference on Acoustics, Speech and Signal Processing*, May 1977, pp. 764–767.

[25] H. Murthy, F. Beaufays, L. P. Heck and M. Weintraub, Robust text-independent speaker identification over telephone channels. *IEEE Trans. Speech and Audio Processing*, **7**(5), 554–568, 1999.

[26] H. Gish, M. Krasner, W. Russell and J. Wolf, Methods and experiments for text-independent speaker recognition over telephone channels. *IEEE International Conference on Acoustics, Speech, and Signal Processing*, Tokyo, 1986, pp. 865–868.

[27] R. Teunen, B. Shahshahani and L. P. Heck, A model-based transformational approach to robust speaker recognition. *Proc. ICSLP*, 2000.

[28] R. Auckenthaler, M. Carey and H. Lloyd-Thomas, Score normalization for text-independent speaker verification systems. *Digital Signal Processing*, **10**, 42–54, 2000.

[29] S. Furui, Cepstral analysis technique for automatic speaker verification. *IEEE Trans. Acoustics, Speech and Signal Processing*, **29**(2), 254–272, 1981.

[30] L. R. Rabiner and B.-H. Juang, *Fundamentals of Speech Recognition*. Prentice Hall, Englewood Cliffs, NJ, 1993.

[31] A. Martin and M. Przybocki, The NIST speaker recognition evaluation – an overview. *Digital Signal Processing*, **10**, 1–18, 2000.

[32] B. Peskin, L. Gillick, Y. Ito, S. Lowe, R. Roth, F. Scattone, J. Baker, J. Baker, J. Bridle, M. Hunt and J. Orloff, Topic and speaker identification via large vocabulary speech recognition, *Arpa Workshop on Human Language Technology*, Princeton, NJ, 1993, pp. 119–124.

[33] M. Newman, L. Gillick, Y. Ito, D. McaAllaster and B. Peskin, Speaker verification through large vocabulary continuous speech recognition. *IEEE International Conference on Acoustics, Speech, and Signal Processing*, 1996, pp. 2419–2422.

[34] O. Kimball, M. Schmidt, H. Gish and J. Waterman, Speaker verification with limited enrollment data. *Proc. Eurospeech '97*, Rhodes, Greece, pp. 967–970.

[35] J. L. Gauvain, L. F. Lamel and B. Prouts, Experiments with speaker verification over the telephone, *Proc. Eurospeech'95*, Madrid 1995.

[36] G. Doddington, Some experiments in ideiolectal differences among speakers, January, 2001. Available online at `http://www.nist.gov/speech/tests/spk/2001/doc/`

[37] W. D. Andrews, M. A. Kohler, J. Campbell, J. Godfrey and J. H. Cordero, Gender-dependent phonetic refraction for speaker recognition. *Proc. IEEE International Conference on Acoustics, Speech, and Signal Processing*, 2002.

[38] W. M. Andrews, M. Kohler and J. Campbell, Acoustic, idiolectical and phonetic speaker recognition. *Proc. 2001: A Speaker Odyssey*, The Speaker Recognition Workshop, Crete, 2001.

[39] SupersSID Team, Johns Hopkins Summer 2002 Workshop. Available online at `http://www.clsp.jhu.edu/ws2002/groups/supersid/SuperSID_Closing_Talk_files/frame.htm`.

[40] H. Gish, Robust discrimination on automatic speaker identification. *IEEE International Conference on Acoustics, Speech, and Signal Processing*, New Mexico, 1990, pp. 289–292.

[41] R. D. Zilca, U. V. Chaudhari and G. N. Ramaswamy, The sphericity measure for cellular speaker verification. *IEEE International Conference on Acoustics, Speech, and Signal Processing*, Orlando, 2002 pp. 1-697–1-670.

Technology Evaluation of Fingerprint Verification Algorithms

6

D. Maio, D. Maltoni, R. Cappelli, J. L. Wayman and A. K. Jain

6.1 Introduction

In the last few years, many academic and industrial research and development groups have created new measurement techniques and new acquisition sensors for automatic fingerprint recognition. Fingerprint-based biometric systems represent a significantly growing commercial segment for pattern recognition applications [9]. Nevertheless, given the lack of standards, in the past most developers have generally performed only internal tests over self-collected databases. Few standardized benchmarks have been available for comparing developments in fingerprint verification. This deficiency has unavoidably led to the dissemination of confusing, incomparable and irreproducible results, sometimes embedded in research papers and sometimes enriching the commercial claims of marketing brochures.

The only public domain data sets have been the (US) National Institute of Standards and Technology (NIST) CD-ROMs [21,22] containing thousands of images scanned from paper cards where fingerprints were impressed by rolling inked fingers from "nail to nail". These images differ significantly from those acquired directly from the finger by "live-scan" optical or solid state sensors. Although these collections of "rolled" images constitute an excellent data set for benchmarking forensic AFIS (Automated Fingerprint Identification Systems) [12] and fingerprint classification development [4, 10], they are not well-suited for testing "online" fingerprint systems [9] commonly used in access control and civilian AFIS applications (i.e. driver's licensing and social service systems). In 1998, NIST released a database containing digital videos of live-scan fingerprint data [23]. As this database was specifically collected for studying the effects of both finger rotation and plastic skin distortion on the online acquisition process [5,6], it models only certain fingerprint variations and it is not applicable to the general evaluation of fingerprint verification algorithms.

The aim of the Fingerprint Verification Competition (FVC), organized by the authors for the first time in 2000, was to attempt to establish a common basis for better understanding, within both academia and industry, the state-of-the-art and the future development needs of fingerprint technology. Analogous efforts have been carried out for other biometric technologies (e.g. face [13, 16], and voice [2, 17]) and for other more classical pattern recognition tasks [1, 8, 18, 19]. With the thought that an international open competition could boost interest and give results larger visibility, the 15th International Conference on Pattern Recognition (ICPR) (ICPR 2000) was chosen as the forum for announcing our results. In late spring 1999, the FVC2000 web site [7] was set up to broadly publicize this event and we directly invited several companies and research groups known to us to take part.

The authors believe that the FVC2000 protocol, databases and results have been useful to all practitioners in the field, not only as a benchmark to improve their methods, but also for enabling an unbiased evaluation of algorithms. However, as with all benchmarks, there are limitations to the general applicability of results. From the beginning, we stated that the competition was not meant as an official performance certification of the participant biometric systems, as:

- The databases used in this contest have not been acquired in a real environment and according to a formal protocol [2, 17, 20, 24] (also refer to [25] for an example of performance evaluation on real applications).
- Only parts of the participants' software are evaluated, and this by using images from sensors not native to each system. Fingerprint-based biometric systems often implement proprietary solutions to improve robustness and accuracy (e.g. quality control modules to reject poor quality fingerprints, visual feedback to help users in optimally positioning their fingers, use of multiple fingerprint instances to build more reliable templates etc.), and these contributions are here discounted.

According to the definitions of [17] and [20], FVC2000 should be conceived as a *technology evaluation* (with some analogies with the FERET [16] and the NIST Speaker Verification [2] competitions). In fact, quoting [2]:

> The goal of a technology evaluation is to compare competing algorithms from a single technology. Testing of all algorithms is done on a standardized database collected by a "universal" sensor. Nonetheless, performance against this database will depend upon both the environment and the population in which it was collected. Consequently, the "three bears" rule might be applied, attempting to create a database that is neither too difficult nor too easy for the algorithms to be tested. Although sample or example data may be distributed for developmental or tuning purposes prior to the test, the actual testing must be done on data which has not been previously seen by algorithm developers. Testing is done using "off-line" processing of the data. Because the database is fixed, results of technology tests are repeatable.

FVC2000 received great attention from both academic and commercial organizations. On the one hand, it allowed developers to unambiguously compare their algorithms; on the other, it provided the first overview of the state of the art in fingerprint recognition and shed some light on fingerprint individuality [28]. Specifically, in FVC2000:

- Eleven algorithms were submitted (7 academic, 4 commercial).
- Four databases were collected (one of them was synthetically generated).
- Synthetic fingerprint generation [3, 29] was validated as an effective instrument for comparing algorithms and in-house improvement of methods.
- A CD-ROM containing the four databases and a detailed report was created, more than 80 copies of which have been requested by major institutions and companies in the field. The web site [7] has been visited more than 20,000 times since September 2000.
- Several scientific groups active in the field are currently using FVC2000 databases for their experimentation, allowing them to fairly compare their approaches to published results.
- Some companies which initially did not participate in the competition requested to certify their performance on the FVC2000 benchmark after the competition [7].

The interest aroused by FVC2000, and the encouragement we received, induced us to set up a second competition. In the organization of FVC2002 [30], we took into account advice we received by experts in the field and by reviewers of the FVC2000 paper [31]. By January 10, 2002 (the deadline for FVC2002 registration), we had received 48 registrations (19 academic, 29 industrial), far more than our initial expectation. All the registered participants received the training sets and detailed instructions for the algorithm submission. By March 1, 2002 (the deadline for submission) we had received a total of 33 algorithms from 29 participants (four participants submitted two algorithms). The percentage of withdrawals after registration decreased from 56% in FVC2000 to 31% in FVC2002. The evaluations of the 33 submitted algorithms were presented at the 16th ICPR Conference and are now available online [30].

This chapter is organized as follows. Section 6.2 summarizes the FVC2000 submission rules and Section 6.3 describes the four databases used. In Section 6.4 we present the criteria and the procedures used for performance evaluation. Section 6.5 reports the overall performance of the participating algorithms on each database and concludes with a comparison of the average results. Section 6.6 discusses the FVC2002 databases, the test protocol and the main differences from FVC2000. Finally, in Section 6.7 we draw some concluding remarks and discuss how we intend to continue supporting this initiative in the future.

6.2 FVC2000 Organization and Algorithms Submission Rules

In FVC2000, the "universal" sensor was actually a collection of four different sensors/technologies to better cover the recent advances in fingerprint sensing techniques and to avoid favoring a particular algorithm through the choice of a specific sensor. In fact, of the four databases used in the test, databases 1 and 2 were collected by using two small-size and low-cost sensors (optical and capacitive, respectively). Database 3 was collected by using a higher quality (large area) optical sensor. Images in database 4 were synthetically generated by using the approach described in [3]. Each of the four databases contained 880 fingerprints from 110 different fingers, collected using the "three bears rule" (not too easy, not too hard) [20], based on our prior subjective experiences with fingerprint recognition algorithms. In particular, on the one hand, we discarded fingerprint images we considered completely intractable even for a human expert, while on the other hand we avoided collecting perfect fingerprints which would be very easy for a matching algorithm. Some internally developed algorithms helped us in accomplishing this task. Each database was split into a sequestered "test" set of 800 images (set A) and an open "training" set of 80 images (set B), made available to participants for algorithm tuning. The samples in each set B were chosen to be as representative as possible of the variations and difficulties in the corresponding set A. To this end, fingerprints were automatically sorted by quality as in [15] and samples covering the whole range of quality were included in set B. A final visual inspection of the obtained data sets was carried out to ensure that "dry", "wet", "scratched", "distorted" and "markedly rotated" fingerprints were also adequately represented.

In March 2000, after several months of active promotion, we had 25 volunteering participants (about 50% from academia and 50% from industry), and by the end of April 2000, the training sets were released to the participants.

After the submission deadline (June 2000) for the executables, the number of participants decreased to 11 (most of the initially registered companies withdrew). Nonetheless, the number of participants (see Table 6.1) was more than we had anticipated, so we started working on the submitted executables to complete their evaluation by August 2000.

Once all the executables were submitted, feedback was sent to the participants by providing them the results of their algorithms on training set B (the same data set they had previously been given) to allow them to verify that neither run-time problems nor hardware-dependent misbehaviors were occurring on our side.

Each participant was required to submit two executables in the form of a "win32 console application". According to the given specification, the executables take input from command-line arguments and append the output to a text file. The input includes a database-specific configuration file. Participants were allowed to submit four distinct configuration files – `CF1.cfg`, `CF2.cfg`, `CF3.cfg` and `CF4.cfg` (one for each database) – in order to adjust the algorithm's internal parameters according to each specific database. Configuration files are text files or binary files and their input is the responsibility of the participant's code. Configuration files can also contain pre-computed data to save time during enrollment and matching.

Table 6.1 List of participants: a four digit ID was assigned to each algorithm. (Sagem SA submitted two different algorithms.)

ID	Organization	Type
CETP	CEFET-PR / Antheus Technologia Ltda (Brasil)	Academic
CSPN	Centre for Signal Processing, Nanyang Technological University (Singapore)	Academic
CWAI	Centre for Wavelets, Approximation and Information Processing, Department of Mathematics, National University of Singapore (Singapore)	Academic
DITI	Ditto Information & Technology Inc. (Korea)	Commercial
FPIN	FingerPin AG (Switzerland)	Commercial
KRDL	Kent Ridge Digital Labs (Singapore)	Academic
NCMI	Natural Sciences and Mathematics, Institute of Informatics (Macedonia)	Academic
SAG1	SAGEM SA (France)	Commercial
SAG2	SAGEM SA (France)	Commercial
UINH	Inha University (Korea)	Academic
UTWE	University of Twente, Electical Engineering (Netherlands)	Academic

- The first executable (`ENROLL_XXXX`) enrolls a fingerprint image and produces a template file; the command-line syntax is:

  ```
  ENROLL_XXXX imagefile templatefile configfile outputfile
  ```

  where

`XXXX`	is the participant id
`imagefile`	is the input TIF image pathname
`templatefile`	is the output template pathname
`configfile`	is the configuration file pathname
`outputfile`	is the output text file where a `log string` (of the form `imagefile templatefile result`) must be appended; `result` is "OK" if the enrollment can be performed or "FAIL" if the input image cannot be processed by the algorithm.

- The second executable (`MATCH_XXXX`) matches a fingerprint image against a fingerprint template and produces a similarity score; the command-line syntax is:

  ```
  MATCH_XXXX imagefile templatefile configfile outputfile
  ```

  where:

`XXXX`	is the participant id
`imagefile`	is the input TIF image pathname
`templatefile`	is the input template pathname
`configfile`	is the configuration file pathname
`outputfile`	is the output text file where a `log string` (of the form `imagefile templatefile result similarity`) must

be appended; `result` is "OK" if the matching can be performed or "FAIL" if the matching cannot be executed by the algorithm; `similarity` is a floating-point value ranging from 0 to 1 which indicates the similarity between the template and the fingerprint: 0 means no similarity, 1 maximum similarity.

Two C-language skeletons for `ENROLL_XXXX` and `MATCH_XXXX` were made available online to reduce the participants' implementation efforts. These source files perform all the necessary I/O (including TIF image loading).

We also premised that, for practical testing reasons, we should limit the maximum response time of the algorithms: 15 seconds for each enrollment, 5 seconds for each matching. The test was executed on machines with Pentium III processors running at 450 MHz (under Windows NT 4.0 and Linux RedHat 6.1).

6.3 Databases

Four different databases (hereinafter referred to as DB1, DB2, DB3 and DB4) were collected by using the following sensors/technologies [11] (Fig. 6.1):

- DB1: optical sensor "Secure Desktop Scanner" by KeyTronic
- DB2: capacitive sensor "TouchChip" by ST Microelectronics
- DB3: optical sensor "DFR-90" by Identicator Technology
- DB4: synthetically generated based on the method SFinGe proposed in [3].

Each database is 110 fingers wide (*w*) and 8 impressions per finger deep (*d*) (880 fingerprints in all); fingers numbered from 101 to 110 (set B) were made available to the participants to allow parameter tuning before the submission of the algorithms; the benchmark is then constituted by fingers numbered from 1 to 100 (set A). For a system evaluation, the size of the

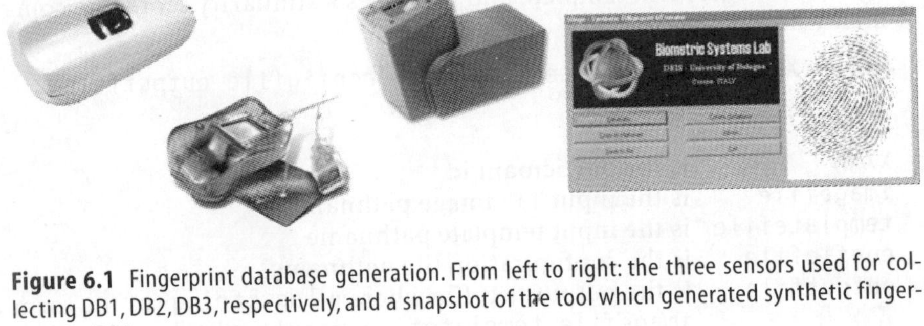

Figure 6.1 Fingerprint database generation. From left to right: the three sensors used for collecting DB1, DB2, DB3, respectively, and a snapshot of the tool which generated synthetic fingerprints in DB4.

Table 6.2 The four FVC2000 databases.

	Sensor type	Image size	Set A (w × d)	Set B (w × d)	Resolution
DB1	Optical sensor	300 × 300	100 × 8	10 × 8	500 dpi
DB2	Capacitive sensor	256 × 364	100 × 8	10 × 8	500 dpi
DB3	Optical sensor	448 × 478	100 × 8	10 × 8	500 dpi
DB4	SFinGe v 2.1	240 × 320	100 × 8	10 × 8	About 500 dpi[1]

[1]In the artificial generation, the resolution is controlled by the average ridge-line inter-distance; this input was estimated from a real 500 dpi fingerprint database.

above four databases is certainly not sufficient to estimate the performance with high confidence. However, in a technology evaluation (like FVC2000) the aim is to capture the variability and the difficulties of the problem at hand and to investigate how the different algorithms deal with them. For this purpose, the sizes of our database are adequate.

Table 6.2 summarizes the global features of the four databases, and Figure 6.2 shows a sample image from each of them.

It is worth emphasizing that the protocol of providing more than one database is not aimed at comparing different acquisition technologies and devices. The results obtained by the algorithms on the different databases cannot be conceived as a quality measure of the corresponding sensors, since the acquisition conditions and the volunteer crew of each database are different.

To summarize, DB1 and DB2 have the following features:

- The fingerprints are mainly from 20 to 30 year-old students (about 50% male).
- Up to four fingers were collected for each volunteer (forefinger and middle finger of both the hands).
- The images were taken from untrained people in two different sessions and no efforts were made to ensure minimum acquisition quality.

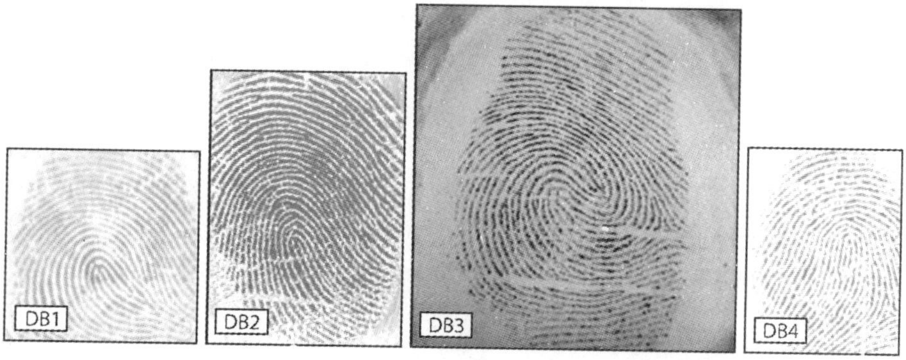

Figure 6.2 Sample images taken from DB1, DB2, DB3 and DB4. In order to show the different image sizes of each database, the four images are displayed at the same scale factor.

- All the images from the same individual were acquired by interleaving the acquisition of the different fingers (e.g. first sample of left forefinger, first sample of right forefinger, first sample of left middle, first sample of right middle, second sample of the left forefinger, ...).
- The presence of the fingerprint cores and deltas is not guaranteed, since no attention was paid to checking the correct finger position on the sensor.
- The sensor platens were not systematically cleaned (as usually suggested by the vendors).
- The acquired fingerprints were manually analyzed to ensure that the maximum rotation is approximately in the range [–15°, 15°] and that each pair of impressions of the same finger has a non-null overlapping area.

Figures 6.3–6.6 show some images from DB1 and DB2.

The database DB3 was collected as follows:

- The fingerprints are from 19 volunteers between the ages of 5 and 73 (55% male).
- One-third of the volunteers were over 55 years of age.
- One-third of the volunteers were under 18 years of age.
- One-sixth of the volunteers were under 7 years of age (children's fingerprints constitute an interesting case study, since the usable image area is small and the ridge-line density is high).
- Two images of up to six fingers (thumb, fore and middle on left and right hands) were taken without interleaving from each volunteer at each session and no efforts were made to ensure a minimum acquisition quality.
- Each volunteer was seen at four sessions, with no more than two sessions on any single day.
- The time gap between the first and last sessions was at least three days and as long as 3 months, depending upon volunteer.

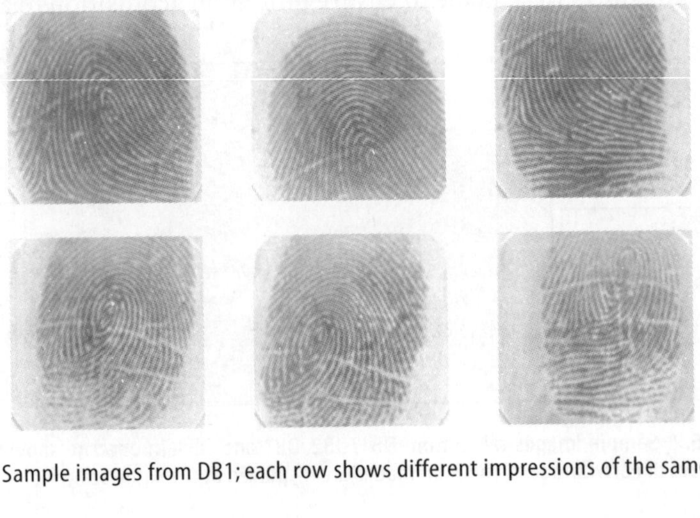

Figure 6.3 Sample images from DB1; each row shows different impressions of the same finger.

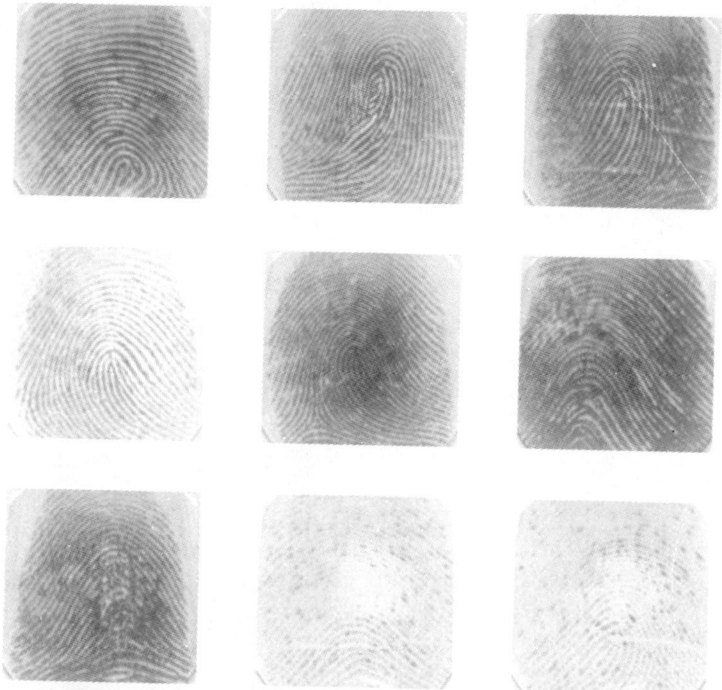

Figure 6.4 Images from DB1; all the samples are from different fingers and are roughly ordered by quality (top left: high quality; bottom right: low quality).

Figure 6.5 Sample images from DB2; each row shows different impressions of the same finger.

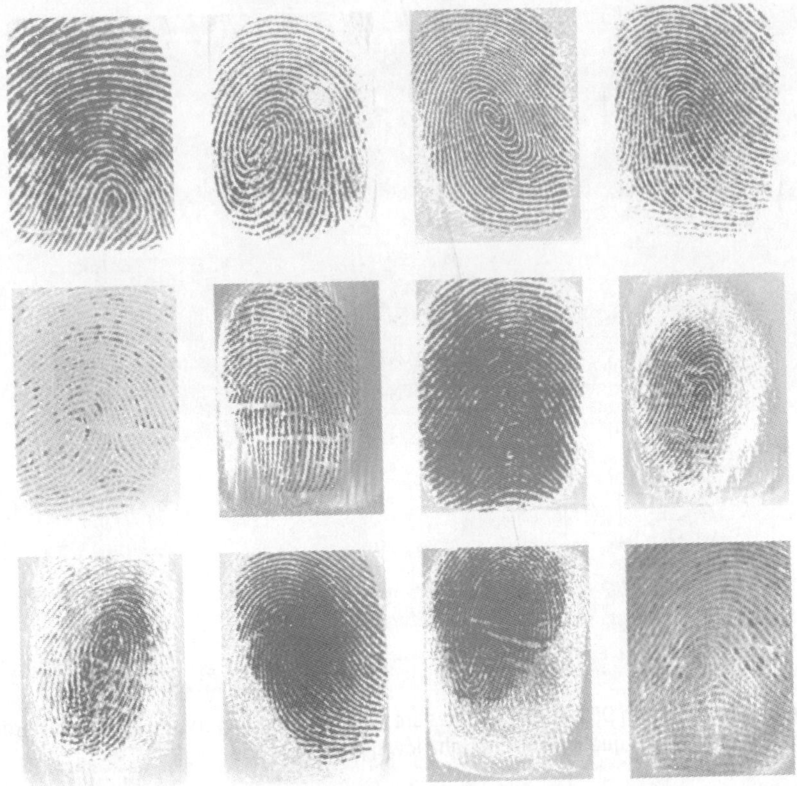

Figure 6.6 Images from DB2; all the samples are from different fingers and are roughly ordered by quality (top left: high quality; bottom right: low quality).

- The sensor plate was systematically cleaned between image acquisitions.
- At one session with each volunteer, fingers were cleaned with rubbing alcohol and dried.
- Some part of the core was apparent in each image, but care was taken to avoid a complete overlap between consecutive images taken during a single session.
- The acquired fingerprints were manually analyzed to ensure that the maximum rotation is approximately in the range [–15°, 15°] and that each pair of impressions of the same finger has a non-null overlapping area.

Figures 6.7 and 6.8 show some sample images taken from DB3.

Collection of DB4 requires some explanation. In general, the use of artificial images for testing biometric systems is not considered to be the "best practice" [20]. Although this may be the case for performance evaluation in real applications, we believe that in a technology evaluation event such as FVC2000, the use of synthetic images has three main advantages:

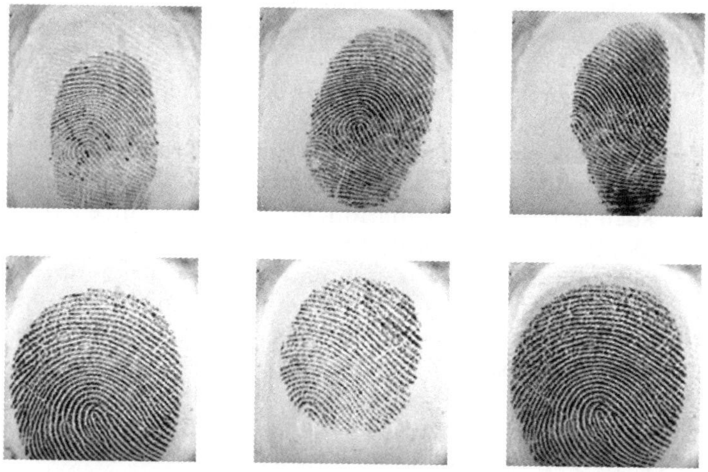

Figure 6.7 Sample images from DB3; each row shows different impressions of the same finger.

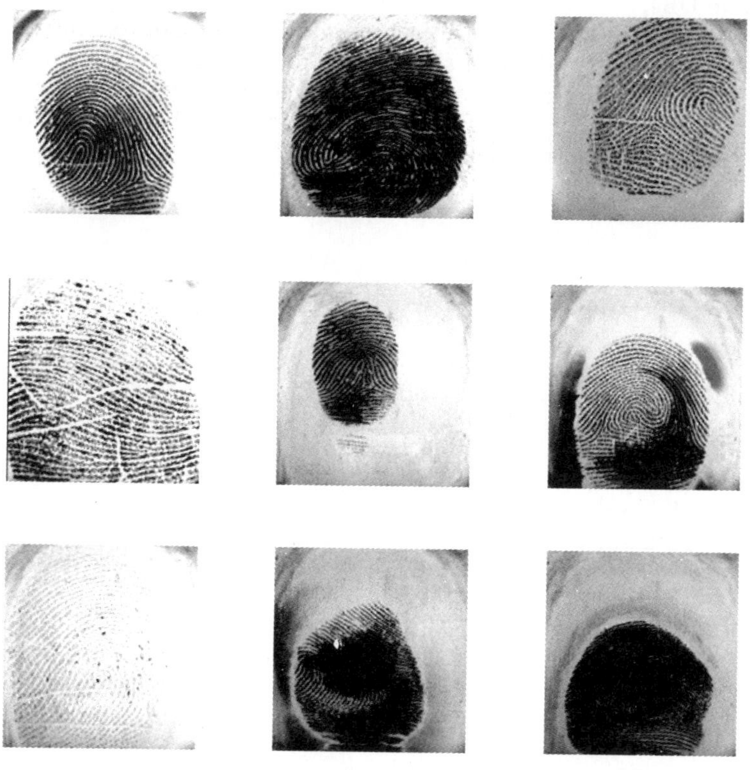

Figure 6.8 Images from DB3; all the samples are from different fingers and are roughly ordered by quality (top left: high quality; bottom right: low quality).

- It supplies images which are native to none of the participant algorithms, thus providing a fair comparison.
- Synthetic fingerprint databases can be created at very low cost. Acquiring a large number of fingerprints for testing purposes may be problematic due to the great amount of time and resources required and also to the privacy legislation which in some countries prohibits the diffusion of such personal information. Furthermore, once a database has been "used", its utility is limited since, for successive testing of algorithms, a new unknown database should be used.
- It is possible to adjust the database difficulty by tuning different kinds of perturbations (e.g. maximum amount of rotation and translation, and the amount of skin distortion).

If the generated artificial images were not a suitable simulation of real fingerprint patterns, the comparisons on the synthetic database would be misleading; furthermore, in order to improve the performance, *ad hoc* algorithms could be designed/tuned according to the same assumptions that model the synthetic generation. However, the presence of three real databases in FVC2000 provides a natural way to check the validity of the results on DB4.

The parameters of the synthetic generator were tuned to emulate a low-cost sensor with a small acquisition area; the maximum rotation and displacement and skin distortion are adjusted to roughly reproduce the perturbations in the three previous databases. Figures 6.9 and 6.10 show some sample images taken from DB4.

Figure 6.9 Sample images from DB4; each row shows different impressions of the same finger.

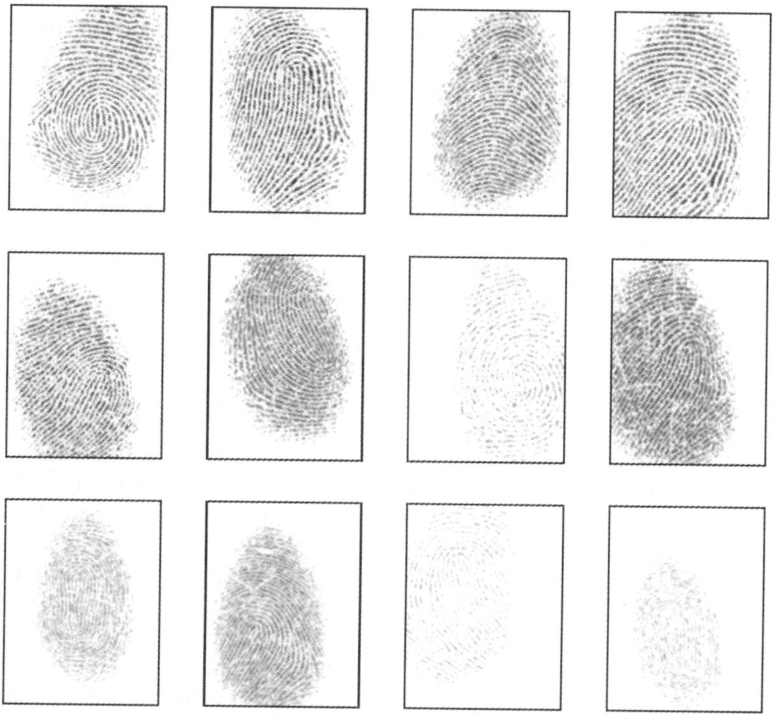

Figure 6.10 Images from DB4; all the samples are from different fingers and are roughly ordered by quality (top left: high quality; bottom right: low quality).

6.4 Performance Evaluation

For each database, we will refer to the jth fingerprint image of the ith finger as F_{ij}, $i = 1..100$, $j = 1..8$ and to the corresponding template (computed from F_{ij}) as T_{ij}.

For each database and for each algorithm:

- The templates T_{ij}, $i = 1..100$, $j = 1..7$ are computed from the corresponding F_{ij} and stored on a disk; one of the following three kinds of rejection can happen for each image F_{ij}:
  1. **F** (Fail): the algorithm declares that it cannot enroll the fingerprint image.
  2. **T** (Timeout): the enrollment exceeds the maximum allowed time (15 seconds).
  3. **C** (Crash): the algorithm crashes during fingerprint processing.

  The three types of rejection, considered as "failure to enroll", are added and stored in $\text{REJ}_{\text{ENROLL}}$.

- Each fingerprint template T_{ij} is matched against the fingerprint images F_{ik} ($j< k \le 8$) and the corresponding *Genuine Matching Scores* gms_{ijk} are stored[1]. The number of matches (denoted as NGRA – Number of Genuine Recognition Attempts) is $((8 \times 7)/2) \times 100 = 2{,}800$ in case $REJ_{ENROLL} = 0$. The failed, timeout (5 seconds) and crash rejections are accumulated into REJ_{NGRA}; no gms_{ijk} is stored in this case.
- Each fingerprint template T_{i1}, $i = 1..100$ is matched against the first fingerprint image from different fingers F_{k1} ($i < k \le 100$) and the corresponding *Impostor Matching Scores* ims_{ik} are stored. The number of matches (denoted as NIRA – Number of Impostor Recognition Attempts) is $((100 \times 99)/2) = 4{,}950$ in case $REJ_{ENROLL} = 0$. The failed, timeout (5 seconds) and crash rejections are accumulated into REJ_{NIRA}; no ims_{ik} is stored in this case.
- The genuine score distribution and the impostor score distribution are computed (actually, the term "distribution" denotes a histogram) and graphically reported to show how the algorithm "separates" the two classes. In fingerprint verification, higher scores are associated with more closely matching images.
- The FMR(t) (False Match Rate) and FNMR(t) (False Non-Match Rate) curves are computed from the above distributions for t ranging from 0 to 1[02]. Given a threshold t, FMR(t) denotes the percentage of $ims_{ik} \ge t$, and FNMR(t) denotes the percentage of $gms_{ijk} < t$. Actually, since FMR and FNMR are used in the contest to compare the performance of different algorithms, FMR and FNMR are "corrected" to keep into account rejections stored in REJ_{NIRA} and REJ_{NGRA}:

$$\mathrm{FMR}(t) = \frac{card\{\mathrm{ims}_{ik} \mid \mathrm{ims}_{ik} \ge t\}}{\mathrm{NIRA}},$$

$$\mathrm{FNMR}(t) = \frac{card\{\mathrm{gms}_{ijk} \mid \mathrm{gms}_{ijk} < t\} + \mathrm{REJ}_{\mathrm{NGRA}}}{\mathrm{NGRA}}$$

where *card* denotes the cardinality of a given set. This correction assumes that a failure to match is always treated by the system as a "non-match" (matching score < 0).

1 If g is matched with h, the symmetric match (i.e. h against g) is not executed.

2 FMR and FNMR are often confused with FAR (False Acceptance Rate) and FRR (False Rejection Rate) respectively, but the FAR/FRR notation is misleading in some applications. For example, in a welfare benefits system, which uses fingerprint identification to prevent multiple payments under false identity, the system "falsely accepts" an applicant if the fingerprint is "falsely rejected" (not matched to the print of the same finger previously stored in the database); similarly, a "false acceptance" causes a "false rejection". Therefore, to avoid this confusion, we distinguish the matching errors made by the algorithm from errors made in the final "accept/reject" decision given the user.

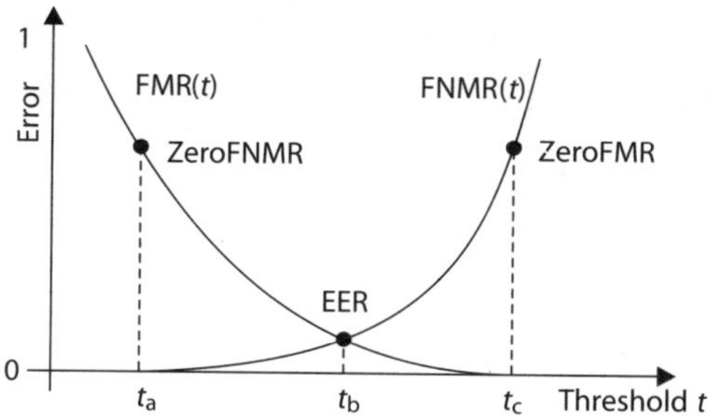

Figure 6.11 An example of FMR/FNMR curves, where the points corresponding to EER, ZeroFMR and ZeroFNMR are highlighted.

- A ROC (Receiving Operating Curve) is obtained, where FNMR is plotted as a function of FMR; the curve is drawn in log–log scales for better comprehension.
- The Equal Error Rate EER is computed as the point where $\text{FNMR}(t) = \text{FMR}(t)$ (see Figure 6.11); in practice, the matching score distributions (histograms) are not continuous and a crossover point might not exist. In this case, we report the interval $[\text{EER}_{low}, \text{EER}_{high}]$. An operational definition of EER is given in Appendix A.
- ZeroFMR is defined as the lowest FNMR at which no False Matches occur and ZeroFNMR is defined as the lowest FMR at which no False Non-Matches occur (Figure 6.11):

$$\text{ZeroFMR}(t) = \min_t \{\text{FNMR}(t) | \text{FMR}(t) = 0\}$$

$$\text{ZeroFNMR}(t) = \min_t \{\text{FMR}(t) | \text{FNMR}(t) = 0\}$$

  Both ZeroFMR and ZeroFNMR may not exist; in such a case we assign to them the value 1.
- The *average enroll time* is calculated as the average CPU time for a single enrollment operation, and *average match time* as the average CPU time for a single match operation between a template and a test image.

6.5 Results

This section reports the performance of the tested algorithms on each of the four databases (Tables 6.3–6.6) and the average results over the four databases (Table 6.7). Figure 6.12 shows the ROC for DB3, which proved to be the most difficult data set. The notation introduced in Section 6.4 is used in both the graphics and tables, with the only exception of reporting $\text{REJ}_{\text{ENROLL}}$ as a percentage value and to collapse both REJ_{NGRA} and REJ_{NIRA} into a single value $\text{REJ}_{\text{MATCH}}$:

Table 6.3 Algorithm performance over DB1 sorted by EER.

Algorithm	EER (%)	REJ_{ENROLL} (%)	REJ_{MATCH} (%)	Average enroll time (s)	Average match time (s)
SAG1	0.67	0.00	0.00	2.48	0.96
SAG2	1.17	0.00	0.00	0.88	0.88
CETP	5.06	0.00	0.00	0.81	0.89
CWAI	7.06	3.71	3.90	0.22	0.32
CSPN	7.60	0.00	0.00	0.17	0.17
UTWE	7.98	0.00	0.00	10.40	2.10
KRDL	10.66	6.43	6.59	1.00	1.06
FPIN	13.46	0.00	0.00	0.83	0.87
UINH	21.02	1.71	5.08	0.53	0.56
DITI	23.63	0.00	0.00	0.65	0.72
NCMI	49.11	0.00	0.12	1.13	1.34

Table 6.4 Algorithm performance over DB2 sorted by EER.

Algorithm	EER (%)	REJ_{ENROLL} (%)	REJ_{MATCH} (%)	Average enroll time (s)	Average match time (s)
SAG1	0.61	0.00	0.00	2.63	1.03
SAG2	0.82	0.00	0.00	0.93	0.93
CSPN	2.75	0.00	0.00	0.17	0.17
CWAI	3.01	1.29	1.29	0.23	0.30
CETP	4.63	0.00	0.09	0.85	0.98
KRDL	8.83	3.29	4.41	1.16	2.88
UTWE	10.65	0.00	0.00	10.42	2.12
FPIN	11.14	0.00	0.00	1.16	1.24
DITI	13.83	0.00	0.00	1.21	1.28
UINH	15.22	0.86	4.08	0.60	0.65
NCMI	46.15	0.00	0.00	1.28	1.57

$$REJ_{MATCH} = \frac{NIRA \cdot REJ_{NIRA} + NGRA \cdot REJ_{NGRA}}{NIRA + NGRA}$$

For a correct interpretation of the results, EER alone is not a sufficient metric; REJ_{ENROLL} should be also taken into account.

For each algorithm, detailed results (including genuine and impostor distributions, FMR and FNMR curves, NGRA, NIRA, ...) are reported in [14]. For each algorithm, detailed results (including genuine and impostor distributions, FMR and FNMR curves, NGRA, NIRA, ...) are reported in Appendix B.

Most of the algorithms submitted to the competition performed well, if we take into account the difficulty of adapting a given algorithm to new types of images. In particular, algorithms SAG1 and SAG2 showed the best accuracy and CSPN exhibited a good trade-off between accuracy and efficiency.

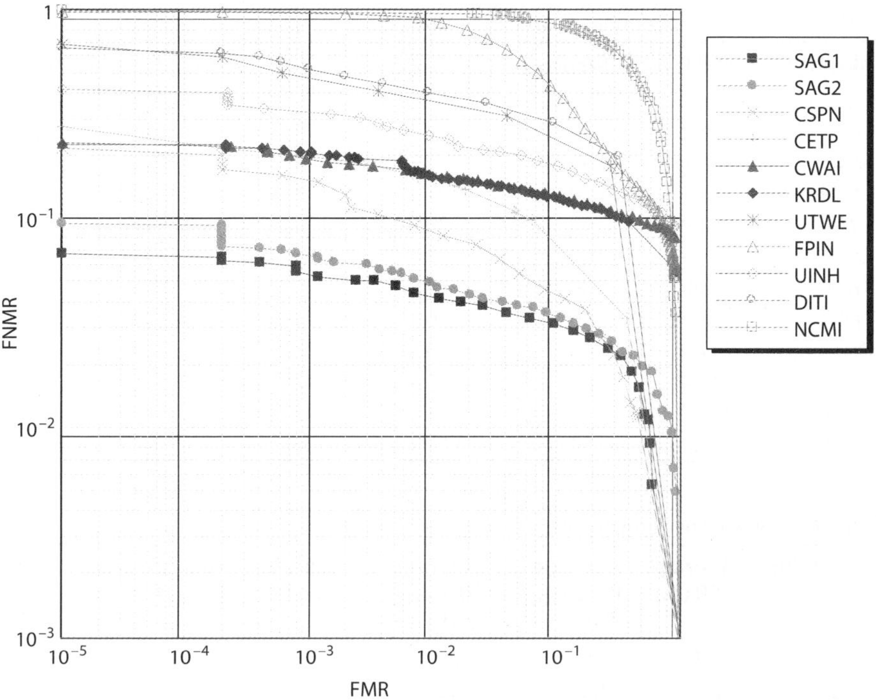

Figure 6.12 ROC curves on DB3. Each point denotes a pair (FMR(t), FNMR(t)) for a given value of t.

Table 6.5 Algorithm performance over DB3 sorted by EER.

Algorithm	EER (%)	REJ_{ENROLL} (%)	REJ_{MATCH} (%)	Average enroll time (s)	Average match time (s)
SAG1	3.64	0.00	0.00	5.70	2.13
SAG2	4.01	0.00	0.00	1.94	1.94
CSPN	5.36	0.57	1.24	0.35	0.36
CETP	8.29	0.00	0.00	1.49	1.66
CWAI	11.94	12.86	8.00	0.46	0.57
KRDL	12.20	6.86	5.12	1.48	1.60
UINH	16.32	10.29	7.64	1.28	1.36
UTWE	17.73	0.00	0.00	10.44	2.31
DITI	22.63	0.00	0.00	2.59	2.67
FPIN	23.18	0.00	0.00	2.13	2.19
NCMI	47.43	0.00	0.01	2.25	2.75

Table 6.7 highlights a significant gap in the performance of the different algorithms and it would be extremely interesting to understand the reasons for such differences. For this purpose, after the presentation of the results, we asked the participants to provide some technical details about their methods, but only a few of them responded (the responses can be found at

Table 6.6 Algorithm performance over DB4 sorted by EER.

Algorithm	EER (%)	REJ_{ENROLL} (%)	REJ_{MATCH} (%)	Average enroll time (s)	Average match time (s)
SAG1	1.99	0.00	0.00	1.90	0.77
SAG2	3.11	0.00	0.00	0.69	0.69
CSPN	5.04	0.00	0.00	0.11	0.11
CWAI	6.30	0.00	0.00	0.16	0.20
CETP	7.29	0.00	0.00	0.65	0.72
KRDL	12.08	10.86	10.24	0.70	0.79
FPIN	16.00	0.00	0.00	0.77	0.80
DITI	23.80	0.00	0.00	0.52	0.60
UTWE	24.59	0.00	0.00	10.42	4.17
UINH	24.77	2.14	4.28	0.42	0.45
NCMI	48.67	0.00	0.25	1.08	1.19

Table 6.7 Average performance over the four databases sorted by average EER.

Algorithm	Average EER (%)	Average REJ_{ENROLL} (%)	Average REJ_{MATCH} (%)	Average enroll time (s)	Average match time (s)
SAG1	1.73	0.00	0.00	3.18	1.22
SAG2	2.28	0.00	0.00	1.11	1.11
CSPN	5.19	0.14	0.31	0.20	0.20
CETP	6.32	0.00	0.02	0.95	1.06
CWAI	7.08	4.46	3.14	0.27	0.35
KRDL	10.94	6.86	6.52	1.08	1.58
UTWE	15.24	0.00	0.00	10.42	2.67
FPIN	15.94	0.00	0.00	1.22	1.27
UINH	19.33	3.75	5.23	0.71	0.76
DITI	20.97	0.00	0.00	1.24	1.32
NCMI	47.84	0.00	0.09	1.44	1.71

the FVC2000 web site [7]). In any case, on the basis of the participant responses and on what we learned from this experience, we can make the following observations:

- A coarse analysis of the errors on genuine attempts showed that most of the errors were made by the algorithms on about 15–20% poor-quality fingerprints in each database. In other words, we could claim that a "20–80 rule" is valid: that is, 20% of the database is responsible for 80% of the errors.
- The most accurate algorithm (SAG1) takes a lot of time for enrollment (3.18 s with respect to a median enrollment time of 1.08 s). This suggests that accurate image enhancement and feature extraction are really important for improving the matching accuracy. Furthermore, feature extraction seems to perform asymmetrically, since the average matching

time (which also includes the feature extraction time for the test image) is substantially lower than a single enrollment time.
- The fastest algorithm (Cspn) extracts minutiae by an adaptive tracing of the gray-level ridges, without *a priori* binarization and thinning (which are time-consuming tasks) [15, 26] and exploits local minutiae arrangement to speed up the initial steps of minutiae matching [27].

Databases DB1 and DB2 proved to be "easier" than DB3, even though the sensor used for DB3 is of higher quality. This means that the acquisition conditions and the volunteer population can have a stronger impact on performance than sensor quality.

The synthetically generated database (DB4) was demonstrated to be adequate for FVC2000 purposes: in particular, from Tables 6.3–6.6, it is evident that the algorithm ranking on DB4 is quite similar to the other databases, proving that no algorithm was favored or penalized by the synthetic images. In particular, if an algorithm performs well on real fingerprints, then it also performs well on synthetic fingerprints, and vice versa. The visual analysis of impostor and genuine distributions (see [14]) definitely supports this claim, since no significant differences are seen between the DB4 graphics and the others.

6.6 Organization of FVC2002

At the end of 2001, when the FVC2002 web site was created [30], we extensively publicized this next competition. To increase the number of companies participating, and therefore to provide a more complete panorama of the state of the art, we decided to allow the participants to remain anonymous. In FVC2002, participants could decide not to publish the name of their organization in case their results were not as they expected.

The FVC2002 announcement clearly stated that, analogously to FVC2000, FVC2002 is not to be viewed as an official certification of fingerprint-based biometric systems, but simply as a technology evaluation [30], where algorithms compliant with a predefined protocol are evaluated on common databases. Neither hardware components nor proprietary modules outside the FVC2002 protocol are tested.

Four new databases were collected and a representative subset of each database was made available to the participants to let them tune their algorithms according to the image size and the variability of the fingerprints in the databases. Four databases constituted the FVC2002 benchmark. Three different scanners and the SFinGe synthetic generator were used to collect fingerprints (see Table 6.8 and Figure 6.13). Figure 6.14 shows an image for each database, at the same scale factor.

At the end of the data collection, we had collected for each database a total of 120 fingers and 12 impressions per finger (1440 impressions) using 30 volunteers. The size of each database used in the FVC2002 test, however, was established as 110 fingers, 8 impressions per finger (880 impressions)

Table 6.8 Scanners/technologies used for the collection of FVC2002 databases.

	Technology	Scanner	Image size – resolution
DB1	Optical	Identix TouchView II	388 × 374 – 500 dpi
DB2	Optical	Biometrika FX2000	296 × 560 – 569 dpi
DB3	Capacitive	Precise Biometrics 100 SC	300 × 300 – 500 dpi
DB4	Synthetic	SFinGE v2.51	288 × 384 – 500 dpi

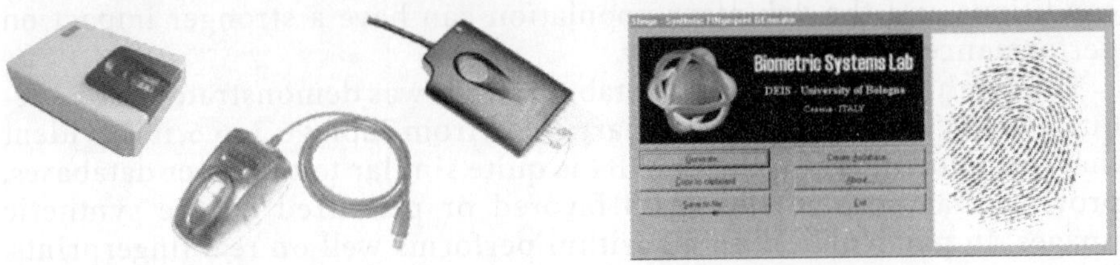

Figure 6.13 From left to right: the three sensors used for collecting DB1, DB2 and DB3, respectively, and a snapshot of the tool which generated synthetic fingerprints in DB4.

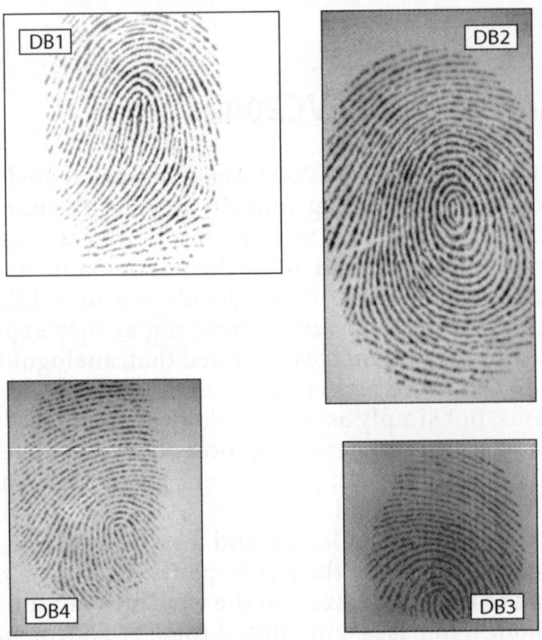

Figure 6.14 One fingerprint image from each database of FVC2002.

(Figure 6.15). Collecting some additional data gave us a margin in case of collection errors, and also allowed us to choose systematically from the collected impressions those to include in the test databases. In the FVC2002 testing protocol, new performance indicators, e.g. FMR100 and FMR1000,

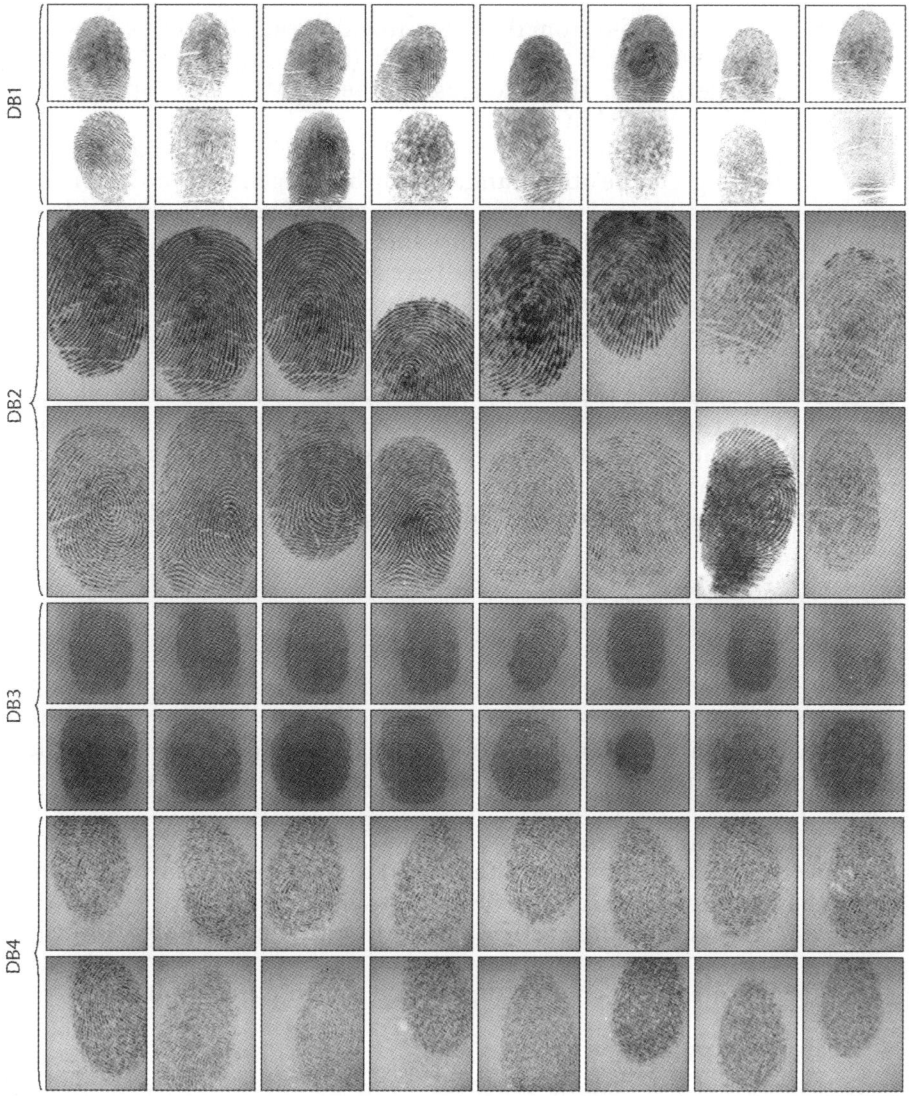

Figure 6.15 Sample images from the four databases; for each database, the first row shows eight impressions of the same finger, the second row reports samples from different fingers, roughly ordered by quality (left: high quality; right: low quality).

which are the values of FNMR for FMR = 1/100 and 1/1000 respectively, were added to those already used in FVC2000. These data are useful to characterize the accuracy of fingerprint-based systems, which are often operated far from the EER point, by using thresholds which reduce FMR at the cost of high FNMR. Failure to enroll errors (FTE) were incorporated into the computation of the false non-match rate (FNMR) and false match rate (FMR) to make the results of the different algorithms directly comparable.

In particular, we clarified from the beginning that each FTE error produces a "ghost" template which does not match (matching score 0) with the remaining fingerprints, thus increasing the FNMR. This approach is consistent with that used in [32].

Ranking the algorithms according to EER (as in FVC2000) may be sometimes misleading. On the other hand, mixing heterogeneous indicators into a unique goodness index is difficult and arbitrary. Therefore, we decided to summarize the results of FVC2002 in a sort of Olympic medal table where three medals (gold, silver and bronze) are assigned to the best three algorithms for each indicator over each database.

The evaluation of the 33 algorithms submitted to the Second International Fingerprint Verification Competition (FVC2002) is available at [30]. A CD-ROM containing the four databases is available to the research community. At the time of this writing, a 2004 version of FVC is being planned, again to have four separate databases: two of these databases will be collected using an optical scanner, one using a thermal swept scanner, and one again synthetically generated. Results will again be posted online [33].

6.7 Conclusions

Once again we would like to remark that the results reported here do not necessarily reflect the performance that the participating algorithms would achieve in a real environment or when embedded into a complete biometric system. In any event, we believe that FVC competition results:

- Provide a useful overview of the state of the art in this field.
- Allow researchers and companies to test their algorithms over common databases collected using state of the art sensors.
- Provide guidance to the participants for improving their algorithms.

In future, we intend to continue supporting this initiative as follows:

- The existing FVC web sites will be maintained to diffuse FVC results, and to promote FVC testing protocol as a standard for technological evaluations.
- Companies and academic research groups will be allowed to test new algorithms or improved versions of existing algorithms on the FVC benchmark databases and to add their results to the FVC web site. New entries will be kept isolated from the original entries, since hereafter the full databases are known in advance, which could allow algorithm tuning to give an unfair advantage to new participants.
- Generating synthetic fingerprint databases for future evaluations will be further investigated.

Appendix A

An operational procedure for computing EER (interval), given a finite number of genuine and impostor matching scores, is reported in the following. Let

$$t_1 = \max_{t \in \{\text{gms}_{ijk}\} \cup \{\text{ims}_{ik}\}} \{t \mid \text{FNMR}(t) \le \text{FMR}(t)\}$$

and

$$t_2 = \min_{t \in \{\text{gms}_{ijk}\} \cup \{\text{ims}_{ik}\}} \{t \mid \text{FNMR}(t) \ge \text{FMR}(t)\}$$

The EER interval is defined as:

$$[\text{EER}_{\text{low}}, \text{EER}_{\text{high}}] = \begin{cases} [\text{FNMR}(t_1), \text{FMR}(t_1)] & \text{if } \text{FNMR}(t_1) + \text{FMR}(t_1) \le \text{FMR}(t_2) + \text{FNMR}(t_2) \\ [\text{FMR}(t_2), \text{FNMR}(t_2)] & \text{otherwise} \end{cases}$$

and EER is estimated as $(\text{EER}_{\text{low}} + \text{EER}_{\text{high}})/2$ (see Figure 6.16).

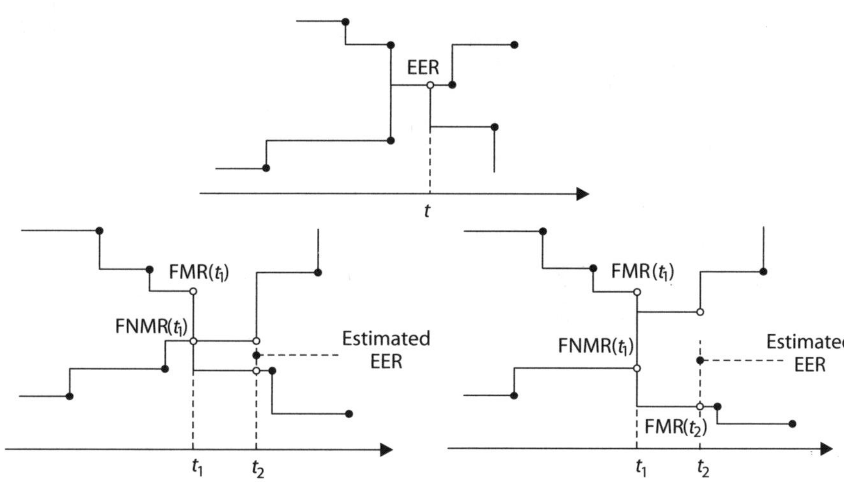

Figure 6.16 Computing the EER interval. At the top an example is given where an EER point exists. Below, two cases are shown where an EER point does not exist and the corresponding intervals are highlighted.

Appendix B

The following pages show the results for each algorithm.

Algorithm CETP on database DB1_A

Average enroll time: 0.81 seconds
Average match time: 0.89 seconds

REJ_{ENROLL}	NGRA	NIRA	REJ_{NGRA}	REJ_{NIRA}
0.00% (F:0 T:0 C:0)	2800	4950	0.00% (F:0 T:0 C:0)	0.00% (F:0 T:0 C:0)

EER	EER*	ZeroFMR	ZeroFNMR
5.06% (4.30%–5.82%)	5.06% (4.30%–5.82%)	18.86%	100.00%

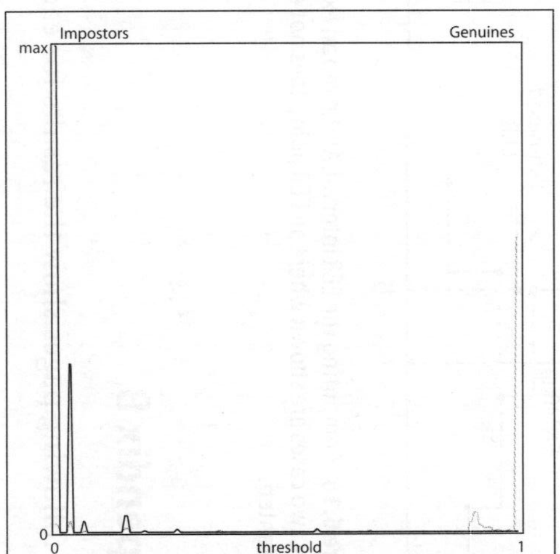

Score distributions

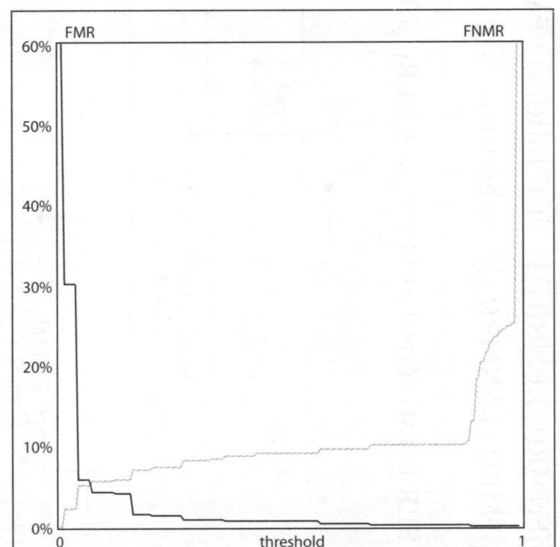

FMR(t) and FNMR(t)

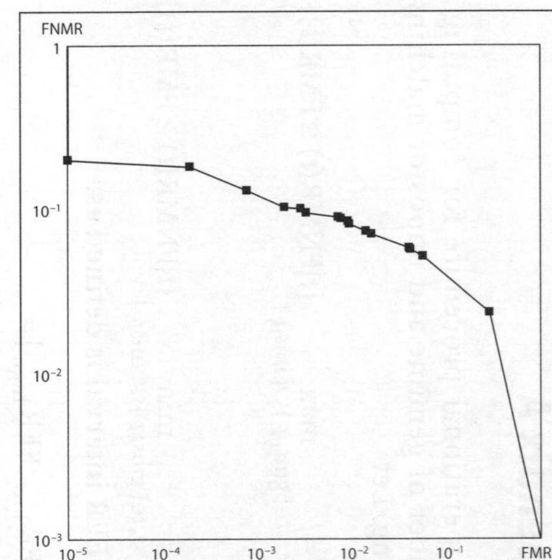

ROC curve

Algorithm CETP on database DB2_A

Average enroll time: 0.85 seconds
Average match time: 0.98 seconds

REJ_{ENROLL}	NGRA	NIRA	REJ_{NGRA}	REJ_{NIRA}
0.00% (F:0 T:0 C:0)	2800	4950	0.25% (F:0 T:0 C:7)	0.00% (F:0 T:0 C:0)

EER	EER*	ZeroFMR	ZeroFNMR
4.63% (3.58%–5.68%)	4.51% (3.58%–5.44%)	11.71%	100.00%

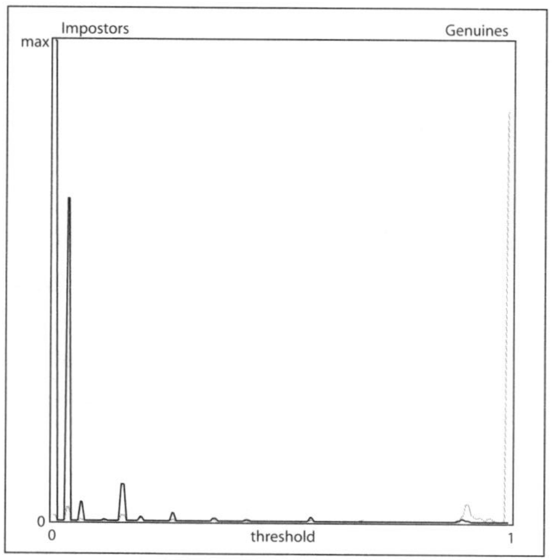

Score distributions

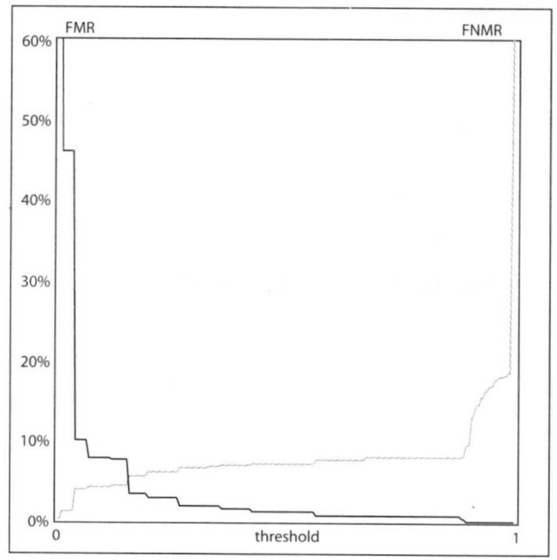

FMR(t) and FNMR(t)

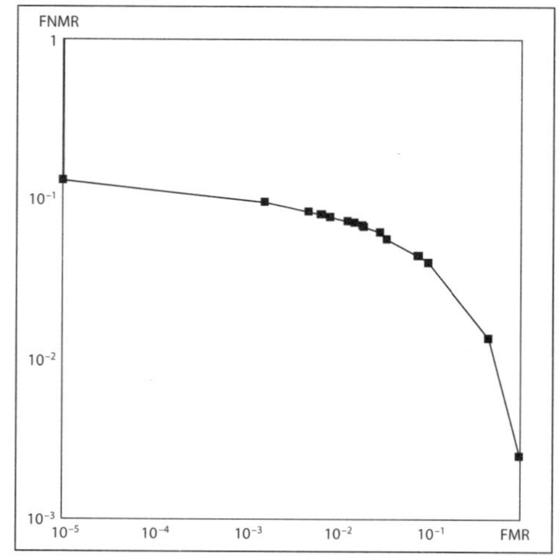

ROC curve

Algorithm CETP on database DB3_A

Average enroll time: 1.49 seconds
Average match time: 1.66 seconds

REJ_{ENROLL}	NGRA	NIRA	REJ_{NGRA}	REJ_{NIRA}
0.00% (F:0 T:0 C:0)	2800	4950	0.00% (F:0 T:0 C:0)	0.00% (F:0 T:0 C:0)

EER	EER*	ZeroFMR	ZeroFNMR
8.29% (6.55%–10.04%)	8.29% (6.55%–10.04%)	22.61%	100.00%

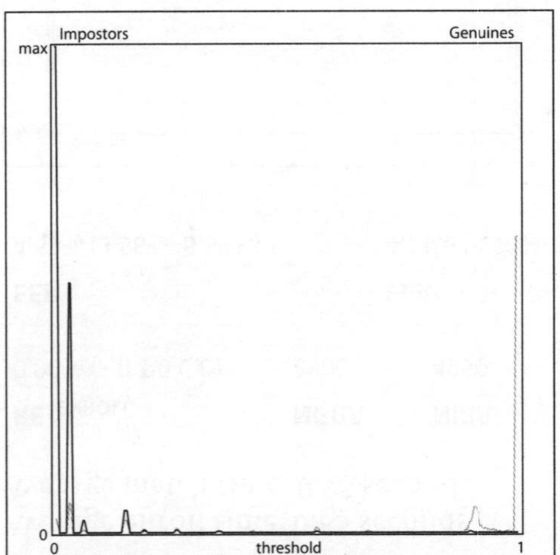

Score distributions

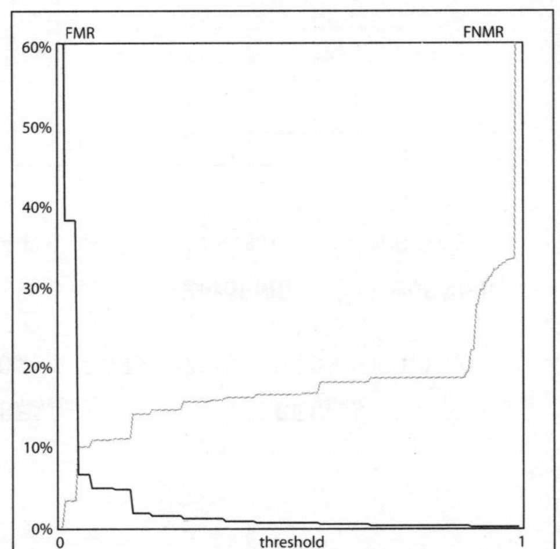

FMR(t) and FNMR(t)

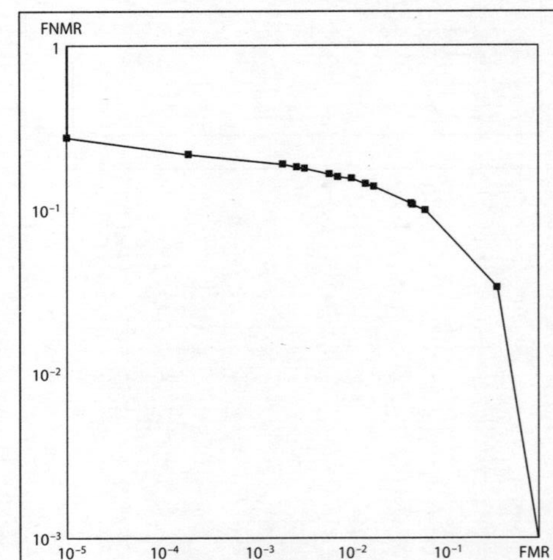

ROC curve

Algorithm CETP on database DB4_A

Average enroll time: 0.65 seconds
Average match time: 0.72 seconds

REJ_{ENROLL}	NGRA	NIRA	REJ_{NGRA}	REJ_{NIRA}
0.00% (F:0 T:0 C:0)	2800	4950	0.00% (F:0 T:0 C:0)	0.00% (F:0 T:0 C:0)

EER	EER*	ZeroFMR	ZeroFNMR
7.29% (7.01%–7.57%)	7.29% (7.01%–7.57%)	29.75%	100.00%

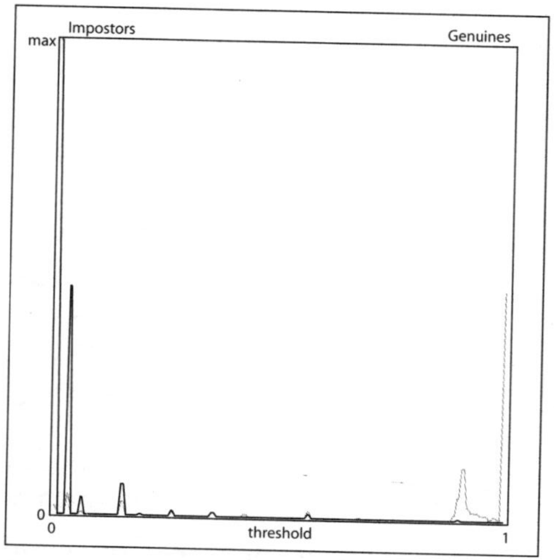

Score distributions

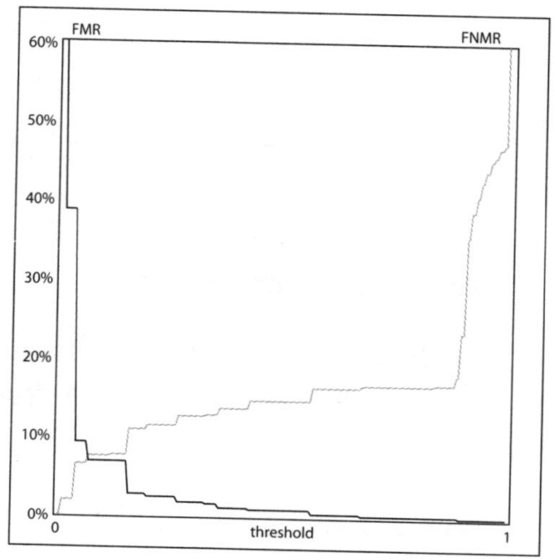

FMR(t) and FNMR(t)

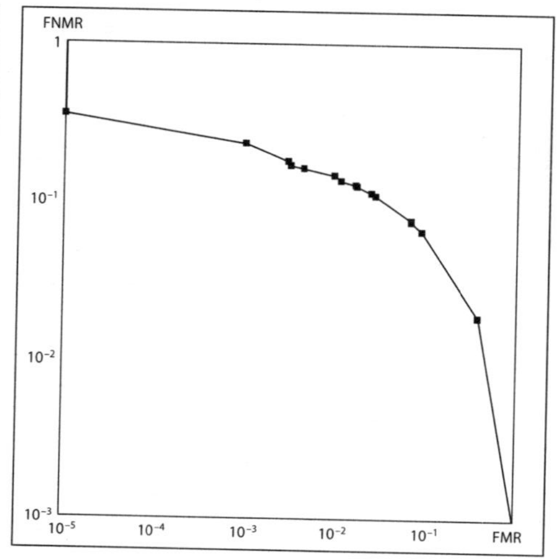

ROC curve

Algorithm CSPN on database DB1_A

Average enroll time: 0.17 seconds
Average match time: 0.17 seconds

REJ$_{ENROLL}$	**NGRA**	**NIRA**	**REJ$_{NGRA}$**	**REJ$_{NIRA}$**
0.00% (F:0 T:0 C:0)	2800	4950	0.00% (F:0 T:0 C:0)	0.00% (F:0 T:0 C:0)

EER	**EER***	**ZeroFMR**	**ZeroFNMR**
7.60% (7.60%–7.61%)	7.60% (7.60%–7.61%)	22.46%	100.00%

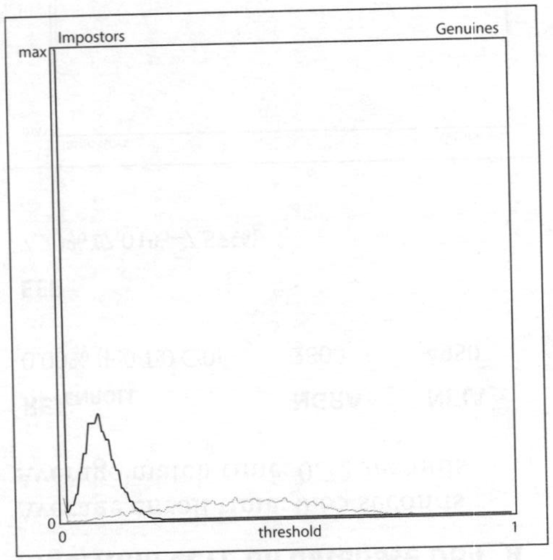

Score distributions

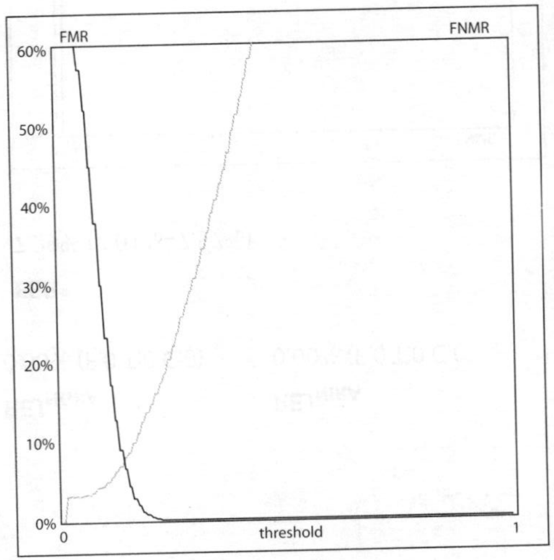

FMR(*t*) and FNMR(*t*)

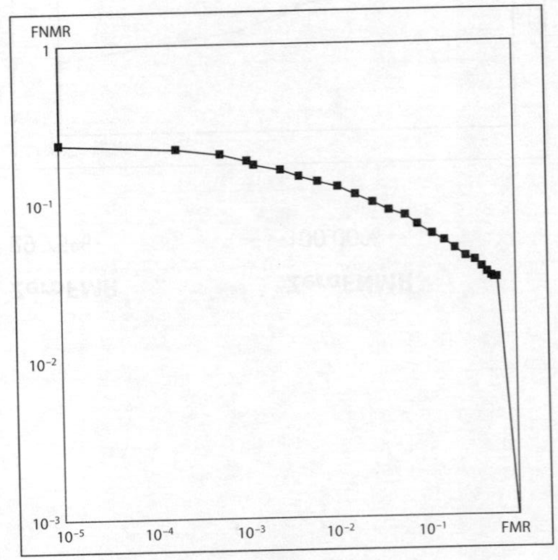

ROC curve

Algorithm CSPN on database DB2_A

Average enroll time: 0.17 seconds
Average match time: 0.17 seconds

REJENROLL	NGRA	NIRA	REJNGRA	REJNIRA
0.00% (F:0 T:0 C:0)	2800	4950	0.00% (F:0 T:0 C:0)	0.00% (F:0 T:0 C:0)

EER	EER*	ZeroFMR	ZeroFNMR
2.75%	2.75%	10.29%	100.00%

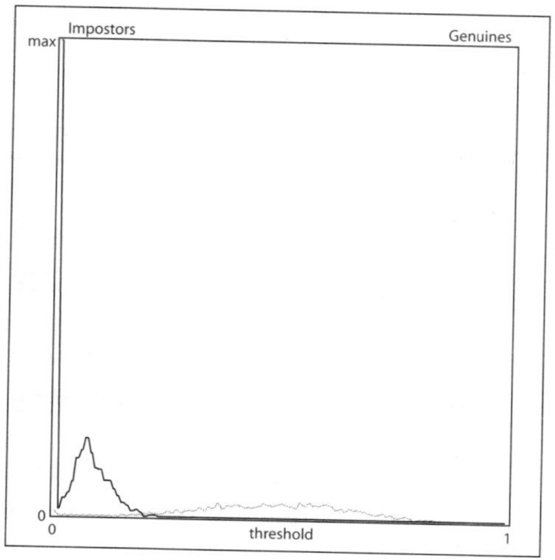

Score distributions

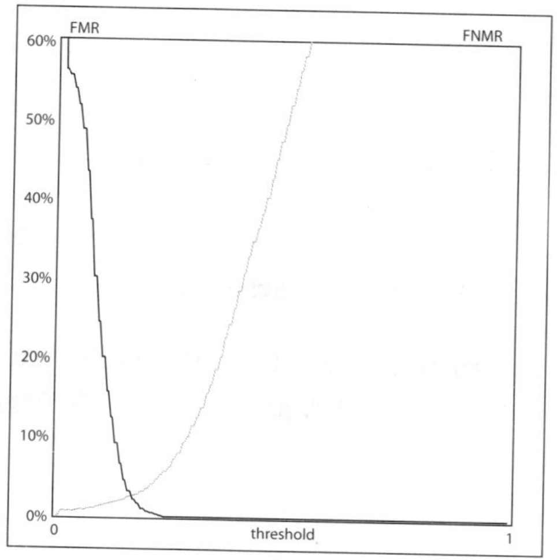

FMR(t) and FNMR(t)

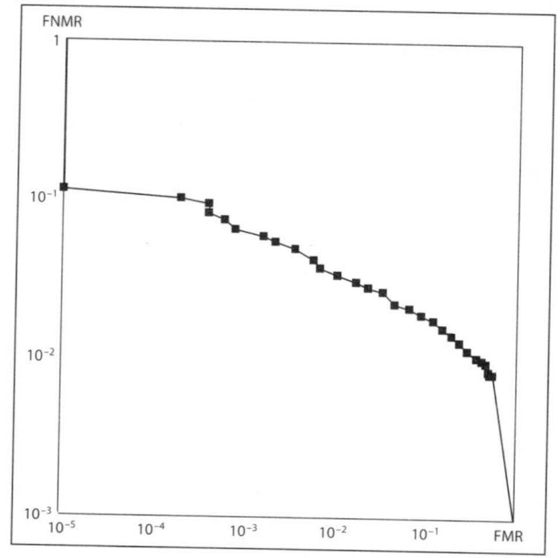

ROC curve

Algorithm CSPN on database DB3_A

Average enroll time: 0.35 seconds
Average match time: 0.36 seconds

REJ_{ENROLL}	NGRA	NIRA	REJ_{NGRA}	REJ_{NIRA}
0.57% (F:0 T:0 C:4)	2777	4945	0.07% (F:0 T:0 C:2)	1.90% (F:0 T:0 C:94)

EER	EER*	ZeroFMR	ZeroFNMR
5.36%	5.33% (5.32%–5.33%)	21.07%	100.00%

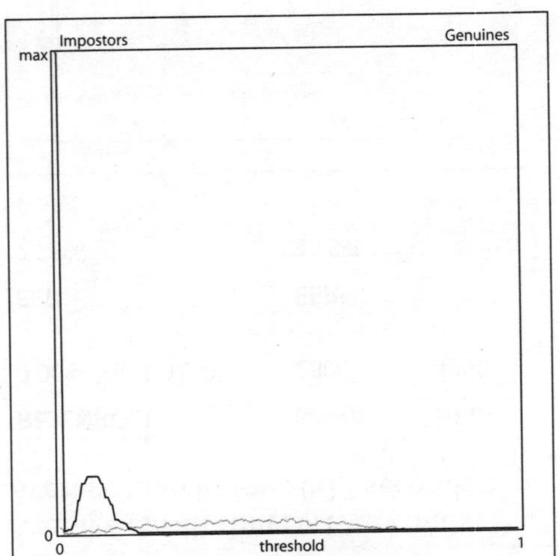

Score distributions

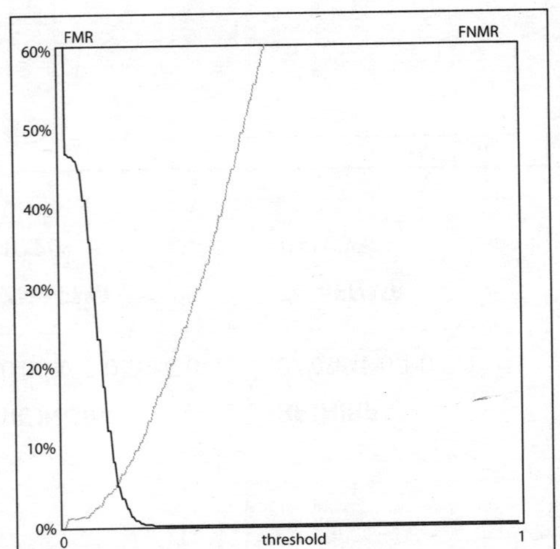

FMR(t) and FNMR(t)

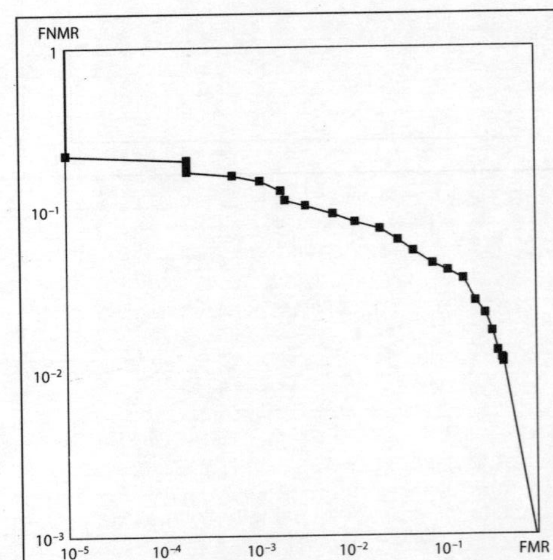

ROC curve

Algorithm CSPN on database DB4_A

Average enroll time: 0.11 seconds
Average match time: 0.11 seconds

REJ_{ENROLL}	NGRA	NIRA	REJ_{NGRA}	REJ_{NIRA}
0.00% (F:0 T:0 C:0)	2800	4950	0.00% (F:0 T:0 C:0)	0.00% (F:0 T:0 C:0)

EER	EER*	ZeroFMR	ZeroFNMR
5.04% (5.04%–5.05%)	5.04% (5.04%–5.05%)	15.54%	100.00%

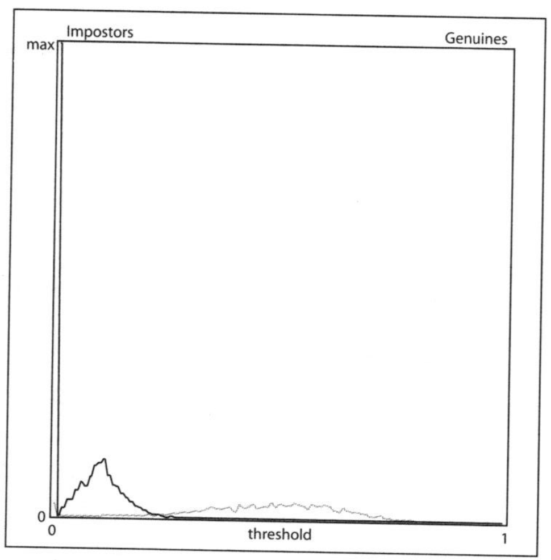

Score distributions

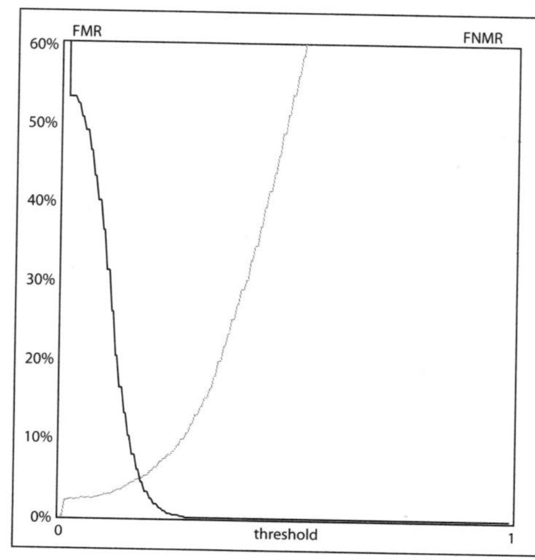

FMR(*t*) and FNMR(*t*)

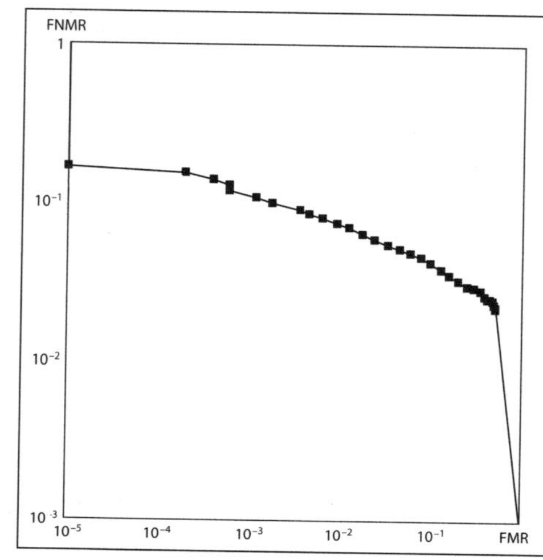

ROC curve

Algorithm CWAI on database DB1_A

Average enroll time: 0.22 seconds
Average match time: 0.32 seconds

REJ_{ENROLL}	NGRA	NIRA	REJ_{NGRA}	REJ_{NIRA}
3.71% (F:26 T:0 C:0)	2717	4944	3.97% (F:108 T:0 C:0)	3.86% (F:191 T:0 C:0)

EER	EER*	ZeroFMR	ZeroFNMR
7.06% (6.61%–7.51%)	4.27% (4.10%–4.45%)	23.15%	100.00%

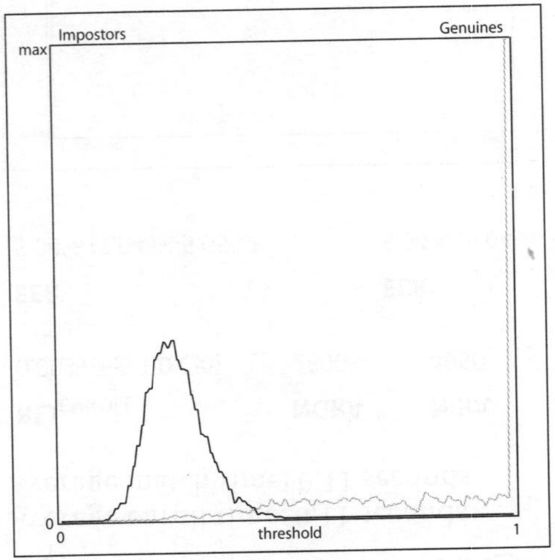

Score distributions

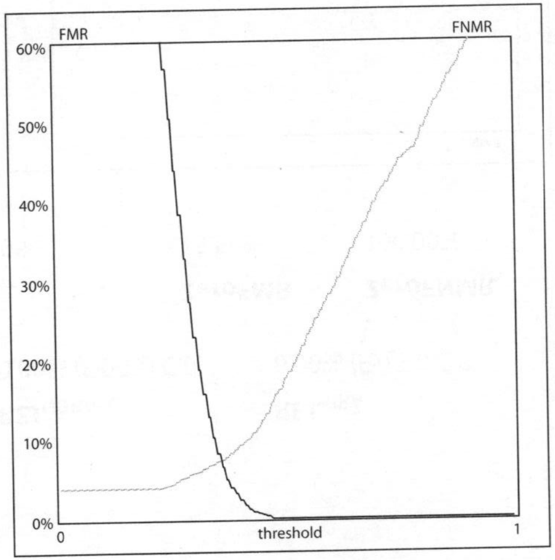

FMR(t) and FNMR(t)

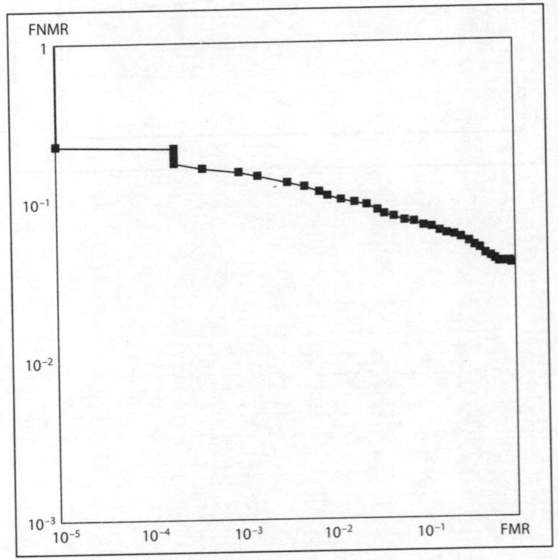

ROC curve

Algorithm CWAI on database DB2_A

Average enroll time: 0.23 seconds
Average match time: 0.30 seconds

REJ_{ENROLL}	NGRA	NIRA	REJ_{NGRA}	REJ_{NIRA}
1.29% (F:9 T:0 C:0)	2768	4916	1.23% (F:34 T:0 C:0)	1.32% (F:65 T:0 C:0)

EER	EER*	ZeroFMR	ZeroFNMR
3.01% (2.66%–3.36%)	2.16% (2.02%–2.30%)	8.74%	100.00%

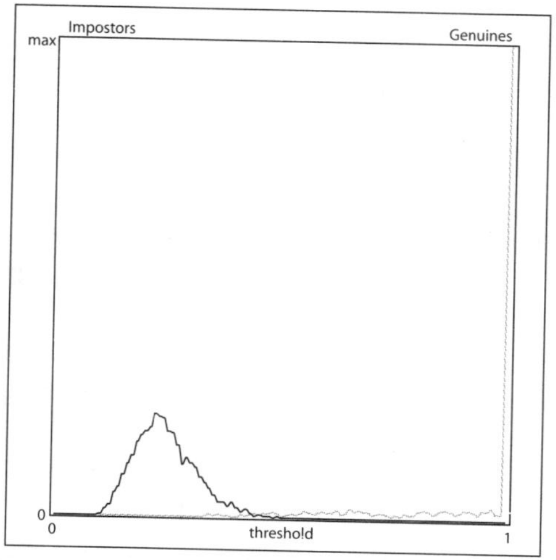

Score distributions

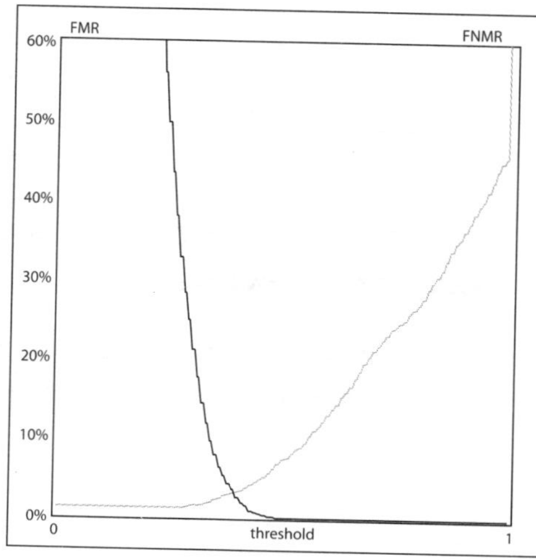

FMR(*t*) and FNMR(*t*)

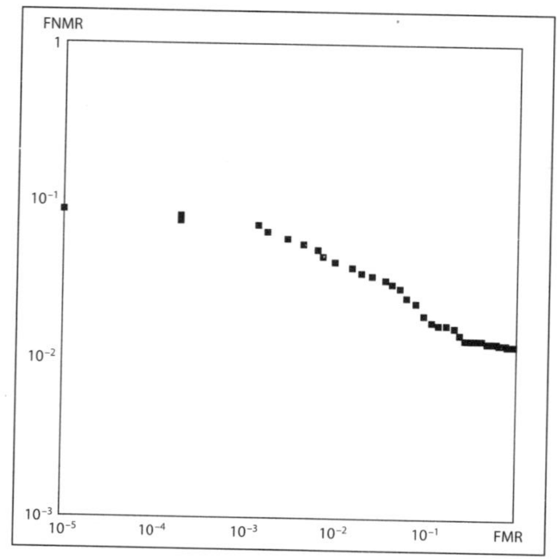

ROC curve

Algorithm CWAI on database DB3_A

Average enroll time: 0.46 seconds
Average match time: 0.57 seconds

REJ_{ENROLL}	NGRA	NIRA	REJ_{NGRA}	REJ_{NIRA}
12.86% (F:90 T:0 C:0)	2475	4252	8.16% (F:202 T:0 C:0)	7.90% (F:336 T:0 C:0)

EER	EER*	ZeroFMR	ZeroFNMR
11.94% (10.87%–13.01%)	5.90% (5.03%–6.78%)	23.43%	100.00%

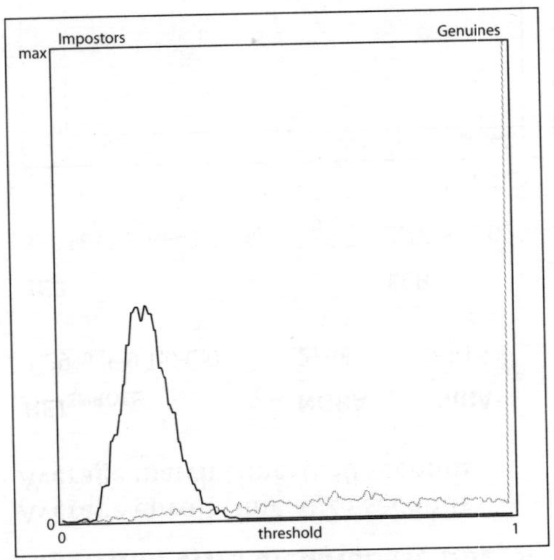

Score distributions

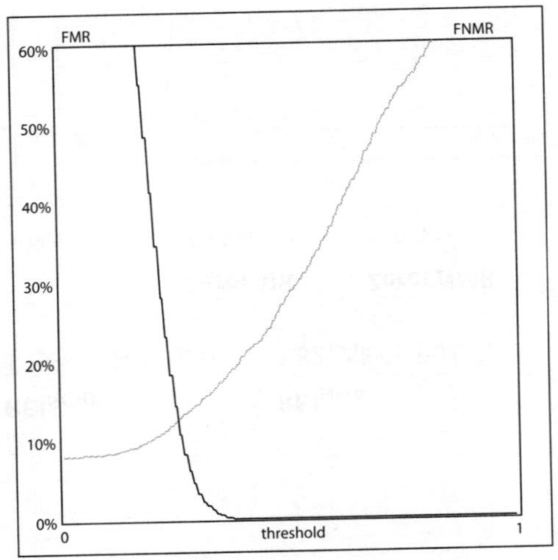

FMR(*t*) and FNMR(*t*)

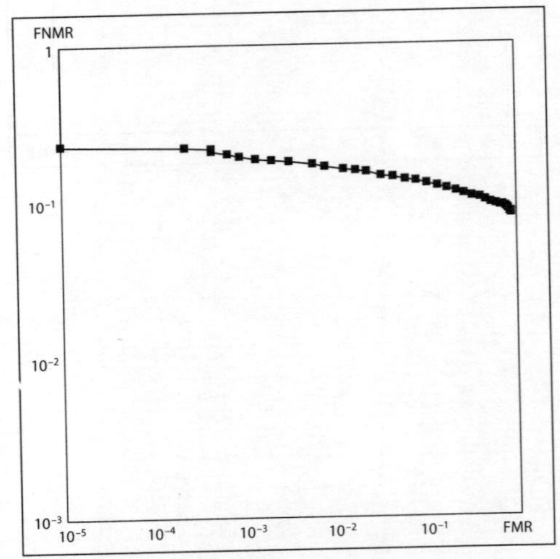

ROC curve

Algorithm CWAI on database DB4_A

Average enroll time: 0.16 seconds
Average match time: 0.20 seconds

REJ_{ENROLL}	NGRA	NIRA	REJ_{NGRA}	REJ_{NIRA}
0.00% (F:0 T:0 C:0)	2800	4950	0.00% (F:0 T:0 C:0)	0.00% (F:0 T:0 C:0)

EER	EER*	ZeroFMR	ZeroFNMR
6.30% (5.74%–6.86%)	6.30% (5.74%–6.86%)	42.18%	78.34%

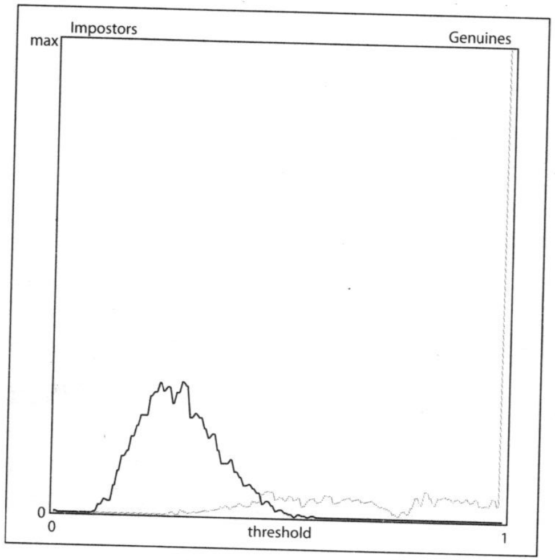

Score distributions

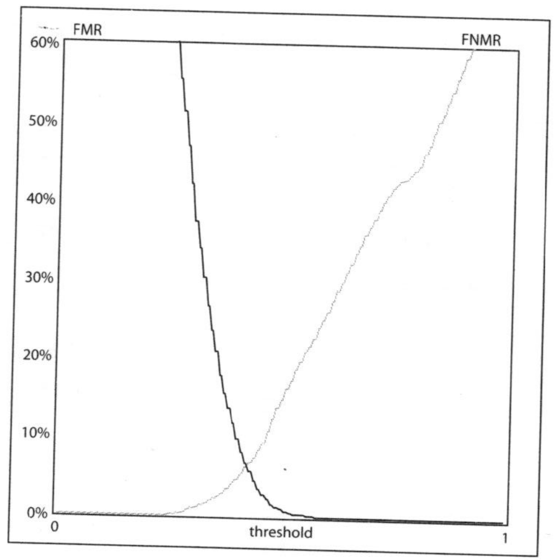

FMR(t) and FNMR(t)

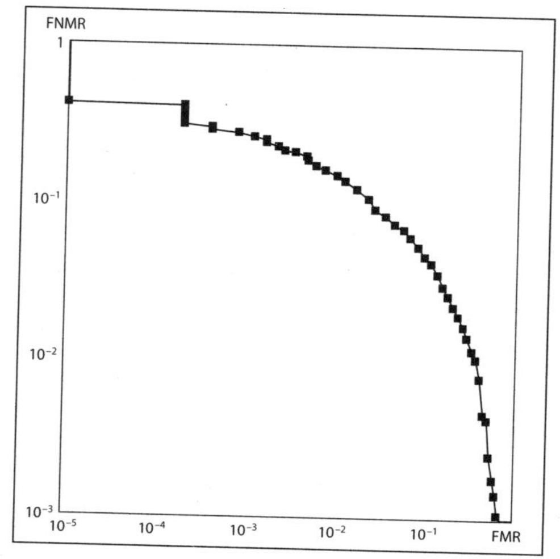

ROC curve

Algorithm DITI on database DB1_A

Average enroll time: 0.65 seconds
Average match time: 0.72 seconds

REJ_{ENROLL}	NGRA	NIRA	REJ_{NGRA}	REJ_{NIRA}
0.00% (F:0 T:0 C:0)	2800	4950	0.00% (F:0 T:0 C:0)	0.00% (F:0 T:0 C:0)

EER	EER*	ZeroFMR	ZeroFNMR
23.63% (23.62%–23.64%)	23.63% (23.62%–23.64%)	50.54%	100.00%

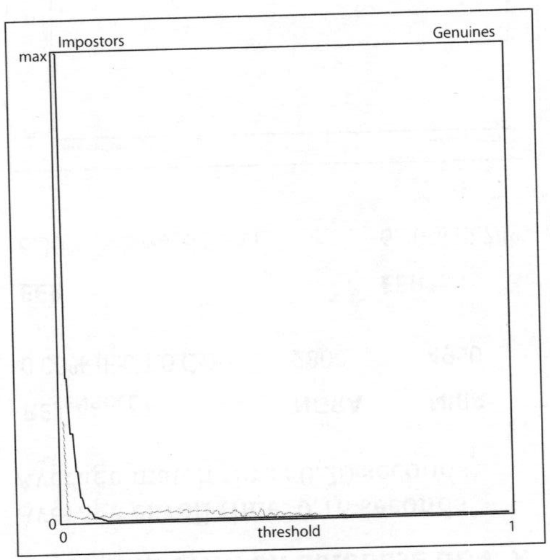

Score distributions

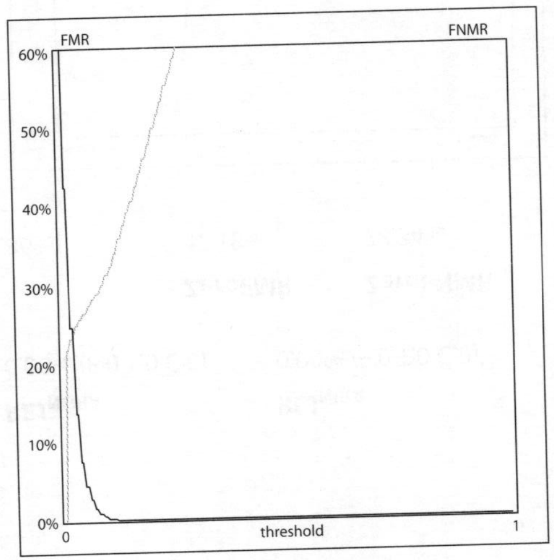

FMR(t) and FNMR(t)

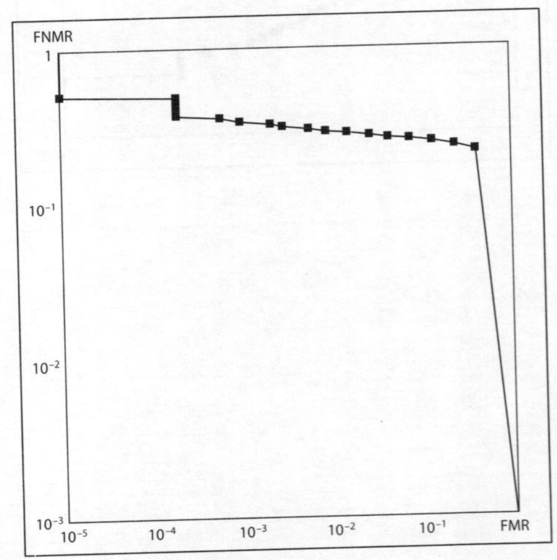

ROC curve

Algorithm DITI on database DB2_A

Average enroll time: 1.21 seconds
Average match time: 1.28 seconds

REJ_{ENROLL}	NGRA	NIRA	REJ_{NGRA}	REJ_{NIRA}
0.00% (F:0 T:0 C:0)	2800	4950	0.00% (F:0 T:0 C:0)	0.00% (F:0 T:0 C:0)

EER	EER*	ZeroFMR	ZeroFNMR
13.83% (13.80%–13.86%)	13.83% (13.80%–13.86%)	37.43%	100.00%

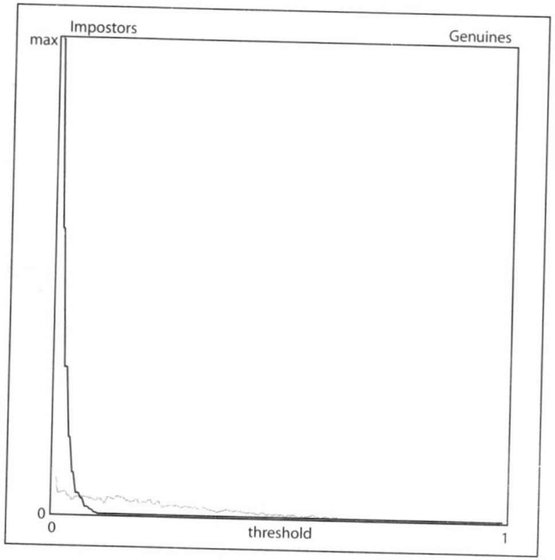

Score distributions

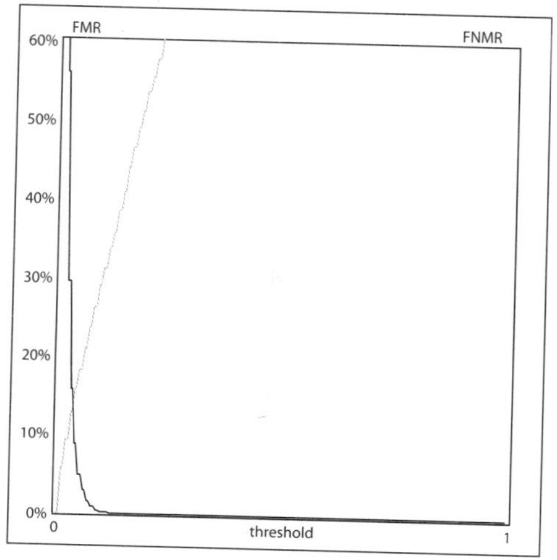

FMR(*t*) and FNMR(*t*)

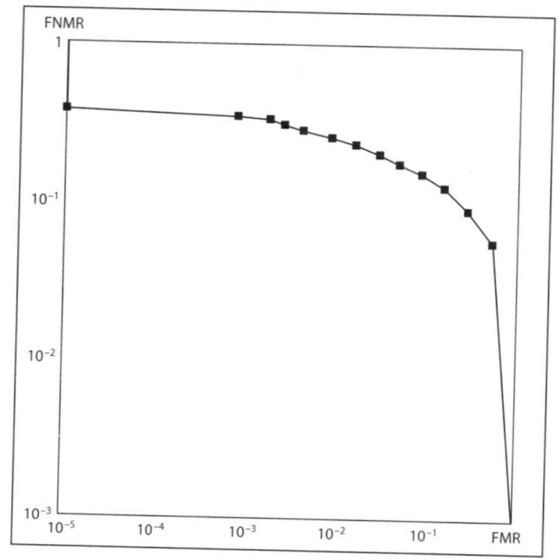

ROC curve

Algorithm DITI on database DB3_A

Average enroll time: 2.59 seconds
Average match time: 2.67 seconds

REJ_{ENROLL}	NGRA	NIRA	REJ_{NGRA}	REJ_{NIRA}
0.00% (F:0 T:0 C:0)	2800	4950	0.00% (F:0 T:0 C:0)	0.00% (F:0 T:0 C:0)

EER	EER*	ZeroFMR	ZeroFNMR
22.63% (22.61%–22.65%)	22.63% (22.61%–22.65%)	65.54%	100.00%

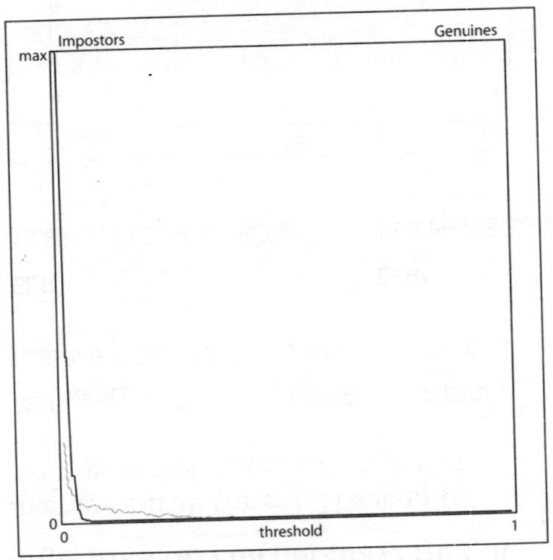

Score distributions

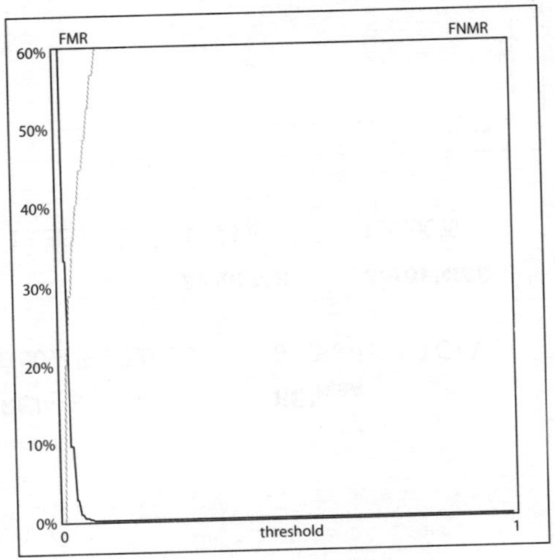

FMR(t) and FNMR(t)

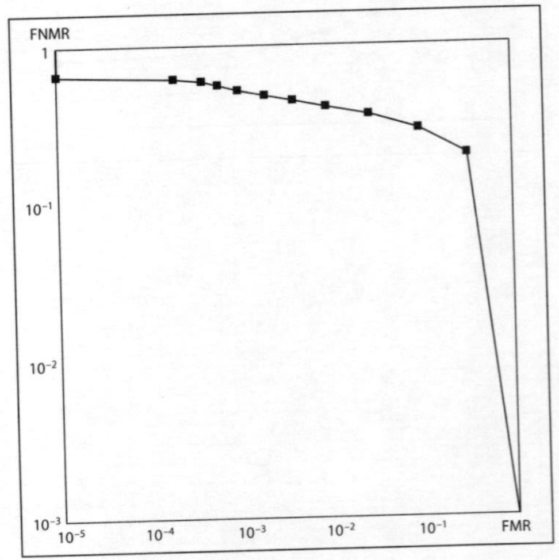

ROC curve

Algorithm DITI on database DB4_A

Average enroll time: 0.52 seconds
Average match time: 0.60 seconds

REJ_{ENROLL}	**NGRA**	**NIRA**	**REJ_{NGRA}**	**REJ_{NIRA}**
0.00% (F:0 T:0 C:0)	2800	4950	0.00% (F:0 T:0 C:0)	0.00% (F:0 T:0 C:0)

EER	**EER***	**ZeroFMR**	**ZeroFNMR**
23.80% (23.74%–23.86%)	23.80% (23.74%–23.86%)	75.36%	100.00%

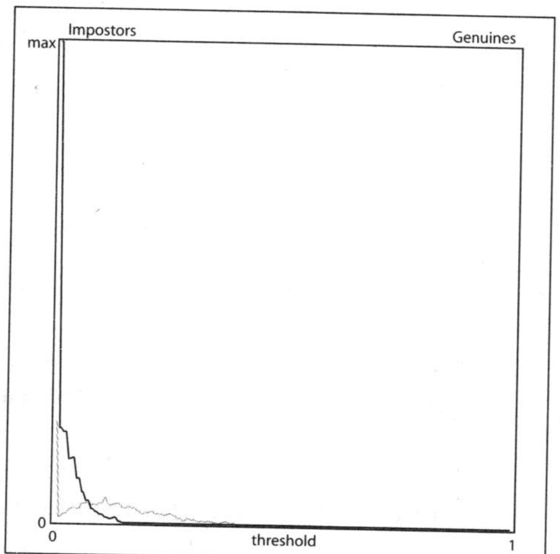

Score distributions

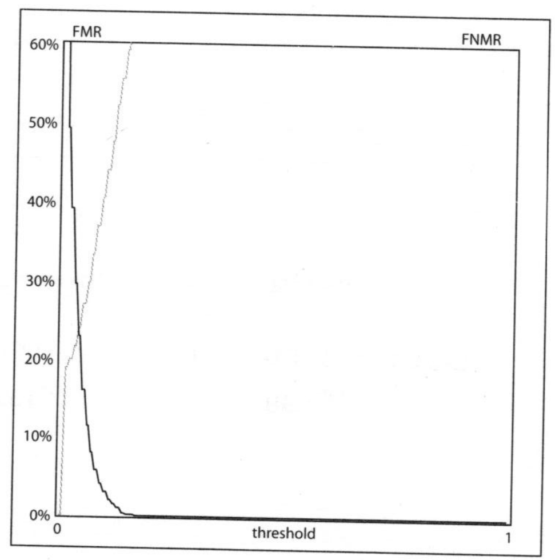

FMR(*t*) and FNMR(*t*)

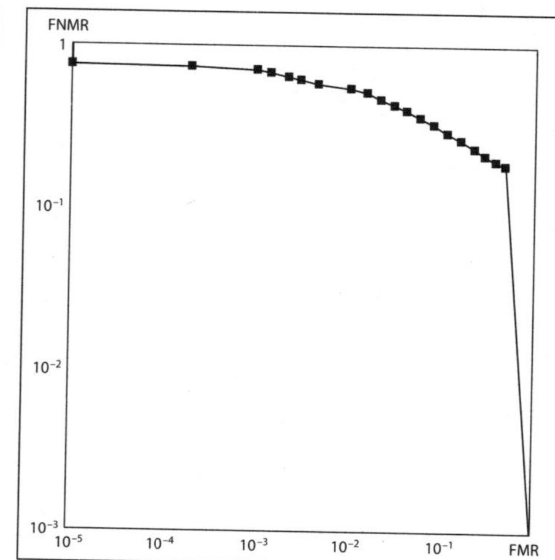

ROC curve

Algorithm FPIN on database DB1_A

Average enroll time: 0.83 seconds
Average match time: 0.87 seconds

REJ_{ENROLL}	NGRA	NIRA	REJ_{NGRA}	REJ_{NIRA}
0.00% (F:0 T:0 C:0)	2800	4950	0.00% (F:0 T:0 C:0)	0.00% (F:0 T:0 C:0)

EER	EER*	ZeroFMR	ZeroFNMR
13.46%	13.46%	96.07%	100.00%

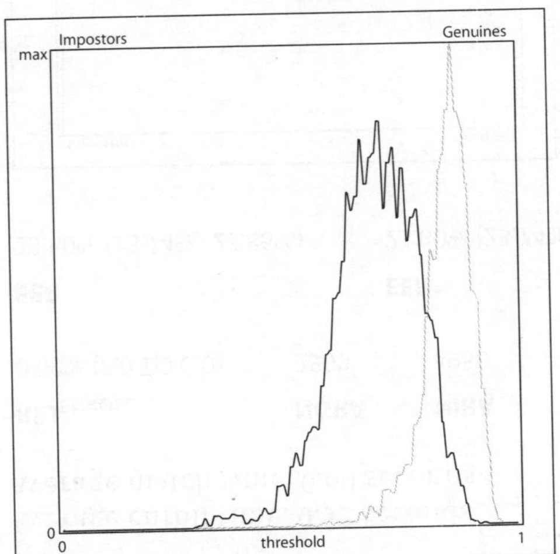

Score distributions

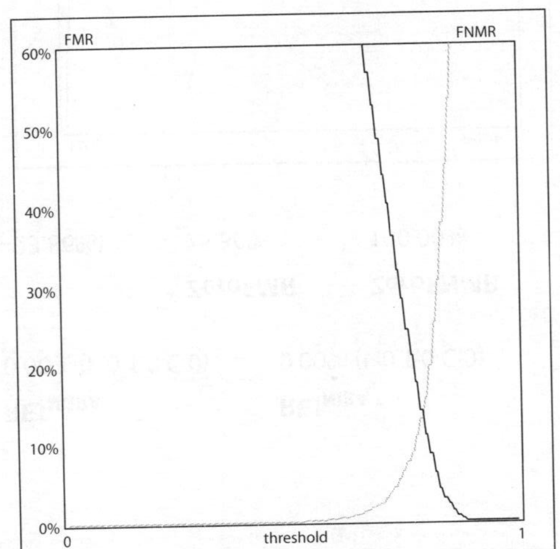

FMR(t) and FNMR(t)

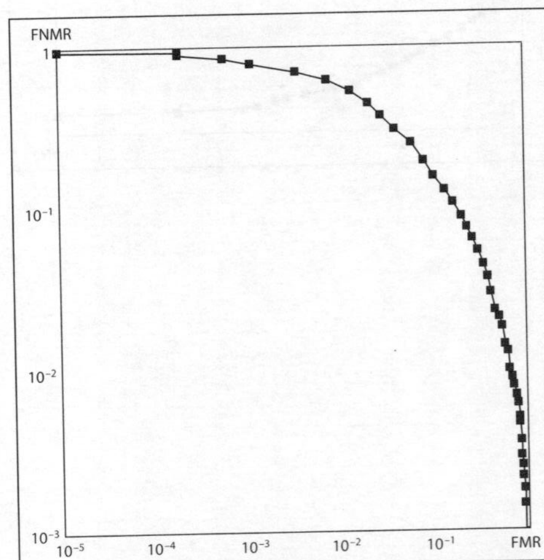

ROC curve

Algorithm FPIN on database DB2_A

Average enroll time: 1.16 seconds
Average match time: 1.24 seconds

REJ_{ENROLL}	NGRA	NIRA	REJ_{NGRA}	REJ_{NIRA}
0.00% (F:0 T:0 C:0)	2800	4950	0.00% (F:0 T:0 C:0)	0.00% (F:0 T:0 C:0)

EER	EER*	ZeroFMR	ZeroFNMR
11.14% (11.13%–11.14%)	11.14% (11.13%–11.14%)	95.61%	99.45%

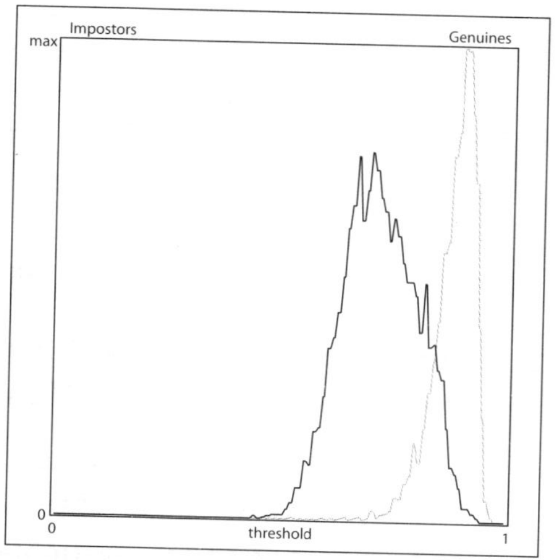

Score distributions

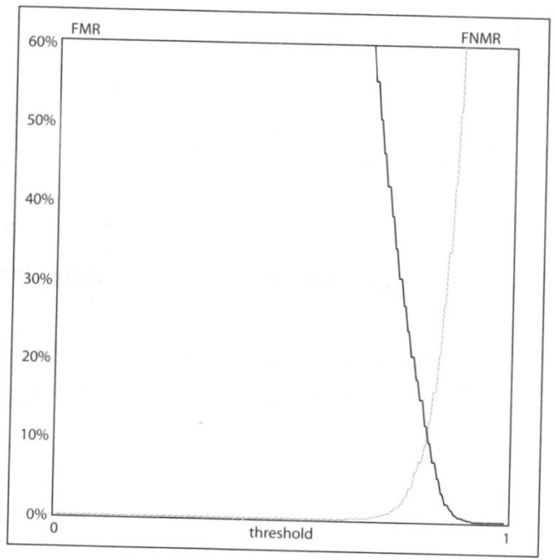

FMR(t) and FNMR(t)

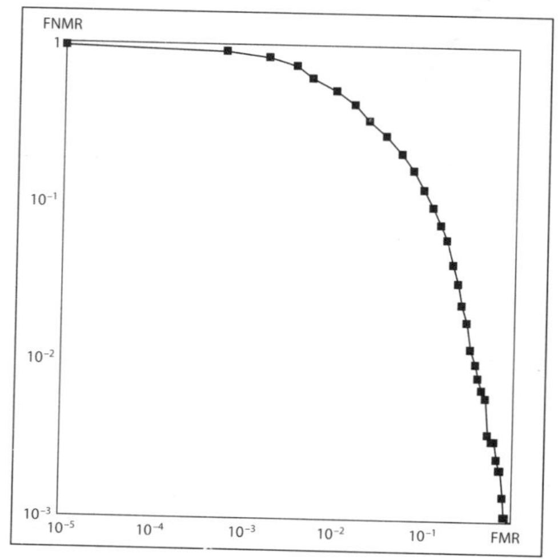

ROC curve

Algorithm FPIN on database DB3_A

Average enroll time: 2.13 seconds
Average match time: 2.19 seconds

REJ_{ENROLL}	NGRA	NIRA	REJ_{NGRA}	REJ_{NIRA}
0.00% (F:0 T:0 C:0)	2800	4950	0.00% (F:0 T:0 C:0)	0.00% (F:0 T:0 C:0)

EER	EER*	ZeroFMR	ZeroFNMR
23.18%	23.18%	98.61%	100.00%

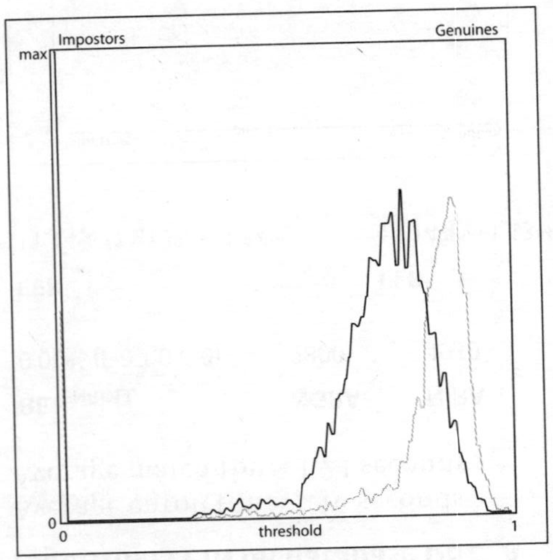

Score distributions

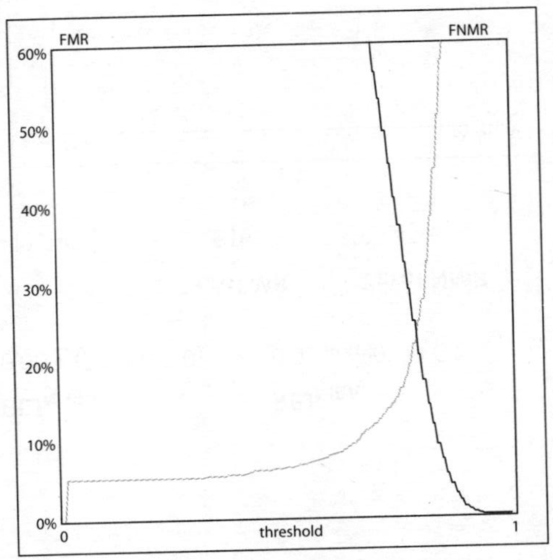

FMR(t) and FNMR(t)

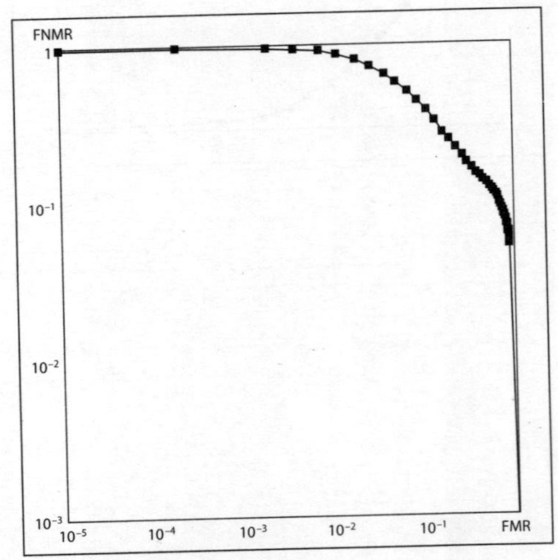

ROC curve

Algorithm FPIN on database DB4_A

Average enroll time: 0.77 seconds
Average match time: 0.80 seconds

REJ_{ENROLL}	NGRA	NIRA	REJ_{NGRA}	REJ_{NIRA}
0.00% (F:0 T:0 C:0)	2800	4950	0.00% (F:0 T:0 C:0)	0.00% (F:0 T:0 C:0)

EER	EER*	ZeroFMR	ZeroFNMR
16.00%	16.00%	97.89%	80.02%

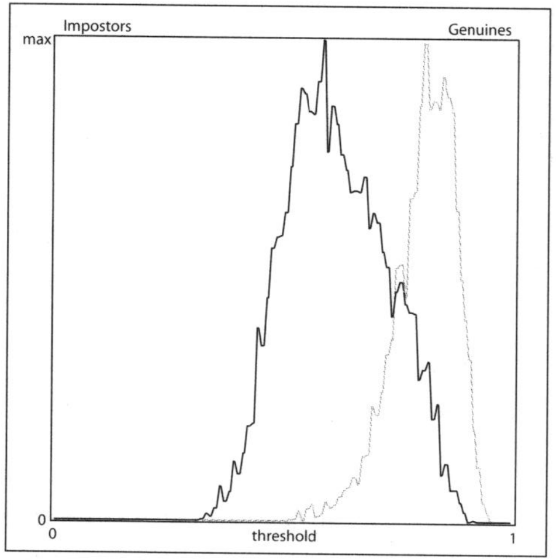

Score distributions

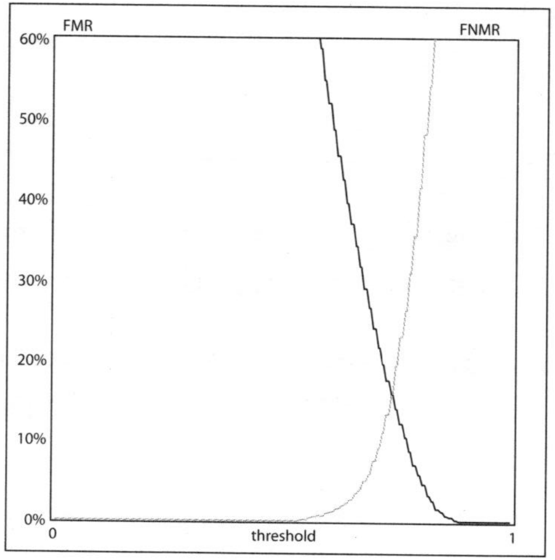

FMR(*t*) and FNMR(*t*)

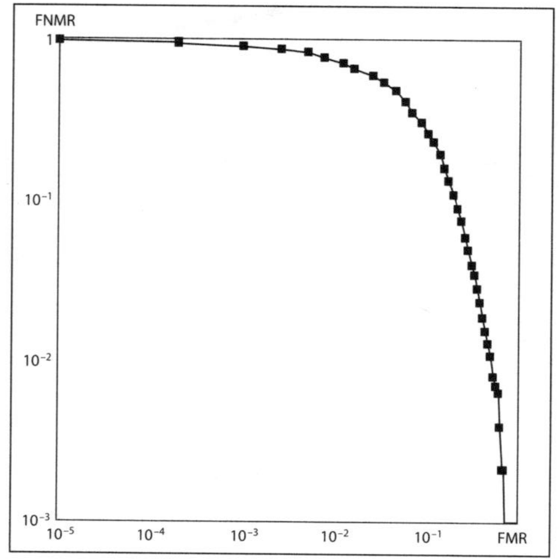

ROC curve

Algorithm KRDL on database DB1_A

Average enroll time: 1.00 seconds
Average match time: 1.06 seconds

REJ_{ENROLL}	NGRA	NIRA	REJ_{NGRA}	REJ_{NIRA}
6.43% (F:43 T:2 C:0)	2644	4841	4.43% (F:113 T:4 C:0)	7.77% (F:376 T:0 C:0)

EER	EER*	ZeroFMR	ZeroFNMR
10.66%	7.35% (7.35%–7.36%)	24.51%	100.00%

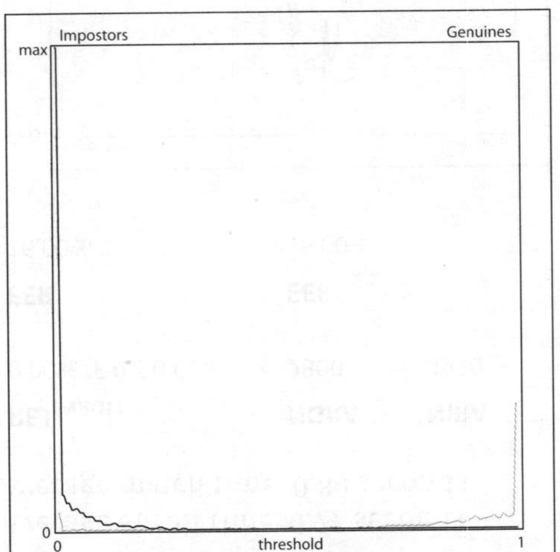

Score distributions

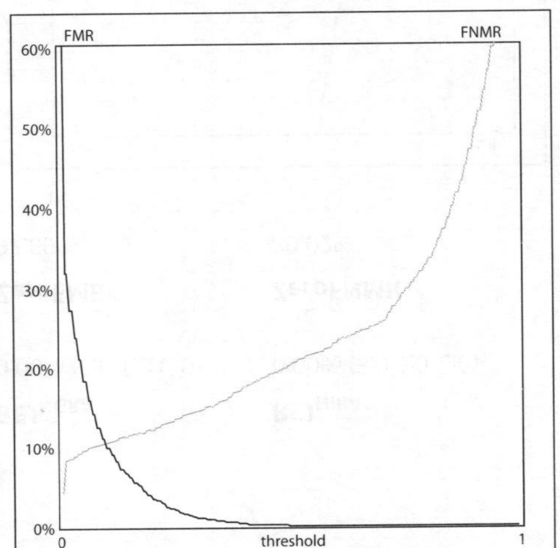

FMR(t) and FNMR(t)

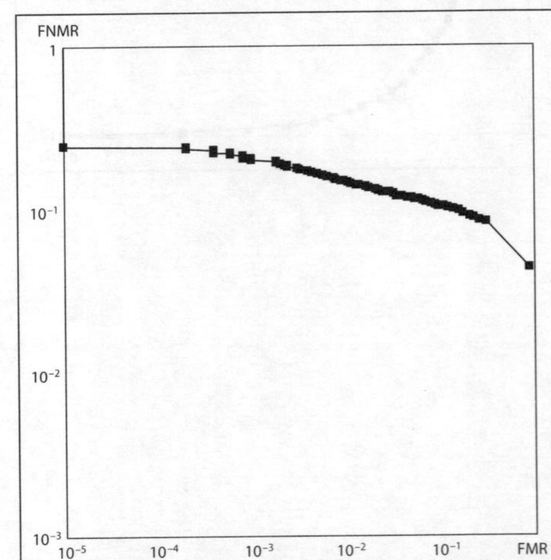

ROC curve

Algorithm KRDL on database DB2_A

Average enroll time: 1.16 seconds
Average match time: 2.88 seconds

REJ_{ENROLL}	NGRA	NIRA	REJ_{NGRA}	REJ_{NIRA}
3.29% (F:22 T:1 C:0)	2702	4942	1.85% (F:46 T:4 C:0)	5.81% (F:286 T:1 C:0)

EER	EER*	ZeroFMR	ZeroFNMR
8.83% (8.82%–8.85%)	7.53% (7.52%–7.54%)	22.13%	100.00%

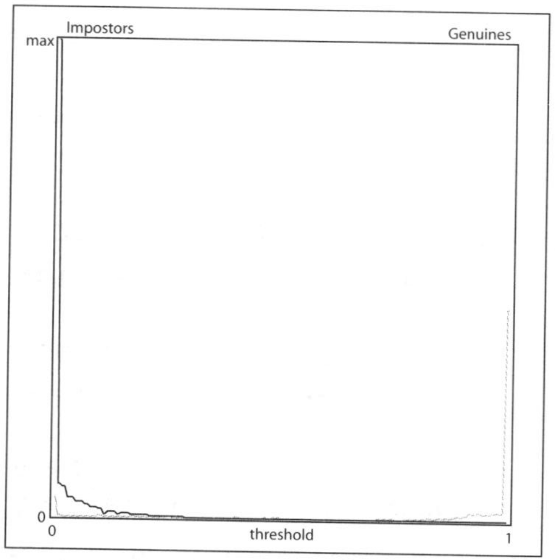

Score distributions

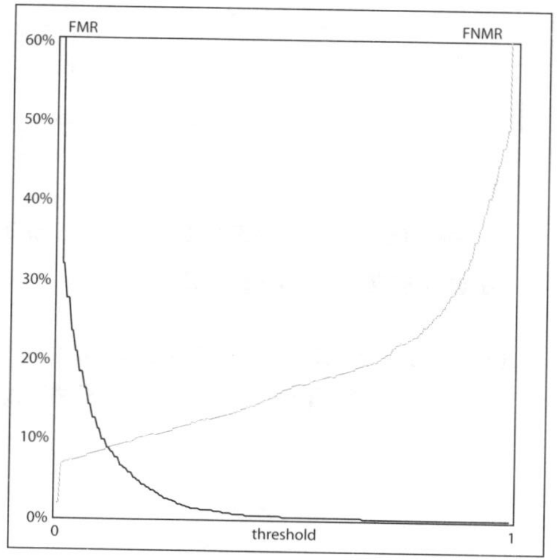

FMR(t) and FNMR(t)

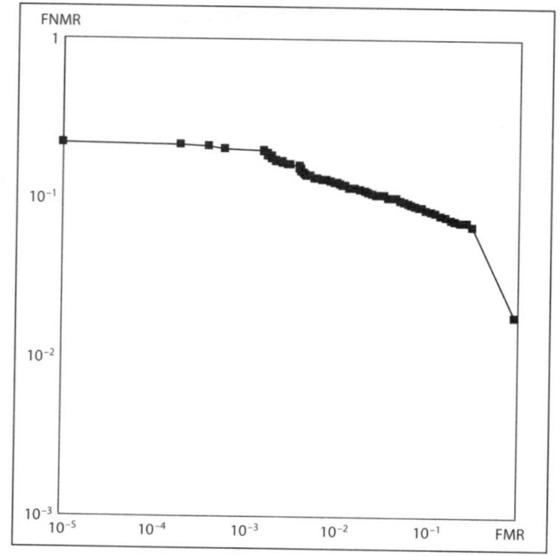

ROC curve

Algorithm KRDL on database DB3_A

Average enroll time: 1.48 seconds
Average match time: 1.60 seconds

REJ_{ENROLL}	NGRA	NIRA	REJ_{NGRA}	REJ_{NIRA}
6.86% (F:48 T:0 C:0)	2637	4691	5.65% (F:149 T:0 C:0)	4.82% (F:226 T:0 C:0)

EER	EER*	ZeroFMR	ZeroFNMR
12.20% (12.19%–12.21%)	8.03% (7.97%–8.08%)	22.83%	100.00%

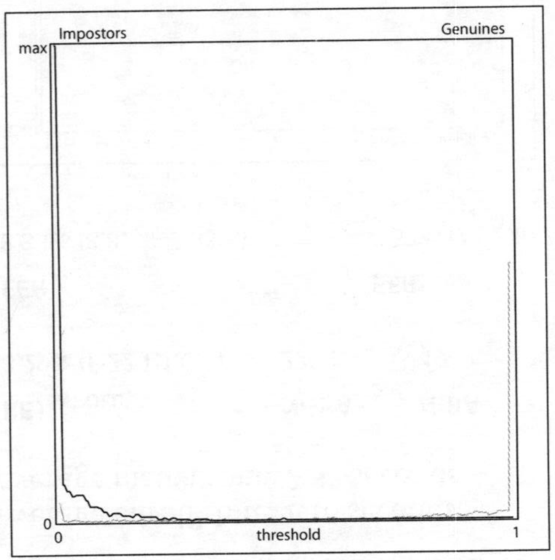

Score distributions

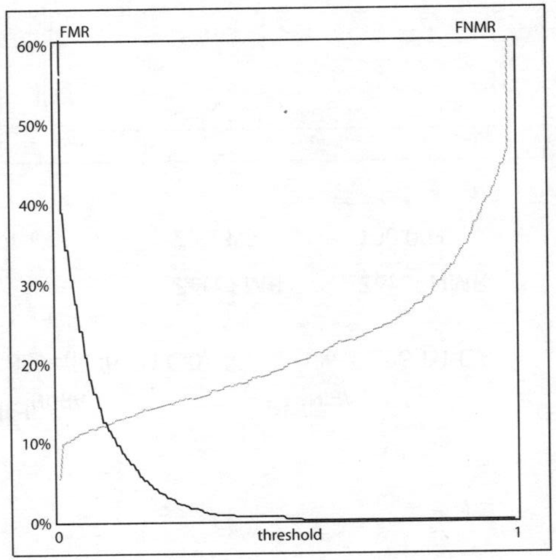

FMR(t) and FNMR(t)

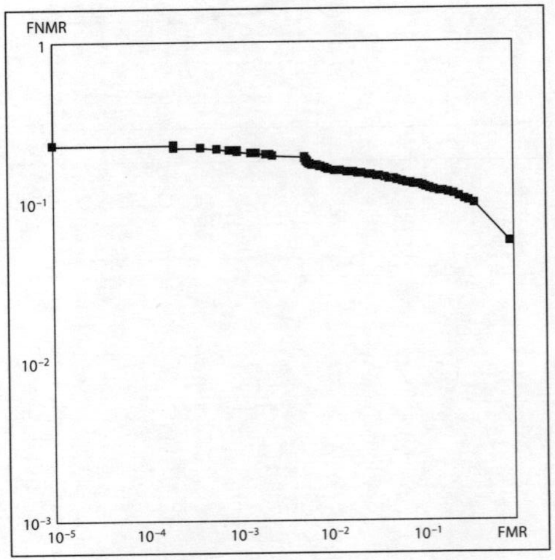

ROC curve

Algorithm KRDL on database DB4_A

Average enroll time: 0.70 seconds
Average match time: 0.79 seconds

REJ_{ENROLL}	NGRA	NIRA	REJ_{NGRA}	REJ_{NIRA}
10.86% (F:76 T:0 C:0)	2488	4477	6.11% (F:152 T:0 C:0)	12.53% (F:561 T:0 C:0)

EER	EER*	ZeroFMR	ZeroFNMR
12.08% (12.06%–12.10%)	7.46% (7.43%–7.49%)	40.19%	100.00%

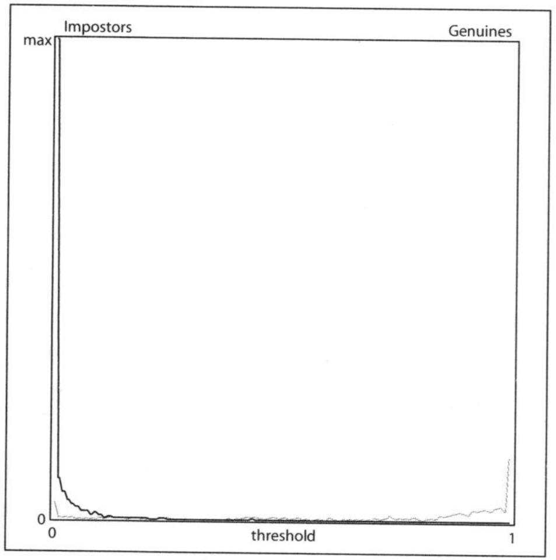

Score distributions

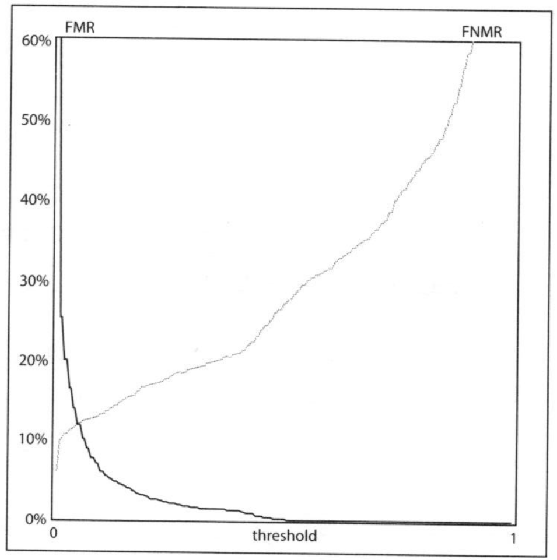

FMR(*t*) and FNMR(*t*)

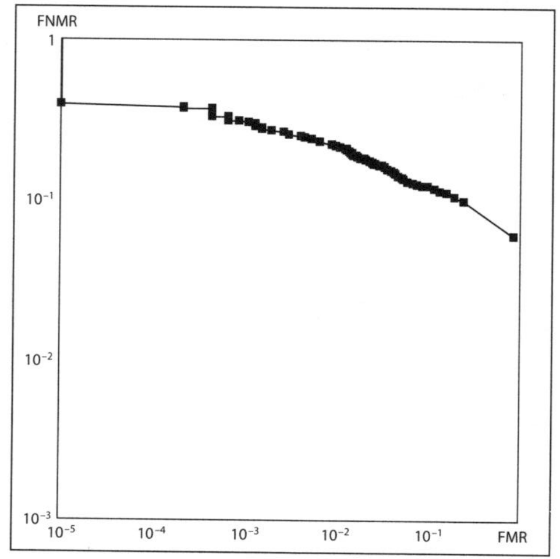

ROC curve

Algorithm NCMI on database DB1_A

Average enroll time: 1.13 seconds
Average match time: 1.34 seconds

REJ_{ENROLL}	NGRA	NIRA	REJ_{NGRA}	REJ_{NIRA}
0.00% (F:0 T:0 C:0)	2800	4950	0.00% (F:0 T:0 C:0)	0.18% (F:9 T:0 C:0)

EER	EER*	ZeroFMR	ZeroFNMR
49.11% (48.82%–49.39%)	49.15% (48.82%–49.48%)	100.00%	99.82%

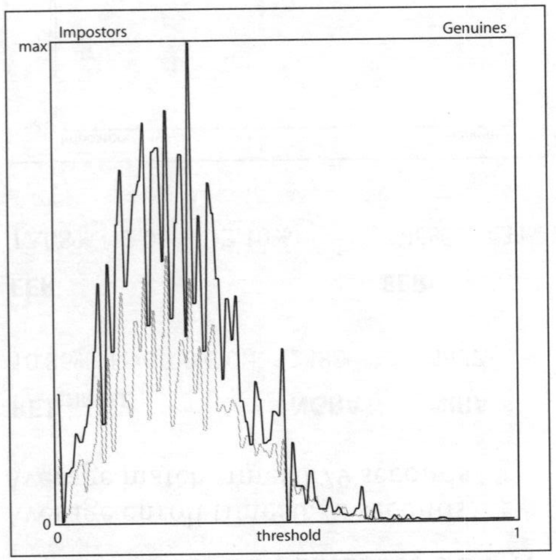

Score distributions

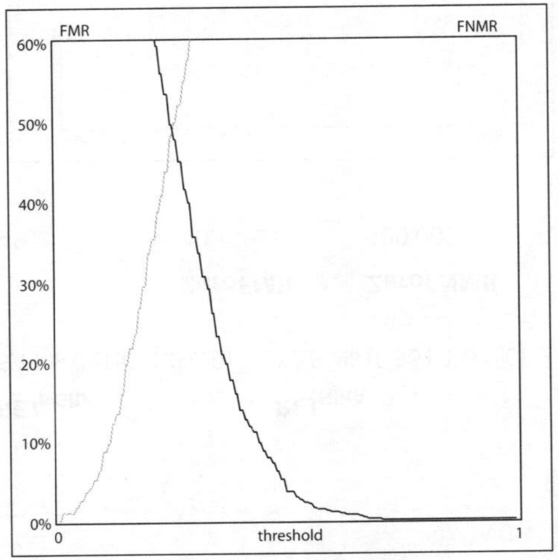

FMR(*t*) and FNMR(*t*)

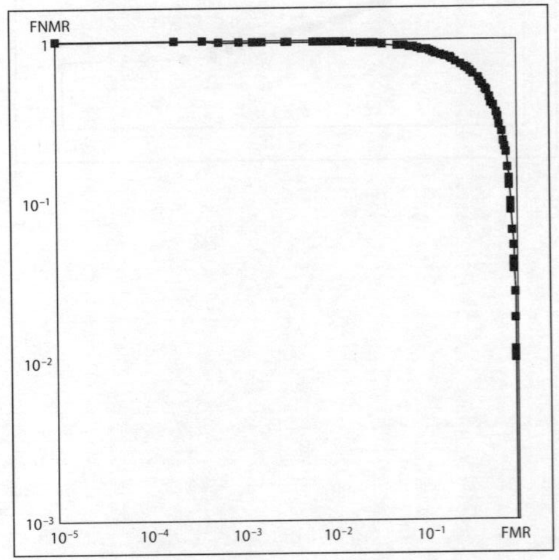

ROC curve

Algorithm NCMI on database DB2_A

Average enroll time: 1.28 seconds
Average match time: 1.57 seconds

REJ_{ENROLL}	NGRA	NIRA	REJ_{NGRA}	REJ_{NIRA}
0.00% (F:0 T:0 C:0)	2800	4950	0.00% (F:0 T:0 C:0)	0.00% (F:0 T:0 C:0)

EER	EER*	ZeroFMR	ZeroFNMR
46.15% (45.82%–46.48%)	46.15% (45.82%–46.48%)	100.00%	100.00%

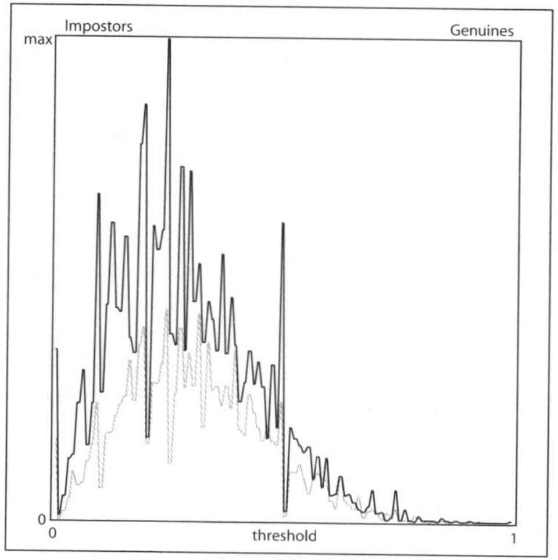

Score distributions

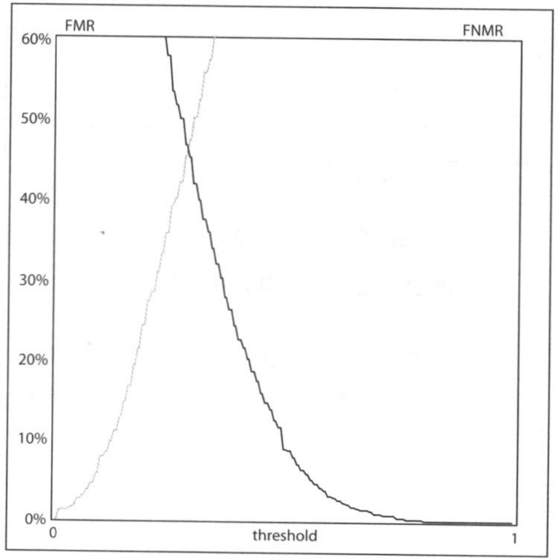

FMR(t) and FNMR(t)

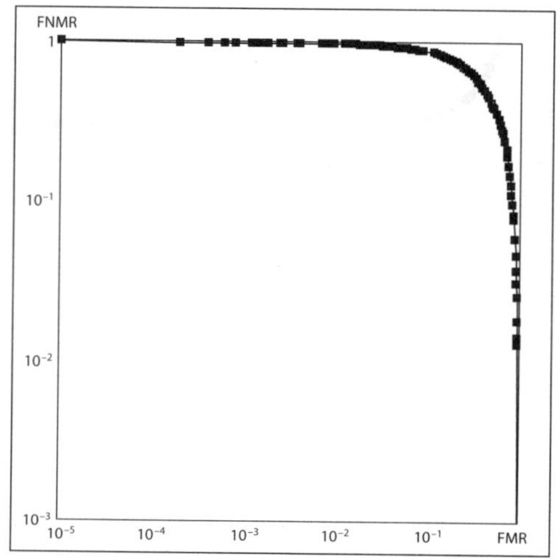

ROC curve

Algorithm NCMI on database DB3_A

Average enroll time: 2.25 seconds
Average match time: 2.75 seconds

REJ_{ENROLL}	NGRA	NIRA	REJ_{NGRA}	REJ_{NIRA}
0.00% (F:0 T:0 C:0)	2800	4950	0.00% (F:0 T:0 C:0)	0.02% (F:1 T:0 C:0)

EER	EER*	ZeroFMR	ZeroFNMR
47.43% (45.76%–49.11%)	47.44% (45.77%–49.11%)	100.00%	99.98%

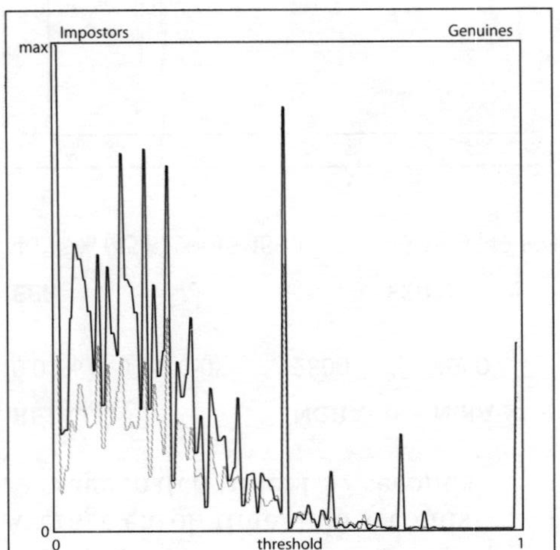

Score distributions

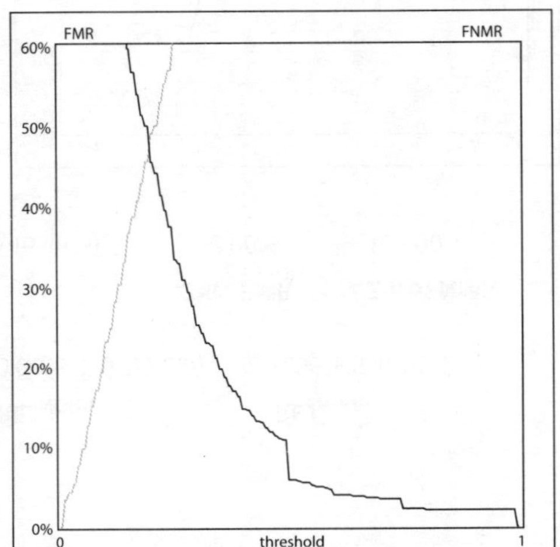

FMR(t) and FNMR(t)

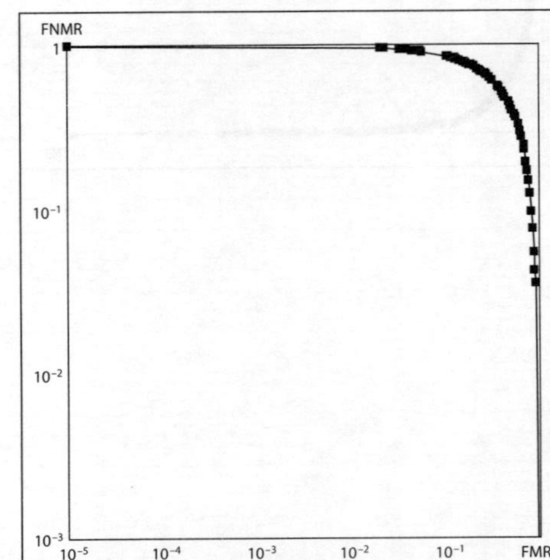

ROC curve

Algorithm NCMI on database DB4_A

Average enroll time: 1.08 seconds
Average match time: 1.19 seconds

REJ_{ENROLL}	NGRA	NIRA	REJ_{NGRA}	REJ_{NIRA}
0.00% (F:0 T:0 C:0)	2800	4950	0.18% (F:5 T:0 C:0)	0.28% (F:14 T:0 C:0)

EER	EER*	ZeroFMR	ZeroFNMR
48.67% (48.67%–48.68%)	48.77% (48.73%–48.80%)	99.96%	100.00%

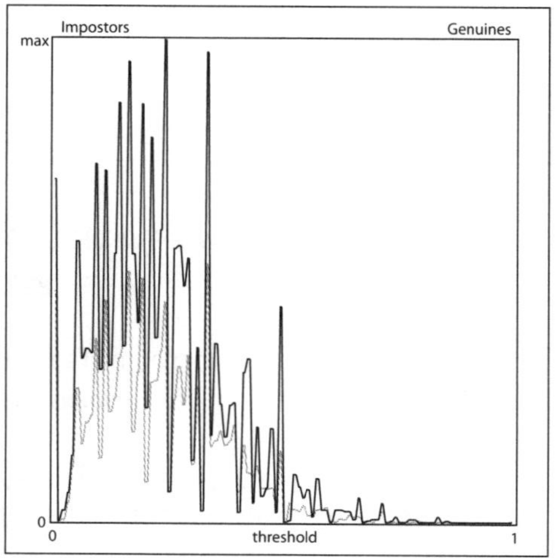

Score distributions

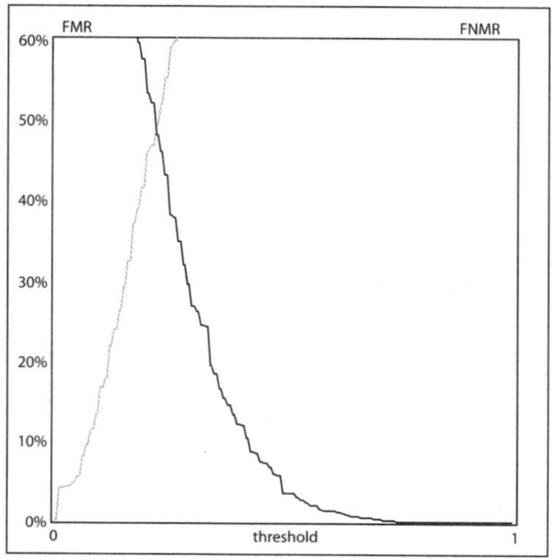

FMR(*t*) and FNMR(*t*)

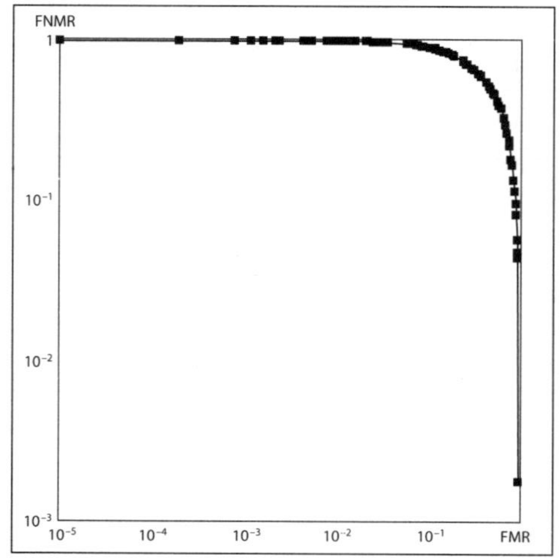

ROC curve

Algorithm SAG1 on database DB1_A

Average enroll time: 2.48 seconds
Average match time: 0.96 seconds

REJ_{ENROLL}	NGRA	NIRA	REJ_{NGRA}	REJ_{NIRA}
0.00% (F:0 T:0 C:0)	2800	4950	0.00% (F:0 T:0 C:0)	0.00% (F:0 T:0 C:0)

EER	EER*	ZeroFMR	ZeroFNMR
0.67% (0.67%–0.68%)	0.67% (0.67%–0.68%)	2.11%	53.13%

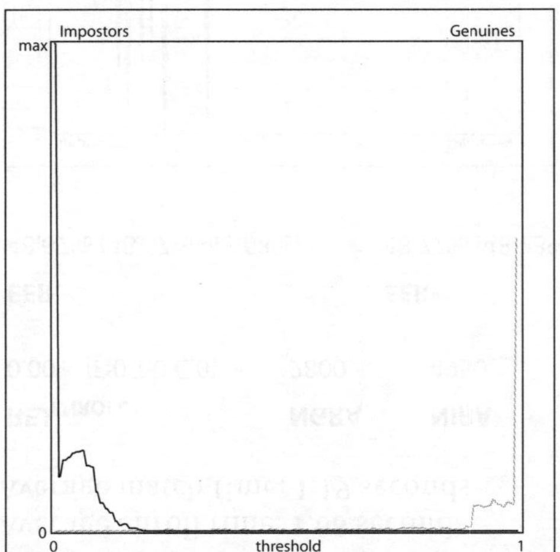

Score distributions

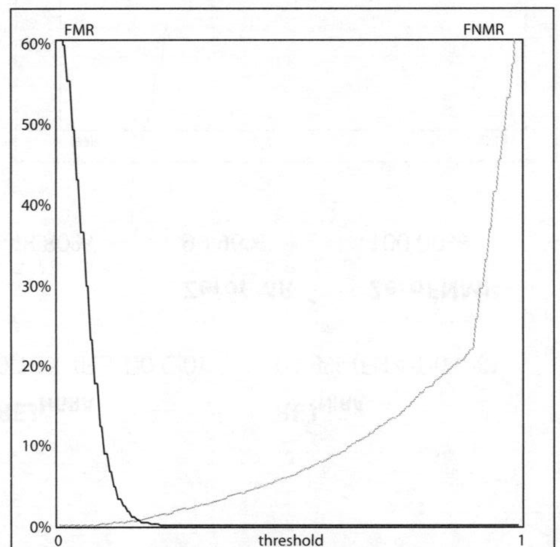

FMR(t) and FNMR(t)

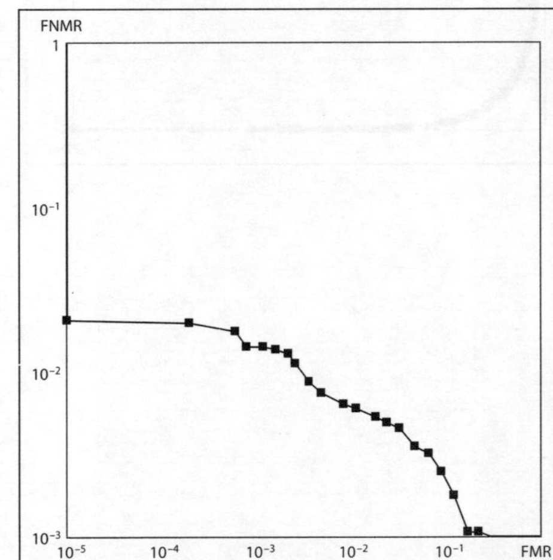

ROC curve

Algorithm SAG1 on database DB2_A

Average enroll time: 2.63 seconds
Average match time: 1.03 seconds

REJ_{ENROLL}	NGRA	NIRA	REJ_{NGRA}	REJ_{NIRA}
0.00% (F:0 T:0 C:0)	2800	4950	0.00% (F:0 T:0 C:0)	0.00% (F:0 T:0 C:0)

EER	EER*	ZeroFMR	ZeroFNMR
0.61%	0.61%	1.36%	50.69%

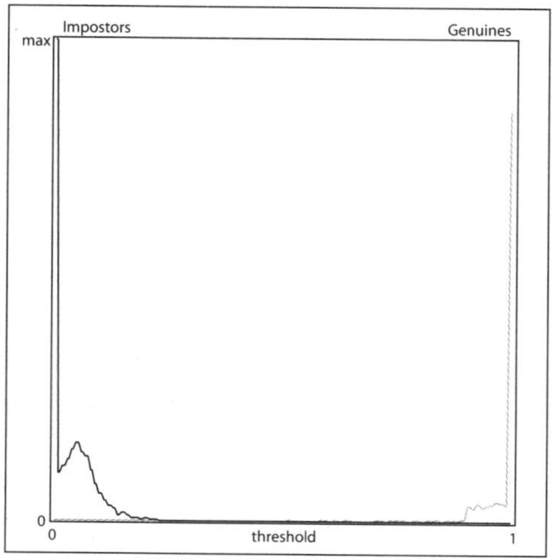

Score distributions

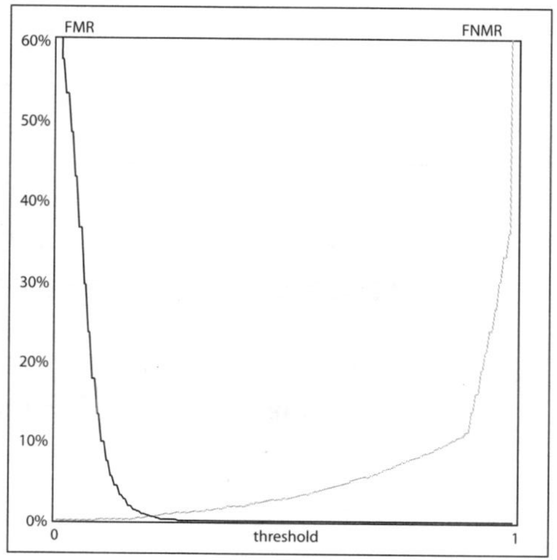

FMR(t) and FNMR(t)

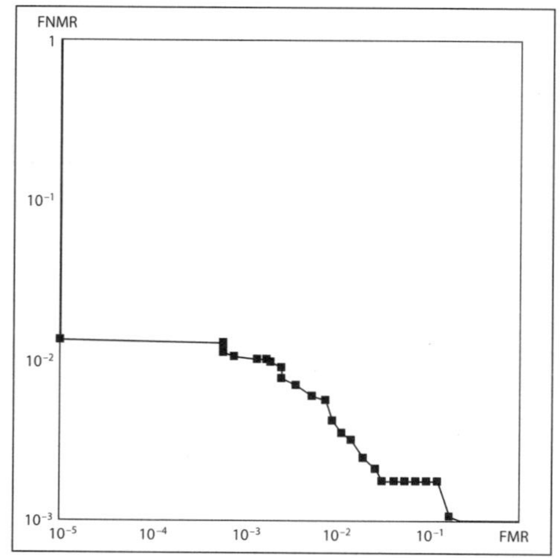

ROC curve

Algorithm SAG1 on database DB3_A

Average enroll time: 5.70 seconds
Average match time: 2.13 seconds

REJ_{ENROLL}	NGRA	NIRA	REJ_{NGRA}	REJ_{NIRA}
0.00% (F:0 T:0 C:0)	2800	4950	0.00% (F:0 T:0 C:0)	0.00% (F:0 T:0 C:0)
EER	**EER***		**ZeroFMR**	**ZeroFNMR**
3.64%	3.64%		6.82%	100.00%

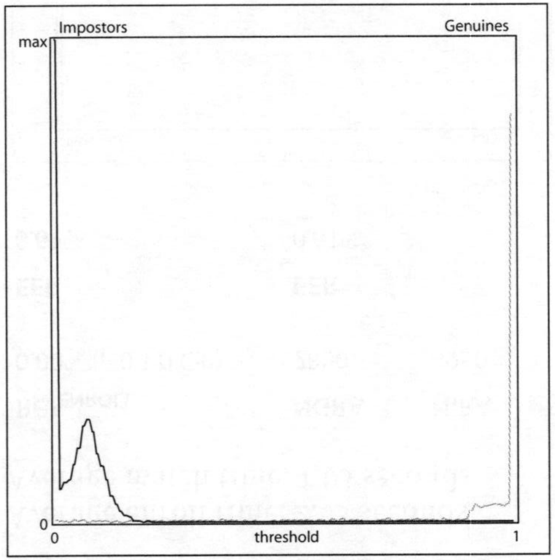

Score distributions

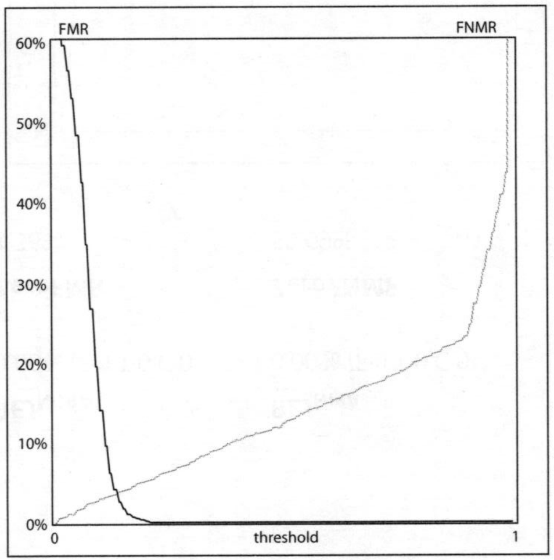

FMR(t) and FNMR(t)

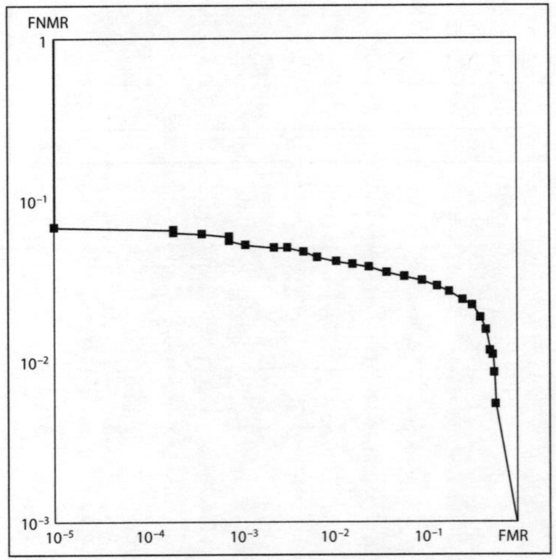

ROC curve

Algorithm SAG1 on database DB4_A

Average enroll time: 1.90 seconds
Average match time: 0.77 seconds

REJ_{ENROLL}	NGRA	NIRA	REJ_{NGRA}	REJ_{NIRA}
0.00% (F:0 T:0 C:0)	2800	4950	0.00% (F:0 T:0 C:0)	0.00% (F:0 T:0 C:0)

EER	EER*	ZeroFMR	ZeroFNMR
1.99% (1.98%–2.00%)	1.99% (1.98%–2.00%)	6.71%	100.00%

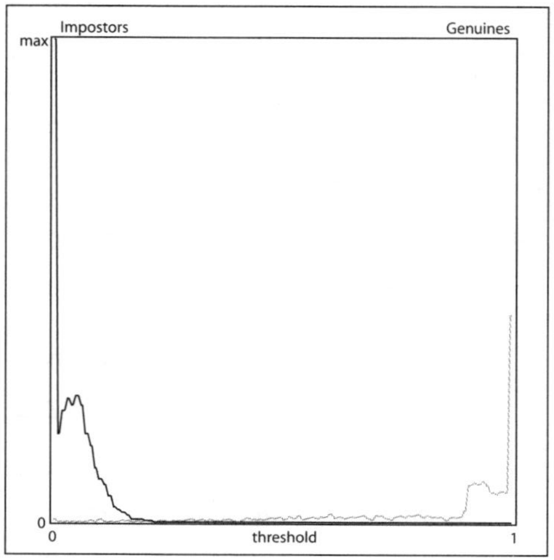

Score distributions

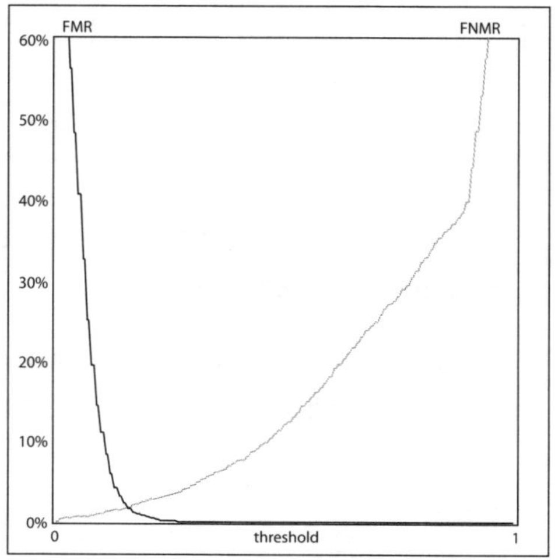

FMR(*t*) and FNMR(*t*)

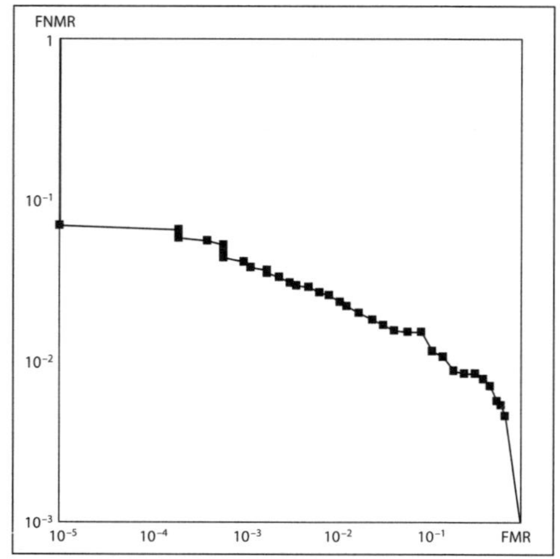

ROC curve

Algorithm SAG2 on database DB1_A

Average enroll time: 0.88 seconds
Average match time: 0.88 seconds

REJ_{ENROLL}	NGRA	NIRA	REJ_{NGRA}	REJ_{NIRA}
0.00% (F:0 T:0 C:0)	2800	4950	0.00% (F:0 T:0 C:0)	0.00% (F:0 T:0 C:0)

EER	EER*	ZeroFMR	ZeroFNMR
1.17% (1.15%–1.18%)	1.17% (1.15%–1.18%)	3.07%	74.57%

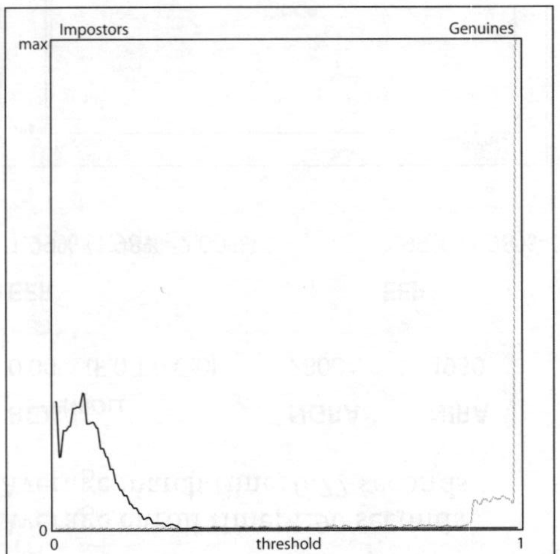

Score distributions

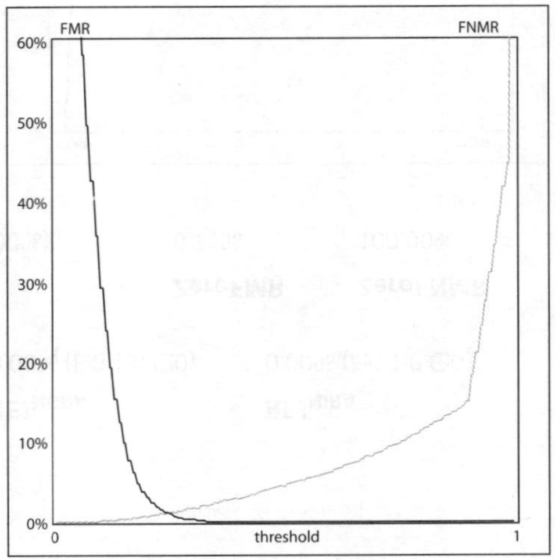

FMR(*t*) and FNMR(*t*)

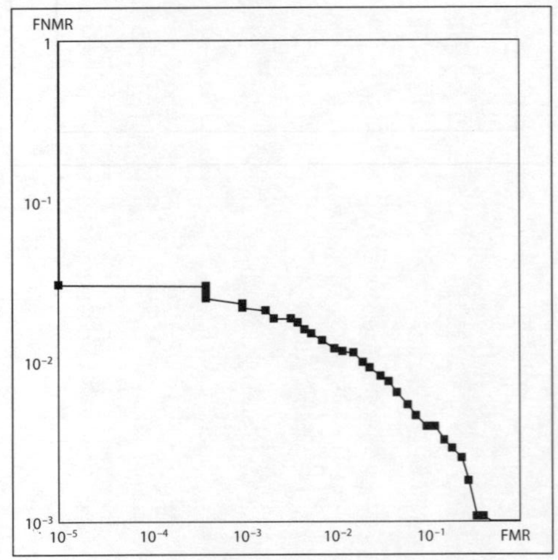

ROC curve

Algorithm SAG2 on database DB2_A

Average enroll time: 0.93 seconds
Average match time: 0.93 seconds

REJ_{ENROLL}	NGRA	NIRA	REJ_{NGRA}	REJ_{NIRA}
0.00% (F:0 T:0 C:0)	2800	4950	0.00% (F:0 T:0 C:0)	0.00% (F:0 T:0 C:0)
EER	**EER***		**ZeroFMR**	**ZeroFNMR**
0.82%	0.82%		2.14%	100.00%

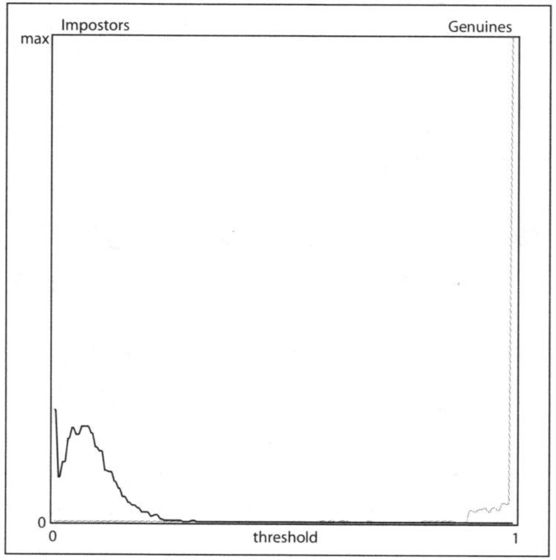

Score distributions

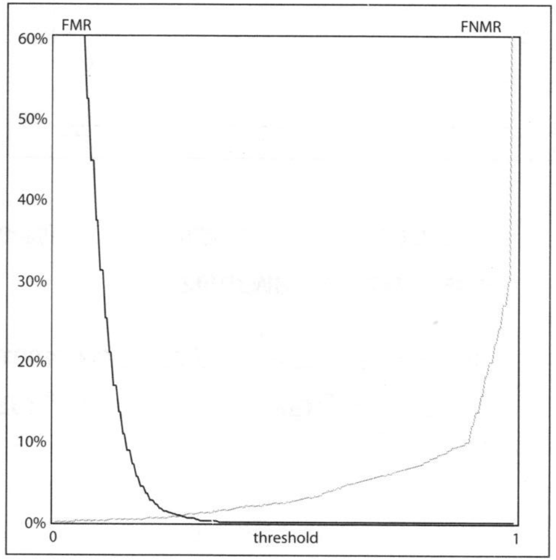

FMR(*t*) and FNMR(*t*)

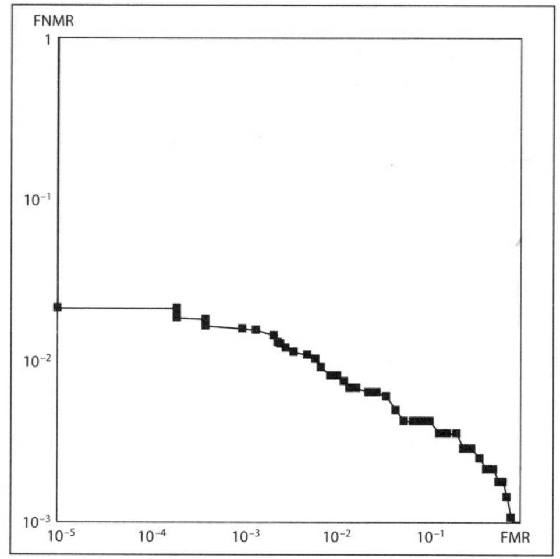

ROC curve

Algorithm SAG2 on database DB3_A

Average enroll time: 1.94 seconds
Average match time: 1.94 seconds

REJ_{ENROLL}	NGRA	NIRA	REJ_{NGRA}	REJ_{NIRA}
0.00% (F:0 T:0 C:0)	2800	4950	0.00% (F:0 T:0 C:0)	0.00% (F:0 T:0 C:0)

EER	EER*	ZeroFMR	ZeroFNMR
4.01% (3.98%–4.04%)	4.01% (3.98%–4.04%)	9.50%	100.00%

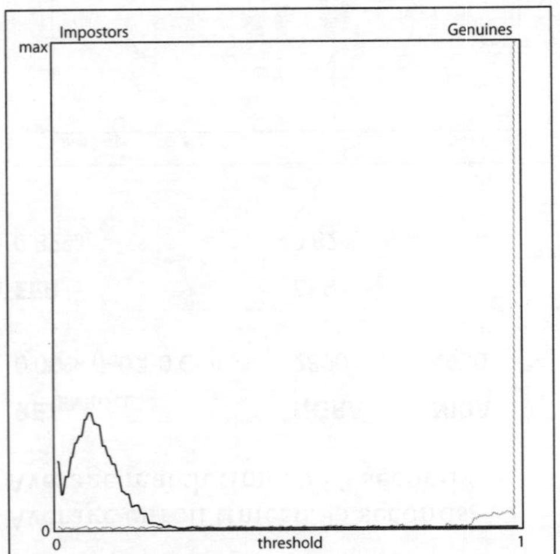

Scoredistributions

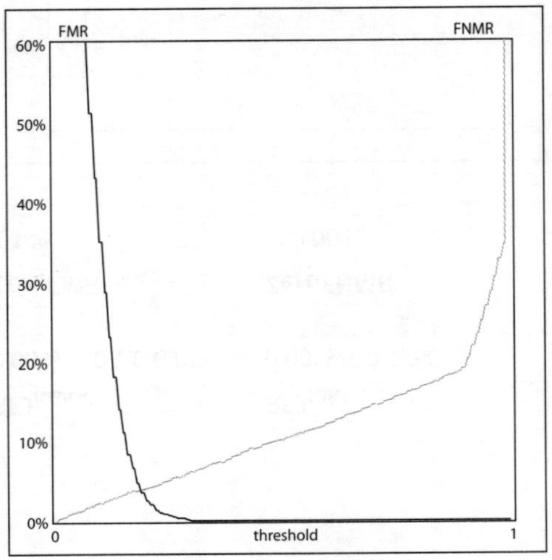

FMR(*t*)andFNMR(*t*)

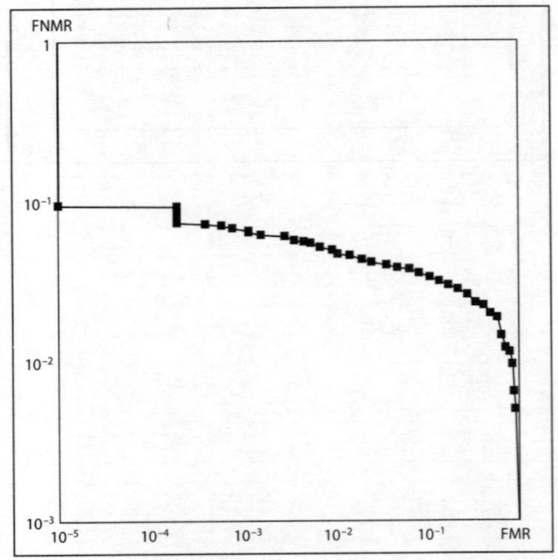

ROCcurve

Algorithm SAG2 on database DB4_A

Average enroll time: 0.69 seconds
Average match time: 0.69 seconds

REJ_{ENROLL}	NGRA	NIRA	REJ_{NGRA}	REJ_{NIRA}
0.00% (F:0 T:0 C:0)	2800	4950	0.00% (F:0 T:0 C:0)	0.00% (F:0 T:0 C:0)

EER	EER*	ZeroFMR	ZeroFNMR
3.11%	3.11%	10.57%	100.00%

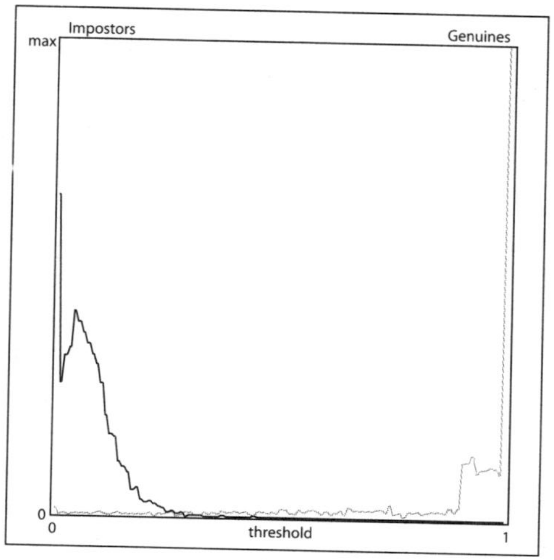

Score distributions

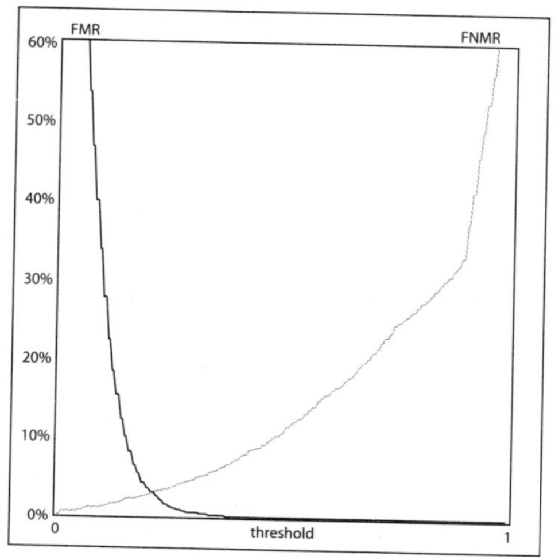

FMR(t) and FNMR(t)

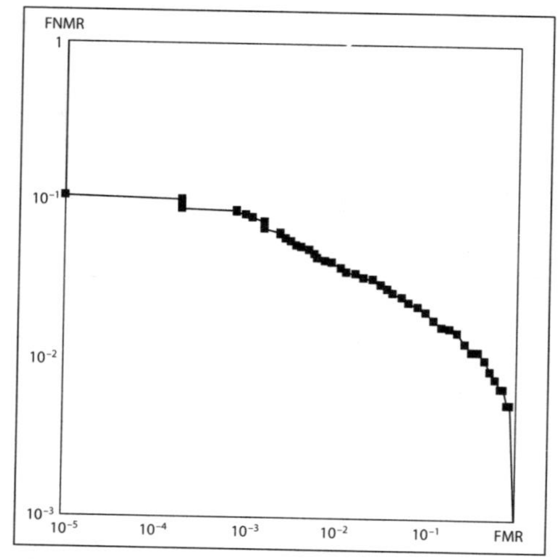

ROC curve

Algorithm UINH on database DB1_A

Average enroll time: 0.53 seconds
Average match time: 0.56 seconds

REJ_{ENROLL}	NGRA	NIRA	REJ_{NGRA}	REJ_{NIRA}
1.71% (F:12 T:0 C:0)	2745	4911	1.71% (F:47 T:0 C:0)	6.96% (F:342 T:0 C:0)

EER	EER*	ZeroFMR	ZeroFNMR
21.02% (20.91%–21.14%)	20.65% (20.64%–20.66%)	82.00%	100.00%

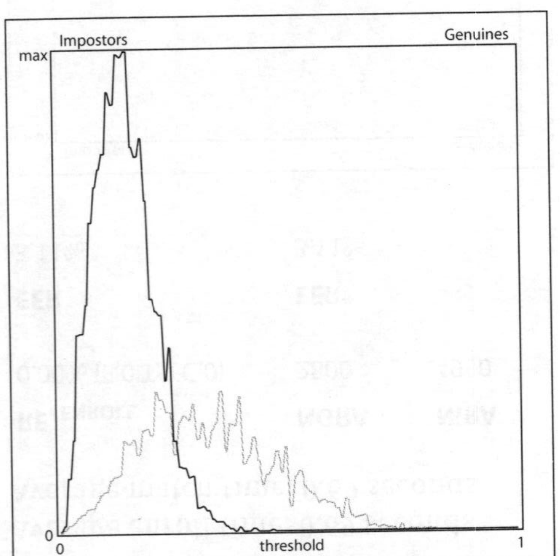

Score distributions

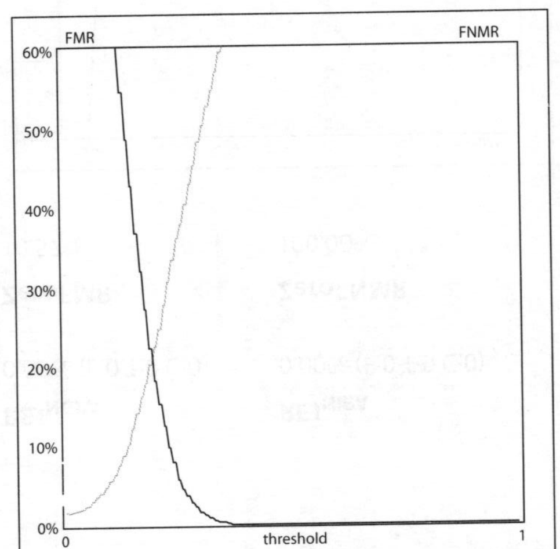

FMR(t) and FNMR(t)

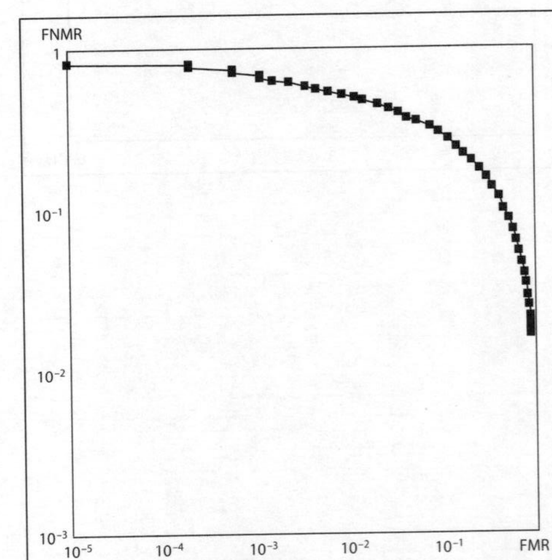

ROC curve

Algorithm UINH on database DB2_A

Average enroll time: 0.60 seconds
Average match time: 0.65 seconds

REJ_{ENROLL}	**NGRA**	**NIRA**	**REJ_{NGRA}**	**REJ_{NIRA}**
0.86% (F:6 T:0 C:0)	2768	4922	1.45% (F:40 T:0 C:0)	5.57% (F:274 T:0 C:0)

EER	**EER***	**ZeroFMR**	**ZeroFNMR**
15.22% (15.21%–15.24%)	14.70%	56.11%	100.00%

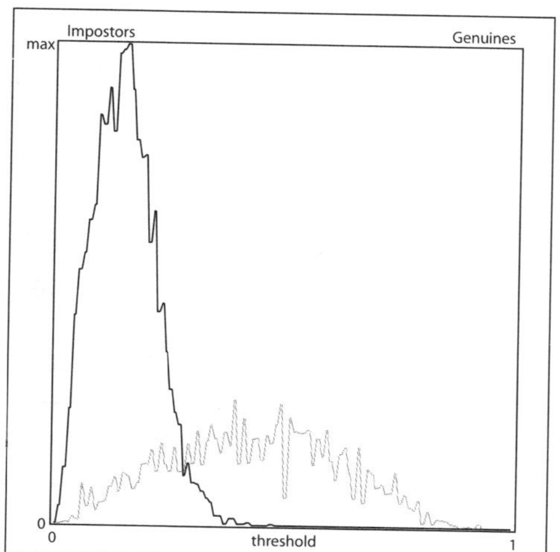

Score distributions

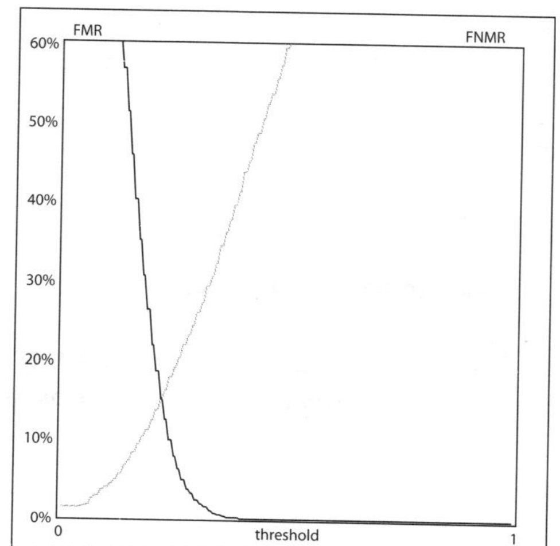

FMR(*t*) and FNMR(*t*)

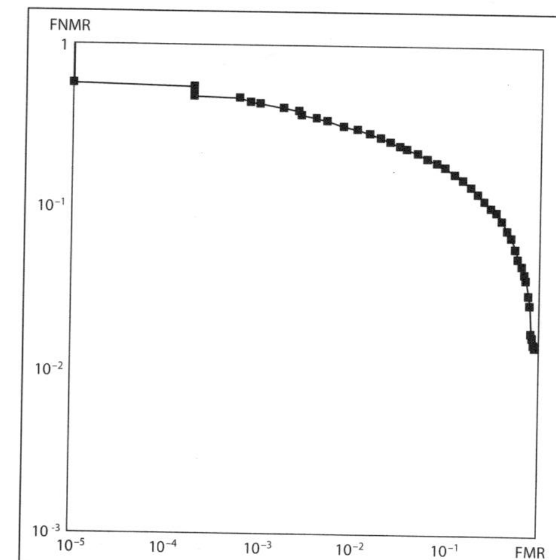

ROC curve

Algorithm UINH on database DB3_A

Average enroll time: 1.28 seconds
Average match time: 1.36 seconds

REJ_{ENROLL}	NGRA	NIRA	REJ_{NGRA}	REJ_{NIRA}
10.29% (F:72 T:0 C:0)	2551	4581	9.21% (F:235 T:0 C:0)	6.77% (F:310 T:0 C:0)

EER	EER*	ZeroFMR	ZeroFNMR
16.32% (16.13%–16.50%)	10.14%	41.24%	100.00%

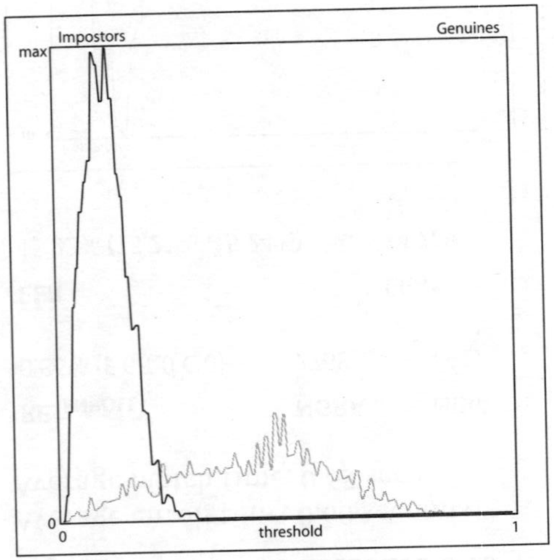

Score distributions

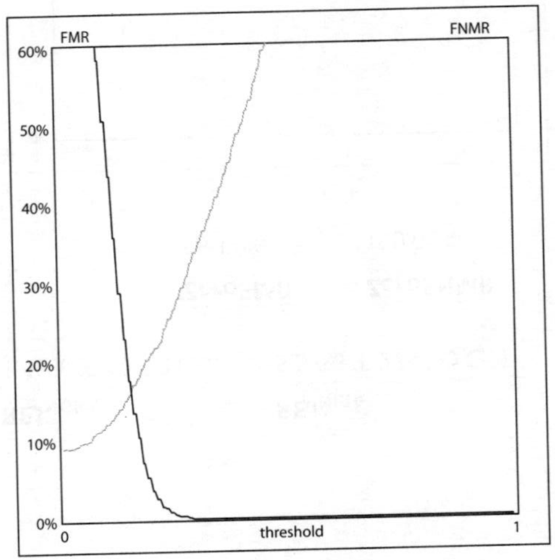

FMR(t) and FNMR(t)

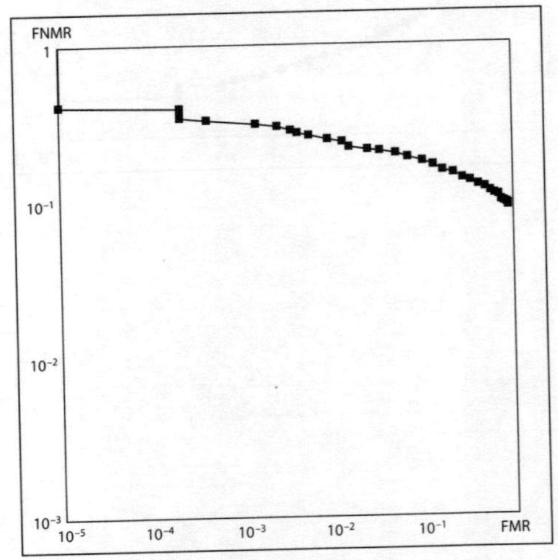

ROC curve

Algorithm UINH on database DB4_A

Average enroll time: 0.42 seconds
Average match time: 0.45 seconds

REJ_{ENROLL}	NGRA	NIRA	REJ_{NGRA}	REJ_{NIRA}
2.14% (F:15 T:0 C:0)	2738	4839	2.81% (F:77 T:0 C:0)	5.10% (F:247 T:0 C:0)

EER	EER*	ZeroFMR	ZeroFNMR
24.77% (24.74%–24.80%)	23.74% (23.71%–23.76%)	97.22%	100.00%

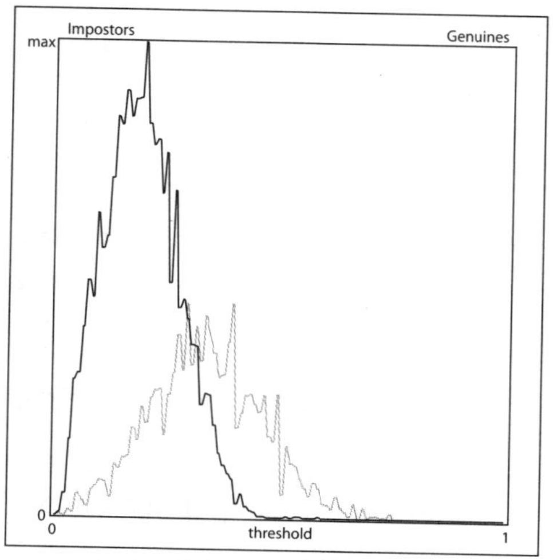

Score distributions

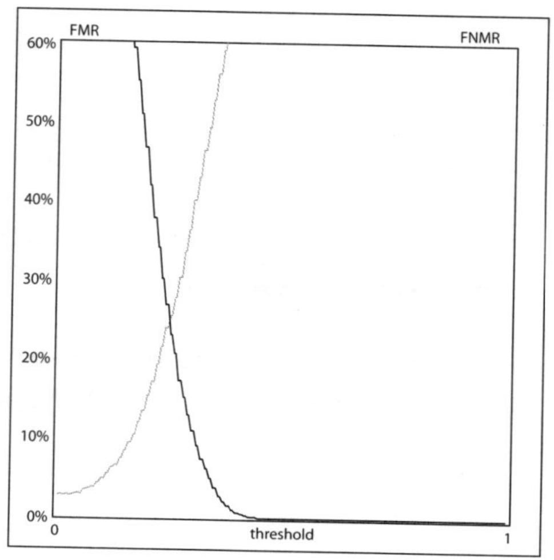

FMR(t) and FNMR(t)

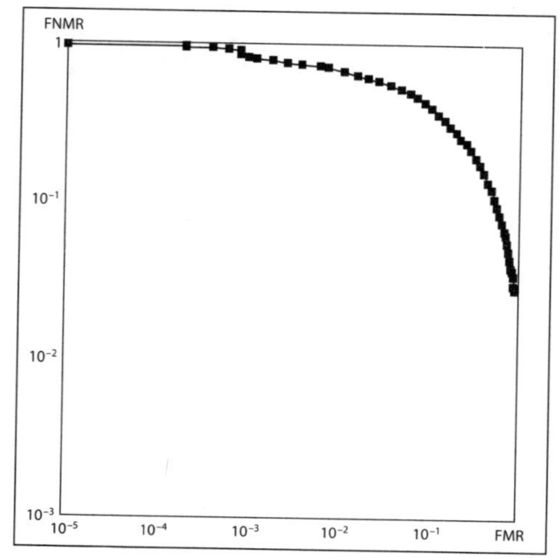

ROC curve

Algorithm UTWE on database DB1_A

Average enroll time: 10.40 seconds
Average match time: 2.10 seconds

REJ_{ENROLL}	NGRA	NIRA	REJ_{NGRA}	REJ_{NIRA}
0.00% (F:0 T:0 C:0)	2800	4950	0.00% (F:0 T:0 C:0)	0.00% (F:0 T:0 C:0)

EER	EER*	ZeroFMR	ZeroFNMR
7.98% (4.00%–11.96%)	7.98% (4.00%–11.96%)	44.00%	100.00%

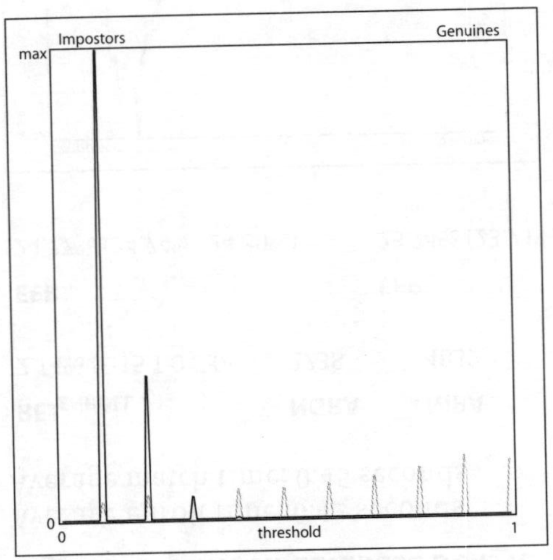

Score distributions

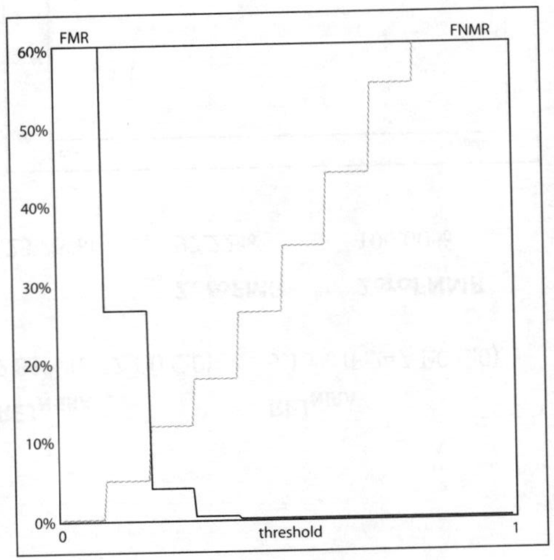

FMR(t) and FNMR(t)

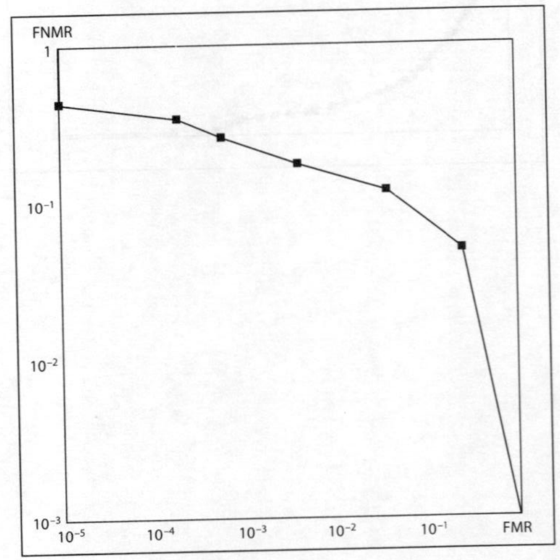

ROC curve

Algorithm UTWE on database DB2_A

Average enroll time: 10.42 seconds
Average match time: 2.12 seconds

REJ_{ENROLL}	**NGRA**	**NIRA**	**REJ_{NGRA}**	**REJ_{NIRA}**
0.00% (F:0 T:0 C:0)	2800	4950	0.00% (F:0 T:0 C:0)	0.00% (F:0 T:0 C:0)

EER	**EER***	**ZeroFMR**	**ZeroFNMR**
10.65% (4.16%–17.14%)	10.65% (4.16%–17.14%)	46.57%	100.00%

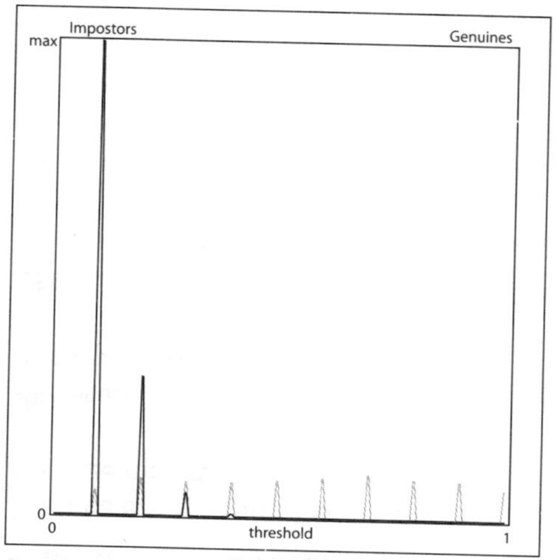

Score distributions

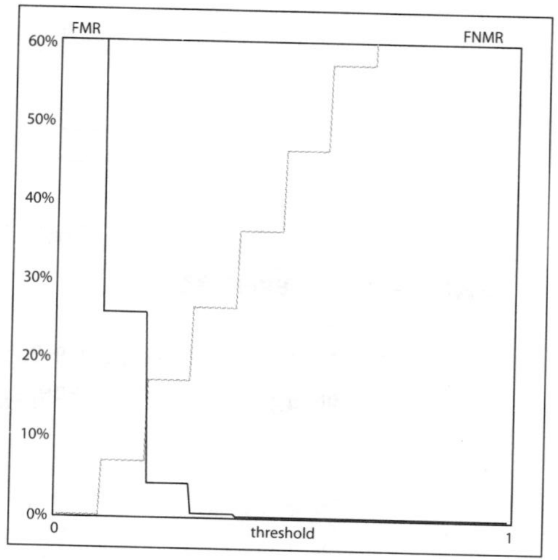

FMR(t) and FNMR(t)

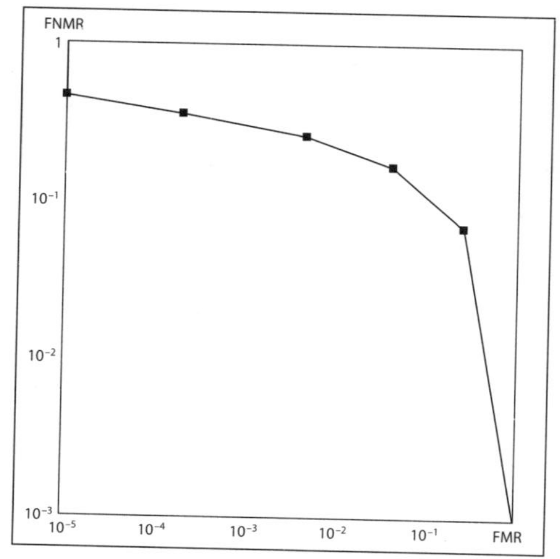

ROC curve

Algorithm UTWE on database DB3_A

Average enroll time: 10.44 seconds
Average match time: 2.31 seconds

REJ_{ENROLL}	NGRA	NIRA	REJ_{NGRA}	REJ_{NIRA}
0.00% (F:0 T:0 C:0)	2800	4950	0.00% (F:0 T:0 C:0)	0.00% (F:0 T:0 C:0)

EER	EER*	ZeroFMR	ZeroFNMR
17.73% (4.06%–31.39%)	17.73% (4.06%–31.39%)	68.82%	100.00%

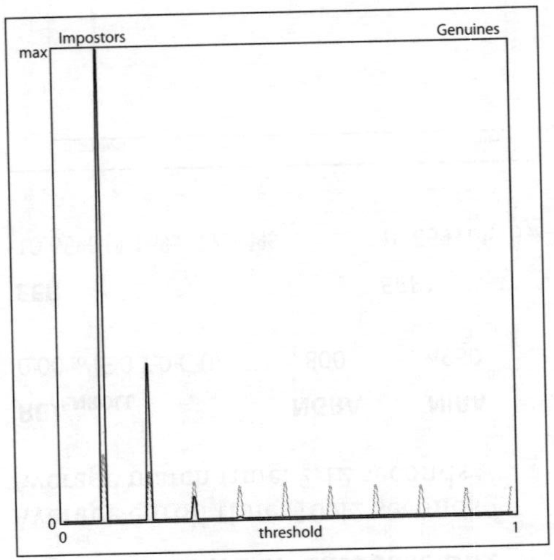

Score distributions

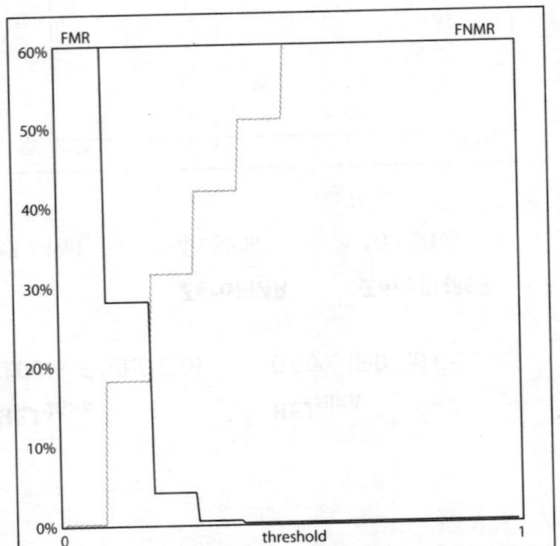

FMR(t) and FNMR(t)

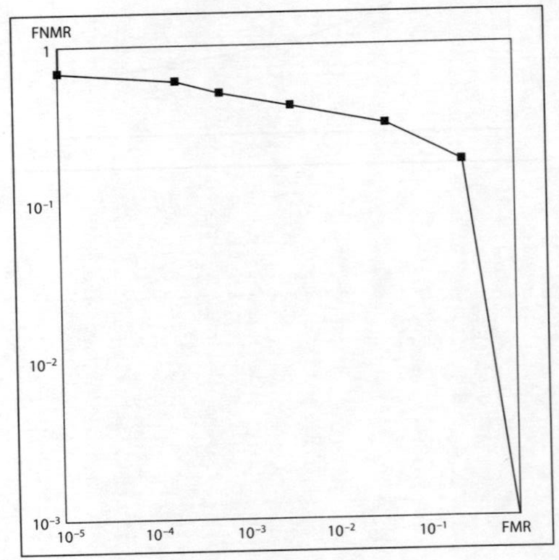

ROC curve

Algorithm UTWE on database DB4_A

Average enroll time: 10.42 seconds
Average match time: 4.17 seconds

REJ_{ENROLL}	NGRA	NIRA	REJ_{NGRA}	REJ_{NIRA}
0.00% (F:0 T:0 C:0)	2800	4950	0.00% (F:0 T:0 C:0)	0.00% (F:0 T:0 C:0)

EER	EER*	ZeroFMR	ZeroFNMR
24.59% (4.79%–44.39%)	24.59% (4.79%–44.39%)	94.29%	100.00%

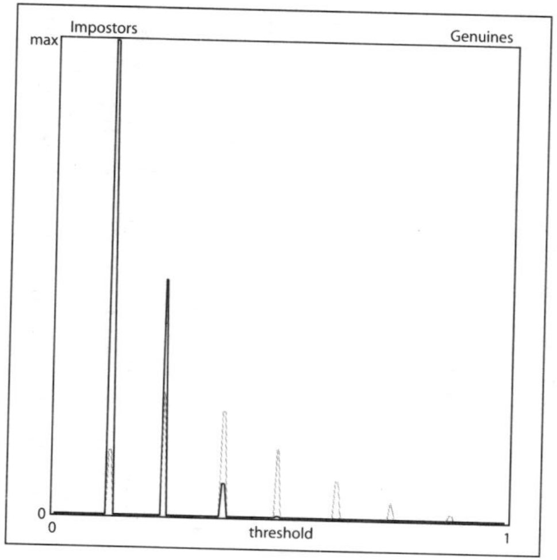

Score distributions

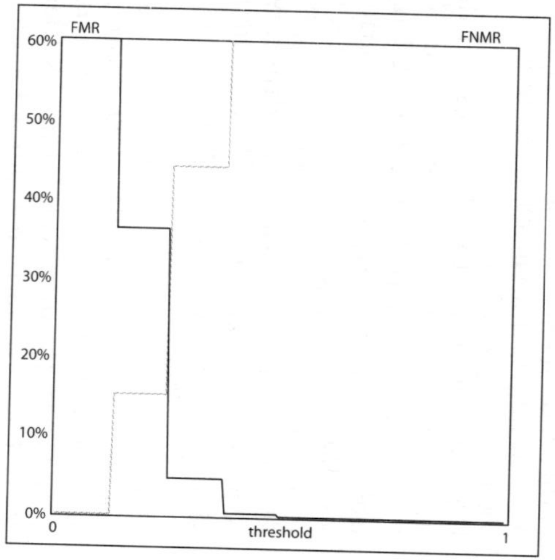

FMR(*t*) and FNMR(*t*)

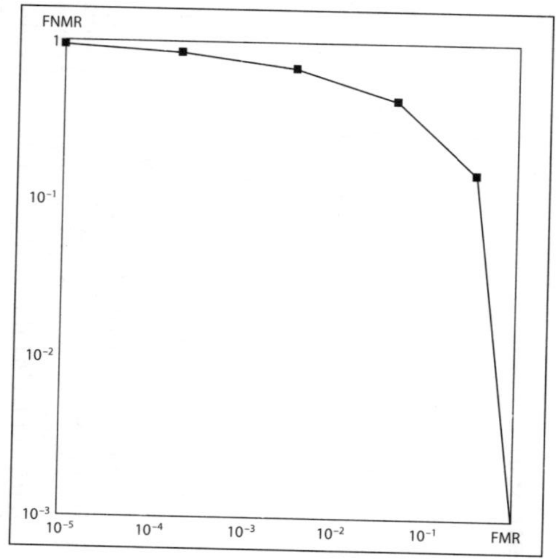

ROC curve

References

[1] J. Barron, D. Fleet and S. Beauchermin, Systems and experiment: performance of optical flow techniques. *Int. J. Computer Vision*, **12**(1), 43–77, 1994.

[2] G. Doddington, M. Przybocki, A. Martin and D. Reynolds, The NIST speaker recognition evaluation – overview, methodology, systems results, perspective. *Speech Communication*, **31**(2–3), 225–254, 2000. Available online at http://www.nist.gov/speech/publications/.

[3] R. Cappelli, A. Erol, D. Maio and D. Maltoni, Synthetic fingerprint-image generation. *Proc. 15th International Conference on Pattern Recognition (ICPR2000)*, Barcelona, September 2000.

[4] R. Cappelli, D. Maio and D. Maltoni, Combining fingerprint classifiers, *Proc. First International Workshop on Multiple Classifier Systems (MCS2000)*. Cagliari, June 2000, pp. 351–361.

[5] R. Cappelli, D. Maio and D. Maltoni, Modelling plastic distortion in fingerprint images, *Proc. 2nd International Conference on Advances in Pattern Recognition (ICAPR2001)*. Rio de Janeiro, March 2001.

[6] C. Dorai, N. K. Ratha and R. M. Bolle, Detecting dynamic behaviour in compressed fingerprint videos: distortion. *Proc. CVPR 2000*, Hilton Head, June 2000, Vol. II, pp. 320–326.

[7] FVC2000 web site: http://bias.csr.unibo.it/fvc2000/.

[8] A. Hoover, R. B. Fisher, G. Jean-Baptiste, X. Jiang, P. J. Flynn, H. Bunke, D. B. Goldgof, K. Bowyer, D. W. Eggert and A. Fitzgibbon, An experimental comparison of range image segmentation algorithms. *IEEE Trans. Pattern Analysis and Machine Intelligence*, **18**(7), 673–689, 1996.

[9] A. K. Jain, R. Bolle and S. Pankanti (eds), *Biometrics – Personal Identification in Networked Society*. Kluwer Academic, 1999.

[10] A. K. Jain, S. Prabhakar and L. Hong, A multichannel approach to fingerprint classification. *IEEE Trans. Pattern Analysis and Machine Intelligence*, **21**(4), 348–359, 1999.

[11] A. K. Jain, S. Prabhakar and A. Ross, Fingerprint matching: data acquisition and performance evaluation. *MSU Technical Report TR99-14*, 1999.

[12] H. C. Lee and R. E. Gaensslen, *Advances in Fingerprint Technology*. Elsevier, 1991.

[13] J. Matas, M. Hamouz, K. Jonsson, J. Kittler, Y. Li, C. Kotropoulos, A. Tefas, I. Pitas, T. Tan, H. Yan, F. Smeraldi, N. Capdevielle, W. Gerstner, Y. Abdeljaoued, J. Bigun, S. Ben-Yacoub andE. Mayoraz, Comparison of face verification results on the XM2VTS database. *Proc. 15th International Conference on Pattern Recognition*, Barcelona, September 2000, Vol. 4, pp. 858–863.

[14] D. Maio, D. Maltoni, R. Cappelli, J. L. Wayman and A. K. Jain, FVC2000: fingerprint verification competition. *DEIS Technical Report BL-09-2000*, University of Bologna. Available online at http://bias.csr.unibo.it/fvc2000/.

[15] D. Maio and D. Maltoni, Direct gray-scale minutiae detection in fingerprints. *IEEE Trans. Pattern Analysis and Machine Intelligence*, **19**(1), 27–40 , 1997.

[16] P. J. Phillips, H. Moon, S. A. Rizvi and P. J. Rauss, The FERET evaluation methodology for face-recognition algorithms. *IEEE Trans. Pattern Analysis and Machine Intelligence*, **22**(10), 1090–1104, 2000.

[17] P. J. Phillips, A. Martin, C.L. Wilson and M. Przybocky, An introduction to evaluating biometric systems. *IEEE Computer Magazine*, **33**, 56–63, 2000. Available online at http://www.frvt.org/DLs/FERET7.pdf.

[18] P. J. Phillips and K. W. Bowyer, Introduction to special section on empirical evaluation of computer vision algorithms. *IEEE Trans. Pattern Analysis and Machine Intelligence*, **21**(4), 289–290, 1999.

[19] T. Randen and J. H. Husoy, Filtering for texture classification: a comparative study. *IEEE Trans. Pattern Analysis and Machine Intelligence*, **21**(4), 291–310, 1999.

[20] A. J. Mansfield and J. L. Wayman, *Best Practices in Testing and Reporting Performance of Biometric Devices*, Version 2.02. UK Government Biometrics Working Group, August 2002. Available online at http://www.cesg.uk.gov/site/ast/biometrics/media/BestPractice.pdf.

[21] C. I. Watson and C. L. Wilson, *NIST Special Database 4, Fingerprint Database*. US National Institute of Standards and Technology, 1992.

[22] C. I. Watson, *NIST Special Database 14, Fingerprint Database*. US National Institute of Standards and Technology, 1993.

[23] C. I. Watson, *NIST Special Standard Reference Database 24, NIST Digital Video of Live-Scan Fingerprint Database*. US National Institute of Standards and Technology, 1998.

[24] J. L. Wayman, Technical testing and evaluation of biometric devices, in A. Jain *et al.* (eds), *Biometrics – Personal Identification in Networked Society*, Kluwer Academic, 1999.

[25] J. L. Wayman, The Philippine AFIS benchmark test, in *National Biometric Test Center Collected Works, 1997–2000*, September 2000. San Jose University.

[26] X. Jiang, W.-Y. Yau and W. Ser, Minutiae extraction by adaptive tracing the gray level ridge of the fingerprint image. *IEEE ICIP'99*, Japan, 1999.

[27] X. Jiang and W.-Y. Yau, Fingerprint minutiae matching based on the local and global structures. *Proc. 15th International Conference on Pattern Recognition*, Barcelona, September 2000.

[28] S. Pankanti, S. Prabhakar and A. K. Jain, On the individuality of fingerprints. *Proc. IEEE CVPR*, Hawaii, December 2001, pp. 805–812.

[29] R. Cappelli, D. Maio and D. Maltoni, Synthetic fingerprint-database generation. *Proc. 16th ICPR*, 2002.

[30] FVC2002 web site: http://bias.csr.unibo.it/fvc2002/.

[31] D. Maio, D. Maltoni, R. Cappelli, J. L. Wayman and A. K. Jain, FVC2000: fingerprint verification competition. *IEEE Trans. Pattern Analysis and Machine Intelligence*, **24**(3), 402–412, March 2002.

[32] A. Mansfield, G. Kelly, D. Chandler and J. Kane, *Biometric Product Testing Final Report*, Issue 1.0. UK National Physical Laboratory, March 19, 2001. Available online at: http://www.cesg.gov.uk/site/ast/biometrics/media/BiometricTestReportpt1.pdf.

[33] FVC2004 web site: http://bias.csr.unibo.it/fvc2004/.

Methods for Assessing Progress in Face Recognition 7

P. Jonathon Phillips, Duane Blackburn, Patrick Grother, Elaine Newton and J. Mike Bone

7.1 Introduction

Evaluations measure the performance and effectiveness of biometrics. Historically, detailed evaluations have been limited to face, fingerprint and speaker recognition. Biometrics, along with character recognition and speech recognition, is one of the few areas in pattern recognition, signal processing and computer vision that have a history of evaluations. Evaluations are effective in these disciplines because performance metrics are easily defined. Regular evaluations have been instrumental in advancing the capabilities of automatic face recognition algorithms.

Evaluations are one of two methods for measuring progress in face recognition that will be addressed in this chapter. The other method is meta-analysis. Meta-analysis is a statistical technique for examining experimental results across multiple papers in a field – for example, face recognition papers published in the scientific literature. From an in-depth analysis of the results from multiple papers, an assessment of progress can be made. Both methods are complementary, with each having a role in measuring progress. Together, evaluations and meta-analysis present a more comprehensive assessment of biometric performance. These methods are also relevant to all areas of biometrics.

A theory and philosophy for evaluation of biometrics is presented in Phillips *et al.* [1], where evaluations are divided into three categories: technology, scenario and operational. We will restrict our discussion to technology evaluations (because there exist a number of well-defined technology evaluations) and accepted protocols for performing them.

The gold standard in technology evaluations is an 'independently administered evaluation'. Independent evaluations are the gold standard because they produce unbiased assessments. In a technology evaluation, face recognition systems are evaluated on the same images and under the same conditions. This allows for the direct comparison among evaluated systems, assessments of individual systems' strengths and weaknesses, and insight into the overall state of the systems' field. Examples of gold standard evaluations are FERET [2, 3], Face Recognition Vendor Test (FRVT) 2000 [35], Fingerprint Verification Competition (FVC) 2000 and 2002

[44–46], and NIST Speaker Recognition Competitions [47, 48]. Key properties of an independent evaluation are: (1) the system developers and testers are separate and independent groups; (2) the systems are evaluated on sequestered data; and (3) all systems are evaluated using the same data.

We examine evaluations by looking at the three FERET evaluations and the Face Recognition Vendor Test (FRVT) 2000. Our analysis shows that periodic evaluations have made significant contributions to advancing the capabilities of face recognition technology over the past decade. In addition, we identify three key areas for future research in automatic face recognition.

An evaluation provides an assessment of the state of the art of a given field at the time of the evaluation. By its nature, an evaluation measures the performance of mature systems. This is because evaluations are normally conducted under competitive conditions where participants have optimized their systems for the specific evaluation. In order to gain a comprehensive assessment of face recognition, one also needs to understand performance trends over time, and what potential breakthroughs are on the horizon. This is found through meta-analysis.

Meta-analysis is a quantitative method for analyzing results from multiple papers on the same subject [28, 29]. Meta-analysis can be performed to consolidate a group of experimental results or to gain deeper insight into methodological techniques in a field. Meta-analysis has been used extensively in medicine, psychology and the social sciences. Its effectiveness in face recognition has also been demonstrated in Phillips and Newton [39], which presents the first meta-analysis in face recognition, biometrics and computer vision. This meta-analysis examines two key issues necessary for the advancement of face recognition. First, is the research community working on the hard issues in the field (e.g. those identified in relevant studies and evaluations such as the FERET evaluations and FRVT 2000)? Second, are the algorithms being developed by the research community significant breakthroughs in face recognition, or are they marginal improvements? If the research community expects to make progress and contribute to advancing face recognition, the answers to both of these questions must be *yes*.

7.2 Face Recognition Evaluations

Independent evaluations have been a critical component of face recognition technology development for the last decade. They have simultaneously measured and driven progress in the field. Independent evaluations allow for an unbiased assessment of the current state of the art, identify the most promising approaches, assess the strengths and weaknesses of both individual approaches and the field as a whole, and identify promising research directions.

7.2.1 Introduction to FERET and FRVT 2000

The three FERET evaluations were the first of their kind conducted in the face recognition community, and proved to be one of the key catalysts in advancing face recognition from its infancy to prototype systems. The FERET evaluations were administered between 1994 and 1996, with the last FERET evaluation measuring the performance of prototype laboratory systems. Between the last FERET evaluation and the beginning of 2000, face recognition matured from prototype laboratory systems to commercially available systems. The Face Recognition Vendor Test (FRVT) 2000 measured the capabilities of these commercial systems as well as progress in the field.

The FERET evaluations and FRVT 2000 were technology evaluations. Technology evaluations measure performance of core face recognition technology, and provide an assessment of the state of the art. Technology evaluations do not directly measure the performance of biometrics systems for general or specific applications. Measuring performance for specific applications is the province of scenario and operational evaluations. Scenario and operational evaluations test biometric systems in field conditions and may take into consideration such factors as the sensors, system integration, human–computer interfaces, operational considerations, and the business model associated with implementing a biometric solution. Technology evaluations provide a more general guide as to which applications a biometric is best suited. A technology evaluation identifies biometric applications that are ready for scenario and operational evaluations, and the systems most likely to succeed for an application. For a detailed discussion of the properties and roles of technology, scenario, and operational evaluations, see Phillips *et al.* [1].

Technology evaluations, such as the FERET evaluations and FRVT 2000, test face recognition algorithms using digital images. The images are acquired prior to the evaluation and sequestered. Sequestering the images enables the algorithms to be tested on images that they have not seen before. This means that the evaluations are repeatable and that all algorithms are tested with the same sets of images. This makes it possible to directly compare performance among the algorithms and the systems that are tested.

The FERET evaluations were one component of a much larger FERET program. The other two components were the FERET database collection effort and FERET algorithm development effort. The goal of the FERET program was to develop and advance face recognition from its infancy in 1993 to a viable technology [42]. To support technology development and evaluation, the FERET database of still facial images was collected. In order to measure progress, a series of three FERET evaluations were administered. The three FERET evaluations took place in August 1994, March 1995 and September 1996.

At the start of the FERET program, face recognition was in its infancy and it was an open question whether or not automatic face recognition was a viable technology. In addition, prior to the first FERET evaluation, there were no independent assessments of the performance of face recognition algorithms. The goals of the FERET evaluations were to show that automatic face recognition was viable, measure progress, identify promising approaches and determine whether key milestones were met in the FERET program.

Table 7.1 Groups that participated in the FERET evaluations. Participants are broken out by each evaluation. In the September 1996 evaluation, two partially automatic systems were tested from both MIT and UMD. To provide benchmark performance scores, the September 1996 evaluation organizers implemented a baseline PCA face recognition algorithm.

		Evaluation		
Version of test	**Group**	**August 1994**	**March 1995**	**September 1996**
Fully automatic	MIT [16, 17]	✓	✓	✓
	Rockefeller [54]		✓	
	Rutgers [36]	✓		
	TASC [7]	✓		
	USC [18]	✓	✓	✓
Partially automatic	Baseline [2]			✓
	Excalibur			✓
	MIT [16, 17]			(2)
	MSU [27]			✓
	Rutgers [36]			✓
	UMD [27]			(2)
	USC [18]			✓

The August 1994 evaluation established the first independent assessment of automatic face recognition performance [3]. It demonstrated that face recognition had the potential to become a viable technology and provided a baseline performance standard for face recognition algorithms. Subsequent FERET evaluations confirmed that automatic face recognition was a viable technology. The August 1994 evaluation measured the performance of four algorithms.

Table 7.1 lists the participants in the FERET evaluations. Baseline performance was established on a gallery of 316 individuals. The evaluation measured the performance of *fully automatic algorithms*. Fully automatic algorithms can automatically locate, normalize, and identify faces from a database. A *partially automatic algorithm* is given the coordinates of both eyes. Thus, partially automatic algorithms do not need to locate the face in an image. The *gallery* contains the set of known individuals. An image of an unknown face presented to the algorithm is called a *probe*, and the collection of probes is called the *probe set*. Since there is only one face in an image, sometimes "probe" refers to the identity of the person in a probe image.

The second FERET evaluation (March 1995) measured progress since the August 1994 evaluation and tested algorithms on larger galleries [3]. The March 1995 evaluation consisted of a single test with a gallery of 817 known individuals. Like the August 1994 evaluation, the March 1995 evaluation tested fully automatic algorithms. The primary emphasis of this test was to measure performance using probe sets that contained duplicate probes on a gallery larger than the galleries in the August 1994 test. A *duplicate* probe is usually an image of a person whose corresponding gallery image was taken on a different day (technically, the probe and gallery images were from different image sets; see the description of the FERET database above).

The third and final FERET evaluation was conducted in September 1996 [2]. This evaluation measured the performance of prototype systems. The September 1996 evaluation tested both fully and partial automatic algorithms. The testing of partially automatic algorithms allowed more groups to participate and produced a more comprehensive assessment of the state of face recognition. Six groups participated (five were from universities). The September 1996 evaluation was administered twice: once in September 1996 and once in March 1997. This evaluation was designed to measure progress over the course of the FERET program. As with the March 1995 test, the emphasis was upon measuring performance on duplicate images and a large gallery of 1196 individuals (a gallery of 1196 was considered very large in 1996). The September 1996 evaluation was open to groups outside of the FERET program, with four of the six participating groups being from outside the FERET research program.

A major innovation of the FERET evaluations was the September 1996 FERET evaluation protocol. Prior to the September 1996 evaluation, generating performance scores for each gallery and probe set required a separate run of an algorithm. The September 1996 FERET evaluation protocol made it possible to compute performance in one run for multiple galleries and probe sets. This provided a significant increase in the ability to measure performance on a wide variety of galleries and probe sets. The multiple galleries and probe sets made it possible to measure performance with advanced statistical techniques. Advanced statistical techniques include computing confidence intervals, performing multi-dimensional analysis, and using resampling techniques [49, 50]. This same protocol was subsequently used in the FRVT 2000.

There was rapid advancement in the development of commercial face recognition systems following the success of the FERET program. This advancement represented not only a maturing of face recognition technology, but also the development of the systems and algorithmic infrastructure necessary to create commercial off the shelf (COTS) systems. Developing systems included converting and porting the code from prototype systems to production-quality code that ran on commercial systems, designing and developing human–computer interfaces for use by non-technical operators, and developing standard interfaces for contending with larger systems and databases. By the beginning of 2000, COTS face recognition systems were readily available. The Face Recognition Vendor Test (FRVT) 2000 was subsequently organized to assess the state of the art in COTS face recognition systems [35].

Participation in the FRVT 2000 was restricted to companies that had commercially available systems. Participants included companies from the USA, Australia and Germany. Five companies participated: Banque-Tec International Pty Ltd, C-VIS Computer Vision und Automation GmbH, Miros, Inc, Lau Technologies and Visionics Corporation.

7.2.2 September 1996 FERET Evaluation Protocol

A design principle and testing protocol describe how evaluations are designed and conducted. Design principles outline the core philosophy and guiding beliefs in designing an evaluation; the evaluation protocol provides the implementation details.

The FERET evaluations and FRVT 2000 design followed the precepts for biometrics evaluations articulated in Phillips *et al.* [1]. Succinctly stated, the precepts are:

1. Evaluations are designed and administered by groups that are independent of algorithm developers and vendors being tested.
2. Test data is sequestered and not seen by the participants prior to an evaluation.
3. The evaluation test design, protocol and methodology are published.
4. Performance results are spread in a manner that allows for meaningful differences among the participants.

Points 1 and 2 ensure fairness in an evaluation. Point 1 provides assurance that the test is not designed to favor one participant over another. Independent evaluations help enforce points 2 and 4. In addition, point 2 ensures that systems are evaluated on their ability to generalize performance to new sets of faces, not the ability of the system to be tuned to a particular set of faces. When judging and interpreting results, it is necessary to understand the conditions under which algorithms and systems are tested. These conditions are described in the evaluation test design, protocol and methodology. Tests are administered using an evaluation protocol that identifies the mechanics of the tests and the manner in which the tests will be scored. In face recognition, the protocol states the number of images of each person in the test, how the output from the algorithm is recorded, and how the performance results are reported. Publishing the evaluation protocol, as recommended in point 3, lets the readers of published results understand how the results were computed. Point 4 addresses the *three bears problem*. If all the scores for all algorithms are too high and within the same error margin, then one cannot distinguish among the algorithms tested. In addition, if the scores are too high in an evaluation, then that is an indication that the evaluation was in reality an exercise in 'tuning' algorithm parameters. If the scores are too low, then it is not possible to determine what problems have been solved.

The goal in designing an evaluation is to have variation among the scores. There are two sorts of variation. The first type is variation among the experiments in an evaluation. Most evaluations consist of a set of experiments, where each experiment reports performance on different problems in face recognition. For example, experiments might look at changes in lighting, or subject pose of a face. The second type of variation is among algorithms for each experiment. Both types of variation are required. The variation in performance among the experiments lets one know which problems are currently sufficiently solved for consideration in operational testing, which problems are research problems, and which problems are beyond the capabilities of the field. The performance variation among algorithms lets one know which techniques are best for a particular experiment. If all the scores for all algorithms across all experiments are virtually the same, then one cannot distinguish among the algorithms.

The key elements that ease adoption of points 3 and 4 can be incorporated into the evaluation protocol. For FERET and FRVT 2000, this was the FERET September 1996 evaluation. This evaluation protocol was designed to assess the state of the art, advance the state of the art, and point to future directions of research. The ability to accomplish these three goals simultaneously was through a protocol whose framework allows for the computation of performance statistics for multiple galleries and probe sets. This allows for the September 1996 evaluation protocol to solve the three bears problem by including galleries and probe sets of different difficulties into the evaluation. This produces a comprehensive set of performance statistics that assess the state of the art and progress in face recognition, and point to future directions of research.

The solution to the three bears problem lies in the selection of images used in the evaluation. The characteristics and quality of the images are major factors in determining the difficulty of the problem being evaluated. For example, if faces are in a predetermined position within the images, the problem is different from that for images in which the faces can be located anywhere within the image. In the FERET database and FRVT 2000 data sets, variability was introduced by the inclusion of images taken at different dates and both outside and indoor locations. This resulted in changes in lighting, scale and background.

The testing protocol is based on a set of design principles. The design principles directly relate the evaluation to the face recognition problem being evaluated. In particular, for FERET and FRVT 2000, the driving applications were searching large databases and access control. Stating the design principles allow one to assess how appropriate the FERET tests and FRVT 2000 are for a particular face recognition algorithm. Also, design principles assist in determining if an evaluation methodology for testing algorithm(s) for a particular application is appropriate.

The FERET evaluation protocol consists of two parts [2]. The first is the rules for conducting an evaluation, and the second is the format of the results that allow for scoring. The last level of detail is the file formats for the images given to algorithms and the file specifications for the output. This level of detail depends on the evaluation being conducted and is beyond the scope of this chapter.

The inputs to an algorithm or system being evaluated are two sets of images: the *target* and *query sets*. Galleries and probe sets are constructed from the target and query sets respectively. The output from an algorithm is a *similarity measure* between all pairs of images from the target and query sets. A similarity measure is a numerical measure of how similar two faces are. Performance statistics are computed from the similarity measures. A complete set of similarity scores between all pairs of images from the target and query set is referred to as a *similarity matrix*. The first rule in the FERET evaluation protocol is that a complete similarity matrix must be computed. This rule guarantees that performance statistics can be computed for all algorithms.

To be able to compute performance for multiple galleries and probe sets requires that multiple images of a person are placed in both the target and query sets. This leads to the second rule: each image in the target and query sets is considered to contain a unique face. In practice, this rule is enforced by giving every image in the target and query set a unique random identifier.

The third rule is that training is completed prior to the start of an evaluation. This forces each algorithm to have a general representation for faces, not a representation tuned to a specific gallery. Also, if training were specific to a gallery, it would not be possible to construct multiple galleries and probe sets from a single run. An algorithm would have to be retrained and the evaluation rerun for each gallery.

Using target and query sets allows us to compute performance for different categories of images. Possible probe categories include (1) gallery and probe images taken on the same day, (2) duplicates taken within a week of the gallery image, and (3) duplicates where the time between the images is at least one year. This is illustrated in the following example. A target and query set consist of images of face taken both indoors and outdoors, with two different facial expressions, and taken on two days. Thus there are eight images of every face. From these target and query sets, one can measure the effects of indoor versus outdoor illumination by constructing a gallery of indoor images with neutral expressions taken on the first day, and the probe set would consist of outdoor images with neutral expressions taken on the first day. Construction of similar galleries and probe sets would allow one to test the effects of temporal changes or expression changes. It is the ability to construct galleries from the target set and probe sets from the query set that allows the FERET September 1996 protocol to perform a detailed analysis.

The FERET September 1996 protocol is sufficiently flexible for subsampling and resampling statistical estimation techniques. For example, one can create a gallery of 100 people and estimate an algorithm's performance of recognizing people in this gallery. Using this as a starting point, we can then create galleries of 200, 300, ..., 1,000 people and determine how performance changes as the size of the gallery increases. Another avenue of investigation is to create *n* different galleries of size 200, and calculate the variation in algorithm performance with the different galleries [37].

The FERET September 1996 evaluation protocol allows for the computation of performance statistics for both identification and verification scenarios. Identification is also referred to as "1 to *n*" and "one to many" matching. Verification is also referred to as authentication and "1 to 1" matching. In an identification application, the gallery (database) consists of a set of known faces. Identification models real-world law enforcement applications where a large electronic mugbook is searched. The input to the system is a probe image of an unknown face. The face recognition system then returns the closest match to the probe in the gallery. A probe is correctly identified if the closest match between the probe and gallery images is the same person. In this chapter we report the fraction of probes in a probe set that are correctly identified. In the general case, the top *n* closest matches in the gallery are reported. The choice of the number of matches reported is dependent of the specific application. The top *n* matches are reported on a cumulative match characteristic. For details on identification performance statistics see Phillips *et al.* [2]; a tutorial on biometric performance measures can be found in Bone and Blackburn [52].

In a verification application, a system is presented with a face and a claimed identity. The system compares the new image to a stored image of the face of the claimed identity. If the match between the two images is sufficiently close, the system accepts the claim; otherwise, it is rejected. Performance for verification is reported on a receiver operating characteristic (ROC) curve. Verification performance can be found elsewhere [2, 35, 40].

7.2.3 Data Sets

The FERET database was designed to advance the state of the art in face recognition, with the images collected to support both algorithm development and the FERET evaluation. During the FERET program approximately one-third of the database was released to researchers for algorithm development with the remaining images sequestered for testing. The images in the development set are representative of the sequestered images. After the conclusion of the FERET and FRVT 2000 evaluations, the entire FERET database was made available to the research community[1].

In the FERET database, the facial images were collected in 15 sessions between August 1993 and July 1996. Collection sessions lasted one to two days. In an effort to maintain a degree of consistency throughout the database, the same physical setup and location were used in each photography session. The setup was a portable studio that consisted of a gray backdrop, a stool for the subject to sit on, and two photographic studio lights on either side of the subject. Thus, subjects were illuminated from both sides. In this chapter, this is referred to as FERET lighting. However, because the equipment in the portable studio had to be reassembled for each session, there was variation from session to session (Figure 7.1).

Images of an individual were acquired in sets of 5 to 11 images, collected under relatively unconstrained conditions. Two frontal views were taken (**fa** and **fb**); a neutral expression was requested for the first image (**fa** image), and a different facial expression was requested for the second frontal image (**fb** image). For 200 sets of images, a third frontal image was taken with a different camera and different lighting (this is referred to as

---

1 See `http://www.nist.gov/humanid/feret/` for details on gaining access to the FERET database.

Figure 7.1 Examples of FERET image and probe categories. (a) Gallery image – **fa** image; (b) FB probe – **fb** from same session as (a); (c) dup I probe – different day from (a); (d) dup II probe – taken at least one year from (a); (e) fc probe – **fc** image from same session as (a).

the **fc** image). The remaining images were collected at various aspects between right and left profile. To add simple variations to the database, photographers sometimes took a second set of images, for which the subjects were asked to put on their glasses and/or pull their hair back. Sometimes a second set of images of a person was taken on a later date; such a set of images is referred to as a *duplicate set*. Duplicates sets result in variations in scale, pose, expression and illumination of the face. Duplicate images refer to both multiple sets of images in a database and to a probe that is from a different set than the gallery image of a person. Usually, a duplicate probe of a person is taken on a different day than the gallery image of that person.

After three years of data collection, the FERET database contained 1564 sets of images consisting of 14,126 total images. The database contains 1199 individuals and 365 duplicate sets of images. For some people, over two years had elapsed between their first and most recent sittings, with some subjects being photographed multiple times (Figure 7.1). The development portion of the database consisted of 503 sets of images and was released to researchers. The remaining images were sequestered.

In the September 1996 FERET evaluation, the target set contained 3323 images and the query set 3816 images. The target set consisted of **fa** and **fb** frontal images. The query set consisted of all the images in the target set plus the **fc**, rotated and digitally modified images. The digitally modified images in the query set were designed to test the effects of illumination and scale. (Results from the rotated and digitally modified images are not reported here.) The similarity matrix for the September 1996 evaluation consisted of 12.5 million similarity scores. Since the target set is a subset of the query set, the test output contains the similarity score between all images in the target set. (Note: having the target set as a subset of the query set does not constitute training and testing on the same images. This is because the face representation is learned prior to the start of the test.) Participants had 72 hours to complete the September 1996 evaluation.

The images in FRVT 2000 came from three sources. In order to measure progress since the September 1996 FERET evaluation and to provide a control, FERET images where included. The second source was digitally modified FERET images. The modifications were designed to measure the impact of compression and changes in resolution on performance. The third source was images from hBase. The hBase is the repository of data collected to support algorithm development and evaluation for the Defense Advanced Research Project Agency's (DARPA) HumanID program [41]. The hBase data included in FRVT 2000 was collected at the National Institute of Standards and Technology (NIST) in 1999 and 2000, and at the Navy's facility at Dahlgren, VA, in 1998 and 1999. The images from the hBase included in FRVT 2000 were sequestered prior to testing.

The images in the Dahlgren 1998 collection were taken indoors using a digital camera and computer-based image capture card. The collection was uncontrolled, and resulted in significant variations in background and lighting conditions.

The remaining Dahlgren and NIST data collections used the same collection protocol. Both indoor and outdoor images were collected. The indoor still image collection was an extension of the FERET image collection protocol. The protocol was extended by adding a digital camera and an additional lighting condition. Six different frontal images of each subject were taken on 35 mm film and with a digital CCD camera. The images collected on 35 mm film were digitized prior to inclusion in the hBase and used in FRVT 2000. Images were collected under three lighting conditions: FERET-style two side lamps, NIST best recommended practice mugshot lighting, and ambient lighting (overhead fluorescent lighting). The overhead lighting is similar to the FERET **fc** images. For each of the three different lighting conditions, two expressions were collected: neutral and alternate facial expressions. The neutral expression corresponds to the FERET **fa** image and the alternate corresponds to the FERET **fb** image. The 35 mm film and digital image for a person for a fixed expression and lighting condition were taken almost simultaneously, within a tenth of a second of each other.

Two images of each person were taken outdoors with a digital CCD camera. One was a frontal image and the other was at a pose angle of 45°. In this chapter, results are only reported for experiments that include the frontal outdoor images. The background was periodically changed and illumination varied because of the diurnal cycle, changes in weather and changes in the background.

The full image data set for FRVT 2000 consisted of 13,872 images of 1,462 individuals. The first part, containing 5,416 images from 1,201 subjects were from the FERET database. The second part consisted of 3,730 images that were digitally generated from FERET images. These images were probe sets in the JPEG compression and resolution experiments. The third part was 4,726 images from 262 subjects from the hBase, the HumanID database. A full description of the FRVT 2000 data set is at Blackburn *et al.* [35, Appendices G and K].

In FRVT 2000, the target and query set were the same set of images. This set contained 13,872 images, and the similarity matrix consisted of 192 million similarity scores. In the FRVT 2000, participants had to perform the 192 million matches in 72 hours. This is compared to 12.5 million matches in the same time period for the September 1996 FERET evaluation.

7.2.4 FERET and FRVT 2000 Results

The FERET evaluations and the FRVT 2000 evaluation are each a sequence of evaluations that increase in difficulty and scope. The August 1994 FERET evaluation established an independent baseline for algorithm performance, with the majority of probes collected on the same day as the gallery image. The FRVT 2000 measured performance for multiple experiments, which provided a detailed assessment of the state of the art in face recognition. This increase in difficulty and scope of the evaluations represents advancement in automatic face recognition technology, greater understanding into the strengths and weakness of the technology, and increased sophistication of evaluation techniques.

In this chapter we look at performance across all three FERET evaluations for four key categories of probes. Detailed descriptions of the FERET results can be found elsewhere [2, 3, 40]. The **fb** probes consisted of images that were taken within five minutes under the same lighting conditions as the gallery image of that person. Scores from the **fb** probes represent an empirical upper bound on performance for algorithms at the time of each evaluation. The **fc** probes consisted of images taken within five minutes but under a different lighting condition from the gallery image of a person. The gallery images were taken under studio lighting conditions, and the **fc** probes were taken with ambient overhead lighting. The **fc** probes examine the sensitivity of algorithms to changes of illumination. The Dup I probes contain faces of people in the gallery that were taken on different days or under different conditions. The Dup I probes measures the effects of temporal differences between gallery and probe images. The Dup II probes are images of faces taken at least one year apart from the corresponding facial image in the gallery. Table 7.2 summarizes the gallery and probe set size for each of the three FERET evaluations. Not all probe categories were included in all evaluations.

Summaries of identification performance for the four categories of probes across the three FERET evaluations are given in Figures 7.2 and Figure 7.3. The x-axis is the FERET evaluation. The y-axis is the probability of correct identification. Figure 7.2 presents the best-performing algorithm in each evaluation. Performance is reported by probe category. For example, the best identification score for the **fb** probe category for the August 1994 evaluation was 0.86. The **fb** (same day) and Dup I (different day) probe categories were in all three evaluations. The **fc** (ambient

Table 7.2 Summary of the gallery and probe sets size for each of the FERET evaluations. Probe set sizes are provide for FB, fc, Dup I and Dup II probe categories.

	August 1994	March 1995	September 1996
Gallery size	317	831	1196
fb probes (same day)	316	780	1195
Dup I probes (different day)	50	463	733
Dup II (one year apart)	–	–	234
fc probes (ambient lighting)	–	–	194

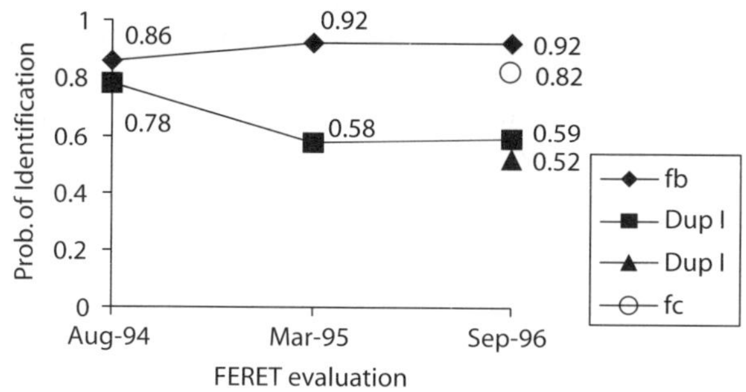

Figure 7.2 FERET performance scores for best algorithm. Performance scores for FB, Dup I, Dup II and fc probe categories are given.

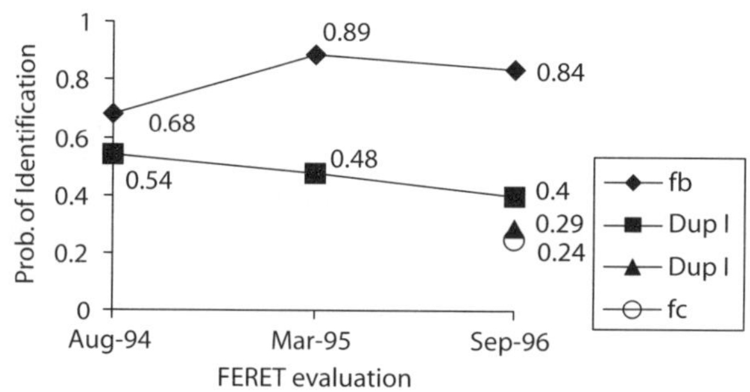

Figure 7.3 Average FERET performance score by evaluation and probe category. Performance scores for FB, Dup I, Dup II and fc probe categories are given.

lighting) and Dup II (images taken over a year apart) probe categories were only included in the September 1996 evaluation. Figure 7.3 presents the average performance for algorithms that took an evaluation. For example, the average identification score for Dup I probes in the September 1996 evaluation was 0.59.

For best performance, there was a steady increase in performance for the **fb** probes across the three evaluations. The increase in performance occurred while the size of the gallery increased, and hence the difficulty of the problem increased. For average performance on the **fb** probes there is an uneven absolute increase in performance. For the best Dup I scores, the performance was the same for the March 1995 and September 1996 evaluations. However, the composition of the Dup I probe set increased in difficulty. The case of the **fc** probes is interesting. For average performance, performance on the **fc** probes was lowest for the four probe categories. However, **fc** performance was significantly higher for the best algorithm.

Each FERET evaluation was more difficult than the preceding evaluations. Algorithm performance did not regress, the evaluations got harder. The two most significant measures of difficulty are gallery size and the number of duplicate probes. In the three FERET evaluations, the size of the gallery increased in each of the evaluations. Therefore the difficulty of the underlining problem being evaluated increased. In each evaluation the number of duplicates increased and the time between acquisition of the gallery image and corresponding duplicate probe increased. These results show that there can be a significant difference in performance between the "average" algorithm and the best algorithm.

The FERET evaluations showed significant advances in face recognition technology development. The advances occurred both in terms of difficulty and size of the problem. The FERET evaluations addressed the following basic problems: the effects on performance of gallery size and temporal variations. The next step in the face recognition evaluation process was to measure performance for a greater number of probe categories. The greater number of probe categories produced a more in depth assessment of the state of the art in face recognition and examined issues directly related to operational considerations.

The FRVT 2000 increased the number of probe categories to eight: compression, distance, expression, illumination, media, pose, resolution and temporal. Unlike the September 1996 FERET evaluation, in FRVT 2000 multiple experiments were performed for each probe category. Each experiment corresponds to a different gallery and probe set. The number and size of the galleries and probe sets varied by category. Three categories – compression, pose and duplicates – contained experiments that used FERET images. For the experiments that use FERET images, the best performance score is reported without attribution to a specific vendor. For experiments that use FRVT 2000 images, results are reported with attribution.

We will summarize the results from the FRVT 2000 by presenting top rank identification scores for key results. Full cumulative match characteristics (CMC) for identification and ROCs for verification, along with complete details, can be found in Blackburn *et al.* [35]. Five vendors participated in FRVT 2000; however, only C-VIS Computer Vision und Automation GmBH, Lau Technologies and Visionics Corporation completed the tests. Results are only presented for the three participants that completed the test.

We started by looking at the performance on the classic face recognition problems identified in FERET: temporal variations (duplicate probes), illumination changes and pose variations. We then presented results for three new categories: effect of image compression on performance, effect on performance on images taken on different media (digital images versus 35 mm film) and effect of size of face on performance.

In FERET, the starting point for measuring performance was the **fb** (same day) probes which provide an empirical upper bound on performance for a system. For **fb** performance, the images of a person are taken in pairs, **fa** and **fb** images, in the same session under the same lighting conditions. The **fa** image was a neutral expression and **fb** was an alternative expression. The **fa** was placed in the gallery and the **fb** was placed in the probe set. In FRVT 2000, the classic **fb** category performance was computed where the **fa** image was in the gallery and the corresponding **fb** was in the probe set. In addition, performance was computed for an **fa** probe set. Here the **fb** image was placed in the gallery, and the corresponding **fa** image was placed in the probe set. For the FRVT 2000 **fb** and **fa** categories, all images were digital

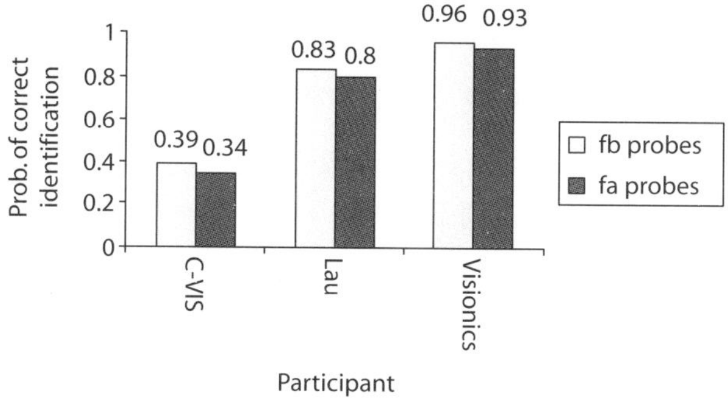

Figure 7.4 FRVT 2000 identification performance for **fa** and **fb** categories.

images with FERET-style lighting from the hBase data set. For the **fa** category, the gallery consisted of 225 **fa** images and the probe set consisted of 228 **fb** images. For the **fa** category, the gallery consisted of 224 **fb** images and the probe set consisted of 228 **fa** images. All images in both categories came from the same set of 225 people. Identification rates for both categories are shown in Figure 7.4.

For the vast majority of applications, the most important issue was the ability of systems to recognize faces when there is a temporal change between the gallery and probe images of a person. In the FERET nomenclature, this is a duplicate probe set. FRVT 2000 measures duplicate performance for two sets of experiments. Performance is computed for the FERET September 1996 Dup I and Dup II probe sets (see Figure 7.5). Figure 7.5 reports the best performance of the FRVT 2000 participants on the Dup I (different day) and Dup II (one year apart) probes. This shows some improvement in performance from FERET on the Dup I probes. For the Dup II probes, an increase from 0.52 on FERET to 0.64 on FRVT 2000.

The second set of duplicate performance scores was computed from images collect at as part of the hBase program. Three experiments report performance for three different galleries and one probe set. The probe set consisted of 467 images with overhead lighting. The galleries consisted of the same people taken on the same day with the neutral expression (one

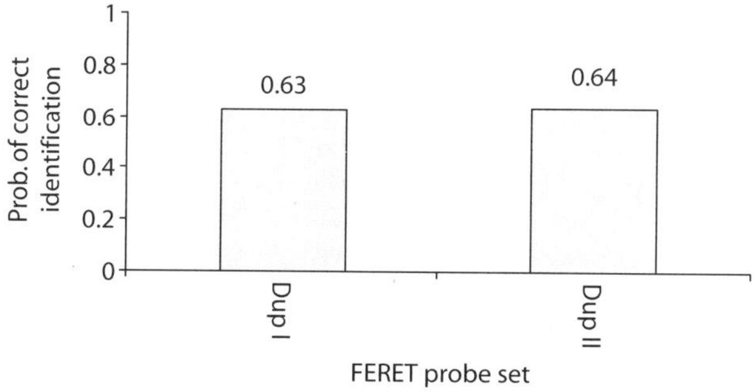

Figure 7.5 FRVT 2000 performance for September 96 FERET Dup I and Dup II probe sets.

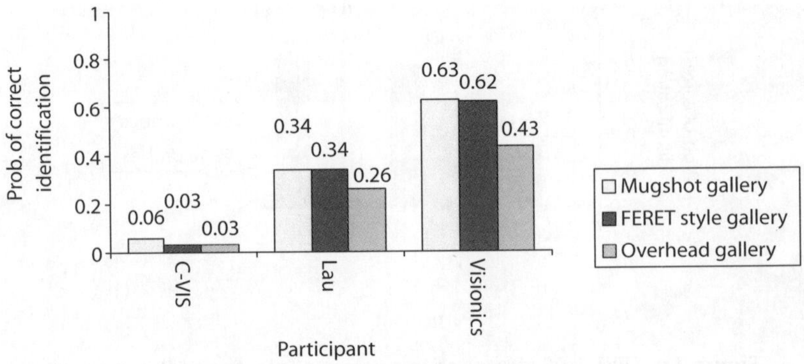

Figure 7.6 FRVT 2000 duplicate performance.

gallery image is missing for one person). The difference among the galleries was the lighting: mugshot, FERET-style and overhead fluorescent. The gallery sizes were 227, 227 and 226, respectively. The time difference between the acquisitions of the gallery and probe images of a person was 11 to 13 months. Performance results are presented in Figure 7.6. For Lau Technologies and Visionics, the performance on the mugshot and FERET style galleries are the virtually the same. For these two vendors, there is a drop in performance for the overhead lighting gallery.

The last major category of probes investigated in the FERET evaluation was the effect of illumination changes on performance. In FERET nomenclature, these was the **fc** (ambient lighting) probes. In FRVT 2000, there were two types of illumination. The first type was indoor images taken with overhead ambient light from fluorescent lights, which are referred to as **fc**-indoor probes. The second type were images taken outdoors, which are referred to as **fc**-outdoor probes. The gallery consisted of images taken indoors under mugshot lighting conditions. All images of a person were taken on the same day within 30 minutes. The images were digital and the neutral expression (**fa**) images were used. The gallery consisted of 227 individuals with one image per person. There were 189 **fc**-indoor probes and 190 **fc**-outdoor probes. The results for the illumination experiments are presented in Figure 7.7. Compared to the same-day (**fa** and **fb**) probe sets in

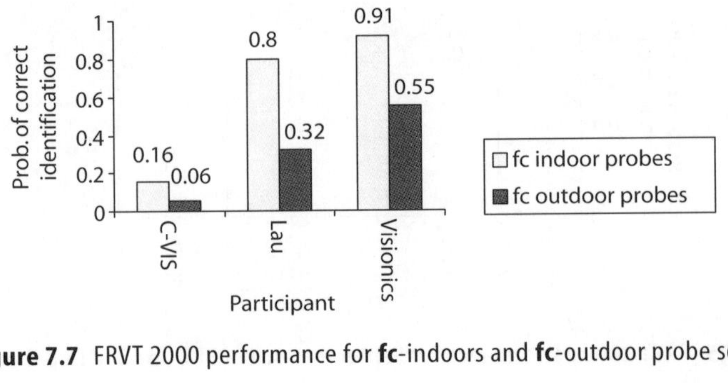

Figure 7.7 FRVT 2000 performance for **fc**-indoors and **fc**-outdoor probe sets.

Figure 7.4, there is a small drop in performance for the **fc**-indoor probes. Compared to the **fc**-indoor probe set, there is a significant drop in performance for the **fc**-outdoor probe set. The results show a significant drop in performance for all systems for the **fc**-outdoor probes, and variable drop in performance for the **fc**-indoor probes.

The majority of face recognition systems are designed to work on frontal facial images. However, the ability to recognize faces from non-frontal poses is important in numerous applications. The effect of pose variation on performance was measured in the August 1994 FERET evaluation, but was not examined in the subsequent FERET evaluations. FRVT 2000 revisited the effects of pose variations. For this experiment, the gallery consisted of 200 frontal images (one image per person) from the FERET database. Performance scores are reported for four probe sets, where the images in each probe set consisted of images at the same angle off frontal. The four angles were ±15°, ±25°, ±40° and ±60°. The rotations result from moving the head right or left in the image plane, with 0° being the frontal image and ±90° being the left and right profile images. The size of each probe set is 400, with two images per person. One image was taken to the right and one to the left at the specified angle. Figure 7.8 shows the results for the system with the best performance results. The results show that performance was not seriously affected for pose angles up to 25°. At 40° performance starts to drop sharply.

In addition to repeating variations on FERET experiments, FRVT 2000 investigated the effects on performance of compression, comparing images taken on 35 mm film and digitally, and the effects of resolution of performance. These categories address system design factors and their effect on vendor performance had not been previously examined.

Compression is of interest because, during transmission or storage, facial images could be compressed. In this experiment, we model the effects on performance of compression on the transmission of the probes. Performance was computed for five levels of JPEG compression: 1:1 (no compression), 10:1, 20:1, 30:1 and 40:1. JPEG compression was selected because it is the *de facto* standard compression technique. A probe set was generated for each compression rate by compressing the probes in the September 1996 FERET Dup I probe set. The gallery consisted of the standard September

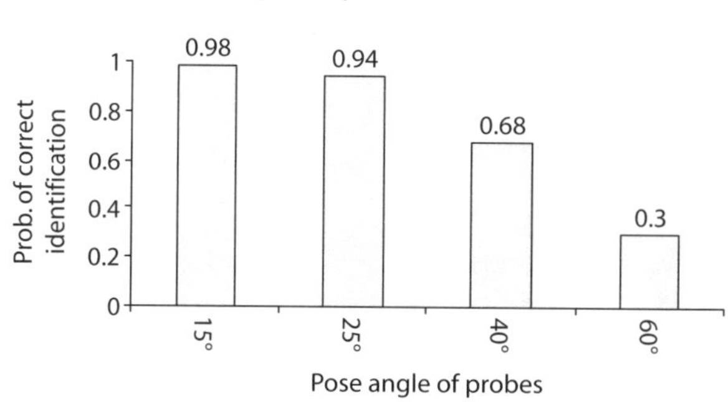

Figure 7.8 FRVT 2000 experiment on pose variation. The gallery consisted of frontal images (0°), and performance is measured for four probes each with a different angle off center.

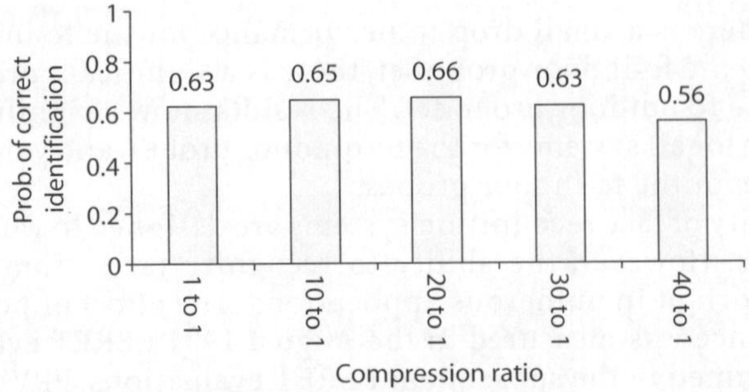

Figure 7.9 FRVT 2000 performance for five levels of JPEG compression.

1996 FERET gallery. The gallery images were not compressed. Figure 7.9 shows results for the algorithm with the best performance. The results show that performance does not deteriorate for compression levels through 30:1.

In FERET and most other face databases used in research, all of the images are collected on the same medium (the medium is the type of material used to record an image). In FRVT 2000, the medium was either 35 mm film or digital electronic storage. In most real-world applications, images will be collected on different media. To examine the effect of shaping media, FRVT 2000 performed two experiments. In both experiments, all images of the same person were taken within a few tenths of a second with the same expression and mugshot lighting. In the first experiment, the gallery consists of 96 individuals taken with 35 mm film, and the probe set consisted of 102 probes taken with a digital camera. In the second experiment, the gallery consisted of digital images of 227 individuals, and the probe set consisted of 99 probes taken with 35 mm film. Results for both experiments are shown in Figure 7.10. The results show that for Lau Technologies and Visionics changing the medium does not significantly affect performance.

One of the critical parameters in designing a system is the number of pixels on a face, or resolution of a face, required to achieve a set performance level. FRVT 2000 ran four experiments to measure the affect of facial resolution on performance. In face recognition, the inter-ocular distance,

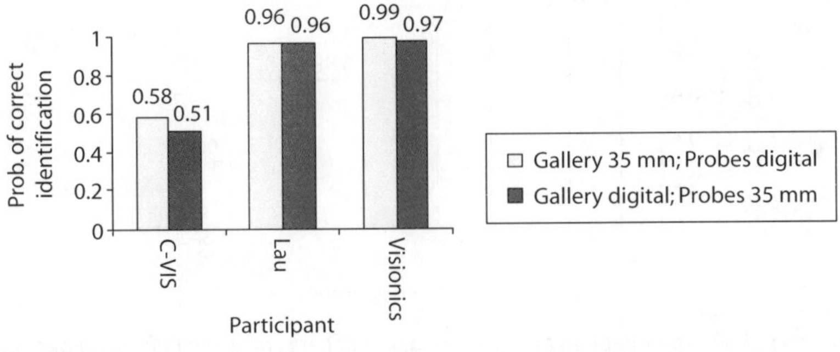

Figure 7.10 Performance results for FRVT 2000 experiments on changes in media.

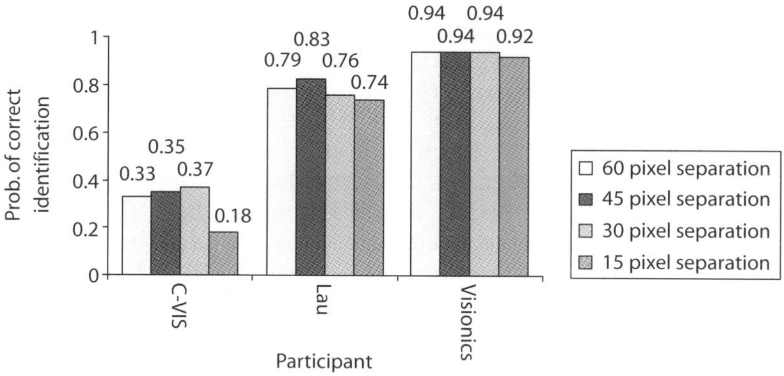

Figure 7.11 Performance results for FRVT 2000 experiments on the effects of resolution.

the number pixels between the centers of the eyes, measures the resolution of a face.

For all resolution experiments, the gallery was the same and consisted of 101 **fa** digital images collected with mugshot lighting (with one image per person in the gallery). The gallery images were not rescaled. The gallery images had a mean eye separation of 138.7 pixels with a range of 88–163 pixels. Four probe sets were generated from the corresponding **fb** image of the **fa** images in the gallery. The four probe sets were generated by digitally rescaling the **fb** images to specified inter-ocular distances. The specified inter-ocular distances were 60, 45, 30 and 15 pixels. All probes in a probe set had the same inter-ocular distance. All probe sets had 102 images (one person had two **fb** images). The results are shown in Figure 7.11. In most cases, performance did not drop significantly throughout the tested range of resolution reduction and, for some vendors, performance increased slightly as resolution was reduced. In fact, for all vendors, performance did not decrease when the inter-ocular distance was decreased from 60 to 45 pixels.

7.2.5 Conclusions Drawn from the FERET Evaluations and FRVT 2000

Comparison of the results from the FERET evaluation and FRVT 2000 shows clear progress in automatic face recognition technology development from 1993 through 2000. This progress was measured from the establishment of the first benchmark performance of face recognition in 1994 (FERET) to measuring performance of COTS systems in 2000 (FRVT 2000). The FERET evaluations and FRVT 2000 both played a critical role in measuring technological advancement and propelling actual technological improvements.

The four evaluations discussed in this chapter clearly identify future research directions, with those directions being consistent across all four evaluations. All the results and experiments in the FERET evaluations and FRVT 2000 fall into four groups, with each group corresponding to a general face recognition problem. The first group consists of experiments where the gallery and probe images were taken indoors and on the same day. This includes **fb** probes (same lighting, different expression) and **fc** probes (different lighting), and resolution and media experiments. The second group consists of experiments where the gallery image is taken indoors with the probe image taken outdoors. In the FRVT 2000, the gallery and probe images were taken on the same day for this group of experiments. The third group consists of experiments where the gallery and probe images are taken on separate days. This includes the FERET Dup I and Dup II categories and temporal variations in FRVT 2000. The fourth group consists of pose variations where gallery and probe images are taken at different poses.

The first group (same day, indoors) represents the easiest category. For all experiments presented in this chapter, identification scores above 80% occurred only for experiments in this category. (Note: not all experiments in this category had scores above 80%. Also, in the pose experiment, the near frontal images taken on the same day had identification rates above 80%.) For FRVT 2000, the best performer for each experiment in this category had an identification rate above 90%. In FRVT 2000, performance on the **fb** probe category was 96% for Visionics, which represents an empirical upper bound on performance at the time of FRVT 2000.

The performance in the other three groups is characterized by identification rates well below 80%. Each of the groups corresponds to a research direction and problem area in face recognition. The second group defines the problem of recognizing faces under temporal variations. This problem has been a classic problem in face recognition and remains an active and important area of research. The vast majority of real-world applications require recognition from images taken on different days. The FERET program and duplicate images set in the FERET database have encouraged research on temporal variations.

The third group of results identify that pose variation is a research area. The August 1994 FERET and FRVT 2000 evaluations measured performance in this area. However, there has not been a strong research emphasis in this area. This lack of emphasis may be attributed to initial research concentrating on frontal face recognition versus recognition with pose variation. The availability of the Carnegie Mellon University (CMU) pose, illumination and expression (PIE) database has contributed to renewed interest in this problem[2] [51].

The last research area in face recognition is recognition of faces in outdoor imagery. This is a problem that had not previously received much attention, but is increasingly important. The greatest challenge in recognition using outdoor imagery is handling illumination variability due to sunlight and the diurnal cycle.

Three of the four groups define interesting areas that need further research. This is from both the technical and application points of view.

---

2 Information on the CMU PIE database can be found at `http://www.hid.ri.cmu.edu/Hid/databases_pie.html`.

However, the "same day, different expression" group does not represent an interesting research area. There are very few, if any, applications where the only difference between gallery and probe images is an expression change. From an algorithmic point of view, to show significant technical progress is this area a much larger database of faces is needed. A larger database of imagery would reduce the identification rate and allow room for improvement to be demonstrated.

FRVT 2000 examined the effects of compression, media changes, and resolution on performance. These issues are important for applications, and their effect on performance needs to be taken into consideration when designing systems. FRVT 2000 showed that for the systems tested, media changes do not affect performance, reasonable compression rates do not affect performance, and resolution above 15 pixels does not affect performance. One interesting observation is that in some cases, compression or a reduction in resolution can increase performance. This improvement in performance most likely occurs because compression and resolution reduction act as bandpass filters [37]. As automatic face recognition continues to advance, it will be necessary to revisit the effects that media changes, compression and resolution have on performance.

FRVT 2000 documented the progress in performance made since the last FERET evaluation. There was improvement in performance on the Dup I and Dup II probe categories. For the best algorithms, performance increased from 0.59 to 0.63 for Dup I and 0.52 to 0.64 for Dup II.

Equally importantly, the FRVT 2000 marks a significant maturing point in automatic face recognition technology. The systems evaluated in FRVT 2000 were all commercial off the shelf systems (COTS) and were required to automatically process 27,744 images and perform 192 million matches in 72 hours. A mere three years after FERET evaluated prototype systems in academia, COTS system were completing evaluations harder than the any of the FERET evaluations. It is significant to note that this occurred only seven years after the start of the FERET program. Progress in face recognition has continued since FRVT 2000. The next significant face recognition evaluation is the Face Recognition Vendor Test 2002, which was being organized at the time this chapter was written[3].

7.3 Meta-Analysis

Independent evaluations are one method of assessing advancement. To get to the point where an algorithm or system is mature enough to participate in an evaluation requires considerable research, engineering and software development. Results from initial research are usually published in the academic literature. Thus, advances in face recognition require that research concentrate on problems that need to be solved, with independent

3 Results of the FRVT 2002 can be found at `http://www.frvt.org/`.

evaluation being very good at identifying these problems. For new techniques, it is necessary to identify which new approaches are potential breakthroughs and which are not. One method of determining whether researchers are concentrating on interesting problems and making progress on these problems is meta-analysis.

7.3.1 Introduction to Meta-Analysis

Meta-analysis is a quantitative method for analyzing the results from multiple papers on the same subject [28, 29]. Meta-analysis can be performed to consolidate a group of experimental results or to gain deeper insight into methodological techniques in a field. Meta-analysis has been used extensively in medicine, psychology and the social sciences.

One type of meta-analysis is a statistical analysis of results from multiple papers on a subject from different research groups. The goal is to take the results of a number of possibly contradictory or inconclusive studies and discover what may be collectively said about a given field. This analysis can provide conclusive results from a series of inconclusive studies or spot trends that cannot be detected from a single experiment. Examples of this are the efficaciousness of Taxol for breast cancer [29], the effectiveness of bilingual education [30] and an assessment of human identification studies in psychology [31].

The key feature of meta-analysis is that it requires results from multiple papers from different groups. This is different from independent evaluations such as FRVT 2000 or FERET. In evaluations, the performance of a set of algorithms is measured on the same images and at the same time. The results are reported in a single paper. Our meta-analysis examines performance through analysis of 24 papers on face recognition. In fact, by looking at performance across multiple evaluations one could perform a meta-analysis on face recognition evaluations.

A second type of meta-analysis examines a field to identify potential methodological problems. Each field has its established conventions for conducting and reporting research results. It is possible that the established conventions will have adverse effects on the field or skew results. In this chapter, we examine the current methods for conducting and reporting results for automatic face recognition algorithms.

In the medicine, the placebo effect is an accepted phenomenon in clinical experiments. However, a recent meta-analysis by Hróbjartsson and Gøtzsche [39] has brought into question this long-accepted idea. In clinical trials, the accepted rule of thumb has been that placebos can improve a wide range of conditions in up to 30–40% of "treated" patients. In this meta-analysis, the authors reexamined clinical trials where patients were randomly assigned either to a no-treatment or to a placebo group. The reexamination found no significant differences in outcomes between the no-treatment and placebo groups.

Two classic studies from medicine further illustrate this category of meta-analysis. The first is the study by Hedges [32] that showed a bias in meta-analyses in medicine because of their tendency to not include unpublished studies. Published studies tend to show greater effectiveness of a new drug or medical regime than unpublished studies. Thus, meta-analyses that excluded unpublished studies would be biased towards showing greater effectiveness of a new drug or regime.

The second is the study by Colditz *et al.* [33] that showed a bias in results from non-randomized experiments in medicine. In a prototypical experiment, a test subject is assigned either to an experimental regime or to a control regime. In a randomized test, subjects are randomly placed in either a treatment (experimental) group or a control group. Colditz *et al.* showed that non-randomized studies report a higher success rate than randomized studies.

Like the two previous examples, our analysis addresses experimental methodological issues and conventions for face recognition algorithms. By performing a meta-analysis, not only can we quantitatively investigate the validity of the reported results, we can also report on the underlying causes and recommend possible solutions.

While the face recognition community has discussed some of the results of this analysis at the philosophical level, none have been studied quantitatively. There is a quip in the face recognition community that researchers always report algorithm performance of 95% and higher (correct identification). At the same time, the FERET evaluations FRVT 2000 show such performance for only one case: images taken on the same day under the same lighting conditions.

We will address the importance of choosing the correct evaluation methodology for conducting experiments; the role of a baseline (or control) algorithm in experiments; and the need to document experimental parameters, design decisions, and performance results.

Automatic face recognition is amenable to meta-analysis for a number of reasons. The first is that this has been a very active area of research for the last decade, so there is a sizable amount of accumulated work in the area. Second, there exists an accepted quantitative performance measure – probability of identification. Third, there exist databases of facial images that are available to researchers and are used to report results in the literature. Fourth, there exist independent measures of performance – the FERET evaluations for example. Fifth, there exists an accepted baseline algorithm that is easily implemented – principal component analysis (PCA)-based algorithms (also known as eigenfaces) [34].

7.3.2 Methodology for Selecting Papers

We selected papers for this study that ran experiments using either the FERET or the AT&T Laboratories-Cambridge (ORL) databases and reported identification performance results for full frontal facial images. The FERET and ORL databases were selected because they are extensively used in the face recognition community[4] [43].

---

4 See `http://www.uk.research.att.com/facedatabase.html` for details on gaining access to the ORL database.

We searched major computer vision and face recognition conference proceedings, journals, edited books and the IEEE Xplore journal and conference archive. This produced 47 papers. We then removed papers that had similar experimental results from the same research group. The similar results occurred because preliminary papers appeared in conferences and the final version appeared in a journal. With one exception, the journal version was selected. The winnowing process produced 24 papers for further analysis. A list of these papers is in the reference section [4–27].

Each paper selected presented a new face recognition algorithm, which we will refer to as an experimental algorithm. The effectiveness of an experimental algorithm is demonstrated by performing experiments on one or more data sets. In addition, some authors report results for more than one variation on the experimental algorithm. If the authors reported performance for a number of variations for an algorithm, we choose the variation with the best overall performance. A number of papers reported performance scores for additional algorithms that served as baselines. If there was only one baseline algorithm, we refer to this as the baseline algorithm for the experiment. In this case, the baseline algorithm was usually a correlation- or PCA-based face recognition algorithm. If there were multiple baseline algorithms, we selected the variation of a PCA-based algorithm with the best performance as the baseline algorithm.

The 24 papers selected yielded 68 performance scores. The 68 scores include multiple experiments in a paper and baseline algorithms. This analysis was performed on the identification error rate, which is one minus the rank one identification rate. A more detailed analysis can be found in Phillips and Newton [39].

We consolidated the results of three sets of papers. The first set of consolidated papers reported results on the ORL database using the same basic evaluation protocol [8–10, 14, 19, 22, 23]. Two of these papers also reported results on the FERET database [10, 23]. The second set were two papers by Moghaddam and Pentland [15, 16] that used the same image sets. The third set consisted of three papers by Liu and Wechsler [11–13] that used the same image sets.

7.3.3 Analysis of Performance Scores – Viewing the Data Through Histograms

We examine the first question, "Are researchers working on interesting problems?", through histograms. Histograms summarize the distribution of performance scores (error rates in the meta-analysis) and allow peaks in the distribution to be easily identified. If the peaks are concentrated with low error rates, then this is evidence that researchers are concentrating on an easy problem that is not interesting. If the peaks are concentrated other places, then researchers are concentrating on hard problems.

We first looked at the distribution of the identification error rates across all experiments and algorithms (experimental and baseline algorithms). Traditionally, researchers have reported identification rate, in other words their success in recognizing faces. In the meta-analysis, we choose to

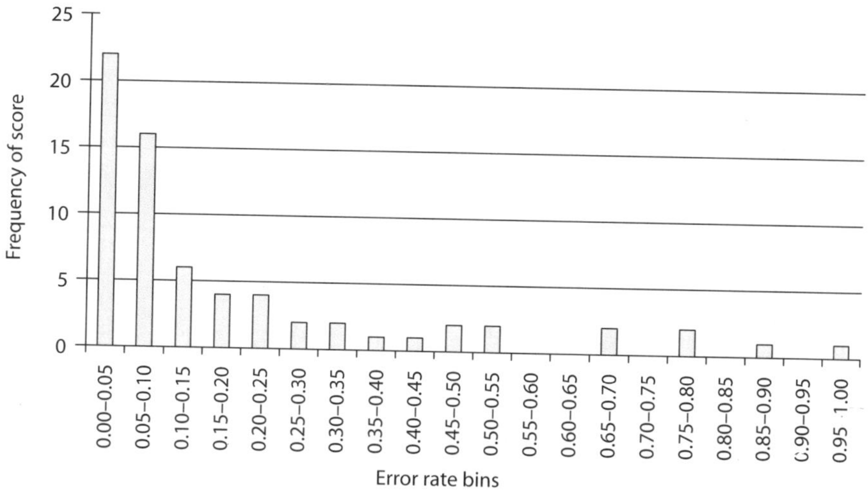

Figure 7.12 Histogram of all experimental and baseline error rate scores. Each bar corresponds to scores greater than the lower interval and equal to or less than the upper interval.

characterize performance by the identification error. The error rate is the percentage of probes that are not correctly identified, which is one minus the top-match identification rate. Figure 7.12 is a histogram of the error rate distribution for algorithms. For the 68 performance scores included in this meta-analysis, 56% (38 out of 68) have an error rate below 0.10.

Next we restricted our attention to the experimental and baseline algorithms according to the exclusion criteria described at the end of the previous section. This yielded 40 experimental algorithms, 33 of which have corresponding baseline algorithms. There are fewer baseline algorithms because seven studies did not use a baseline. Some baseline algorithms correspond to more than one experimental algorithm (e.g. the ORL series has one baseline algorithm for 10 experimental algorithms).

Figure 7.13 shows a histogram of error rates for experimental algorithms in black and baseline algorithms in white. To illustrate the influence of a baseline score, we counted them each time a score served as a baseline (for a total of 33 baselines). For example, for the ORL experiments, we counted the baseline algorithm 10 times. Figure 7.13 shows that 29 of the 40 (73%) experimental algorithms report error rates of 0.10 or less.

We examined the seven experimental algorithms that do not have a baseline score. The error rates for these algorithms are: 0.008, 0.01, 0.02, 0.034, 0.045, 0.046 and 0.28. Their median is 0.034. These scores (1) show that 6 out of 7 experiments have an error rate less than 0.05, (2) contain the best error rate (0.008) for all 40 experimental algorithms in this analysis, and (3) account for one third of the experimental algorithms with error rates below 0.05. Clearly, the results from experimental algorithms without a supporting baseline algorithm are highly biased.

Seven papers report experimental results on the ORL database [8–10, 14, 19, 22, 23]. This produced 11 performance scores: 10 experimental and 1

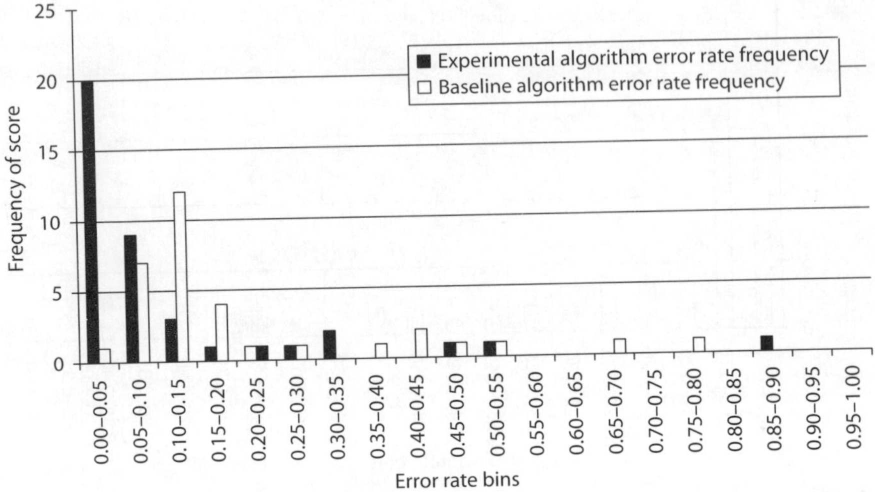

Figure 7.13 Histogram of error rates for experimental and baseline algorithms. Each bar corresponds to scores greater than the lower interval and equal to or less than the upper interval.

baseline. The PCA algorithm in Lawrence *et al.* [8] was the baseline algorithm for the ORL experiments. The baseline error rate is 0.105. The error rate range for the experimental algorithms is between 0.029 and 0.13, with 7 out of 10 performance scores equal to or less than 0.05. This indicates that performance has been saturated using this data set, and the data set does not define a sufficiently challenging problem for automatic face recognition. In the ORL database, all the pictures of a person are taken on the same day. Thus, experiments on the ORL database are equivalent to **fb** (same day, same lighting) experiments on FERET data. Our conclusions on the difficulty of the ORL database are consistent with our findings for **fb** experiments on FERET data: it no longer represents an interesting problem.

7.3.4 Evaluation of Experiments with a Baseline

We next look at the second question: "Is progress being made on interesting face recognition problems?". Ideally this would be accomplished by directly comparing performance of the experimental algorithms. This would require that all the algorithms report performance on the same data sets. Unfortunately, this is not possible. Therefore, to compare performance we have to use indirect methods. This restricts our analysis to experimental algorithms that also report performance for a baseline algorithm. To indirectly compare algorithms, we measure the relationship between the performance scores of the experimental and baseline algorithms on the same experiments.

We assess progress by examining the relationship between experimental and baseline performance scores. A scatter plot is commonly used to show this relationship. Figure 7.14 is a scatter plot of the 33 experimental scores

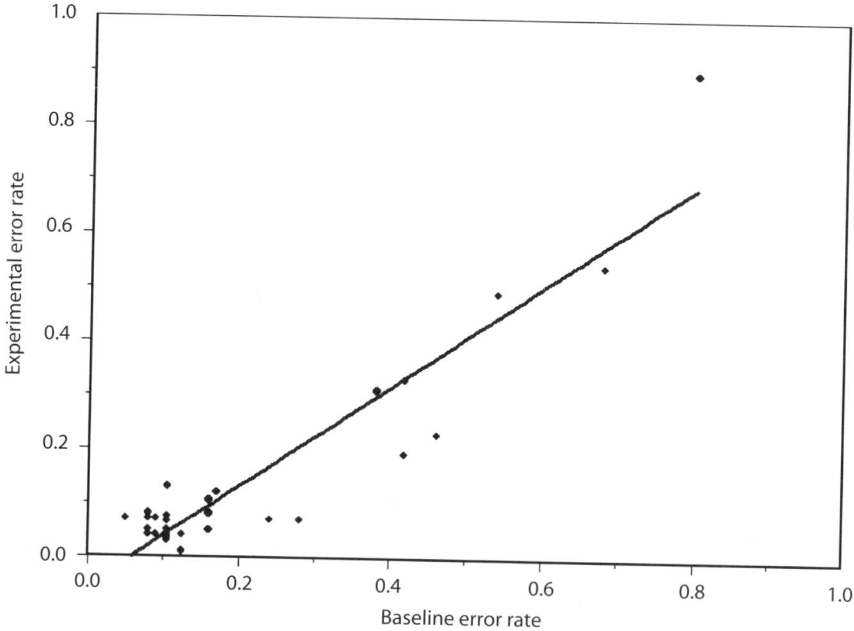

Figure 7.14 Scatter of baseline and experimental error rates for all experiments with a baseline algorithm. The best-fit line shows the correlation between the experimental and baseline scores.

that had corresponding baseline scores. Each point in the scatter plot is an experiment. The x-axis is the baseline score, and the y-axis is the corresponding experimental score. A best-fit line for the data is shown in Figure 7.14 as well. If the points are close to the best-fit line, then there is a correlation between the experimental and baseline scores. In other words, the baseline scores are predictive of the experimental scores. To formalize the analysis of this relationship, we computed the correlation coefficient value r for the 33 pairs of scores. If the experimental and baseline scores were perfectly correlated, then the r-values would be 1.0. The observed correlation value r is 0.932 with a significance level greater than 0.01. This shows strong correlation between the 33 pairs of baseline and experimental scores.

Next we divided the pairs of scores into two groups and examined the relationship between experimental and baseline scores for pairs with baseline scores above 0.20 and below 0.20. Figure 7.15 is a scatter plot of the nine experimental algorithms with baseline scores above 0.20. The correlation coefficient r is 0.953, which has a significance level greater than 0.01. This shows that the baseline scores are predictive of the experimental error rates when the baseline scores are greater than 0.20. In fact, the performance scores of experimental algorithms can be predicted from the baseline performance scores. This implies that all the algorithms are making the same incremental increase in performance over the baseline algorithm.

The correlation coefficient for algorithms with baseline scores less than 0.20 was 0.283. A correlation value of 0.283 is not significant and shows that the experimental and baseline scores are not correlated. There are two

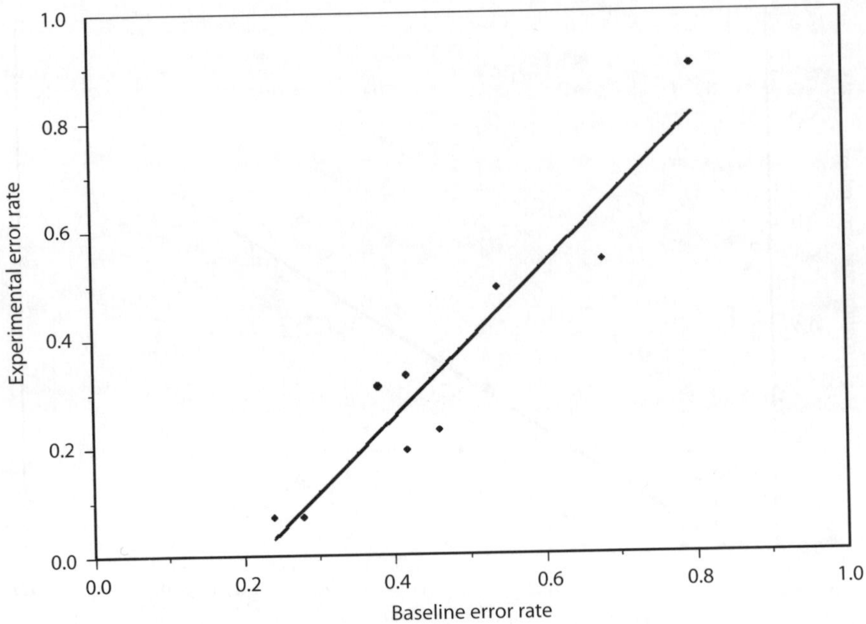

Figure 7.15 Scatter plot for experimental and baseline scores with a baseline error rate > 0.20. The best-fit line shows the correlation between experimental and baseline error rates.

Table 7.3 Summary of results from Figures 7.14 and 7.15.

	All algorithms with a baseline	**Algorithms with baseline error rates >0.20**	**Algorithms with baseline error rates <0.20**	**Figure No.**
No. of experimental algorithms	33	9	24	14
Correlation value *r*	0.932	0.953	0.283	15
Level of significance	<0.01	<0.01	Not significant	N/A

possible explanations for the scores not being correlated: first, that performance is saturated for low error rates with PCA-based algorithms; or second, that PCA-based algorithms are not appropriate baseline algorithms for low error experiments. The results of the scatter plot analysis are summarized in Table 7.3.

7.3.5 Meta-Analysis Conclusions

The meta-analysis discussed in this section looked at two questions: are researchers concentrating on interesting problems in face recognition, and is progress being made on interesting problems?

For the first question, this meta-analysis identifies that the majority of papers in scientific literature concentrated on developing face recognition algorithms to solve problems that are not interesting. These problems are characterized by images being taken on the same day under the same lighting conditions. This problem does not model a real-world application and only provides an upper bound on algorithm performance. Results on experiments that model this problem are a first step in demonstrating the capabilities of an algorithm. In addition to including results on this easy problem, papers need to include experimental results on harder problems. Experimental results on harder problems will give the face recognition community a chance to properly assess whether a new algorithm represents an improvement over existing approaches.

When designing experiments on images collected on the same day with the same light (an easy problem), existing results must be taken into consideration. On the equivalent FRVT 2000 results, **fb** probe sets, the best COTS system obtained an identification rate 0.96 (error rate of 0.04) on a gallery of 225 people. Since these results are on COTS systems, researchers would need to demonstrate that an experimental algorithm performs considerably better than available COTS systems.

There are many researchers who are working on interesting problems. In the papers studied in this meta-analysis, an interesting problem is characterized by images of a person taken indoors on different days. However, is significant progress being made on this problem? The answer from this meta-analysis is no. Meta-analysis results show that baseline and experimental algorithm scores are correlated and that all algorithms are making the same incremental improvement over the baseline. This strong correlation raises three questions for future investigations. First, why is the improvement in performance of experimental algorithms only incremental over the experimental scores? Second, could one detect breakthroughs in face recognition through performance of an experimental algorithm that is not predicted by a baseline score? Third, because of the strong correlation, are all the algorithms using essentially the same information to perform recognition?

The answers to the two questions studied in the meta-analysis suggest that a new methodology for conducting and reporting experiments needs to be adopted by the face recognition community. To foster discussion on a new experimental methodology we make the following recommendations:

- The face recognition community should establish an algorithm implementation as a baseline.
- The face recognition community should establish a set of standard challenge problems.
- Published papers should report results on appropriate challenging problems and, for new data sets, provide new performance baseline results.

A common baseline will allow for the difficulty of data sets and problems to be calibrated. In addition, it will allow for indirect comparison among algorithms. The establishment of a set of challenge problems will allow for the direct comparison among algorithms and a measure of progress. As progress is made in face recognition, the baseline and challenge problems will need to be updated to adjust for improvements and progress in face recognition technology. One possible source of baseline implementations is the baseline suite from Colorado State University [55].

Publishing the performance results of experimental algorithms on challenging problems will fix a base level of performance. Some experimental algorithms will be designed to address new problems that are not part of the standard set of challenge problems. Reporting baseline performance on new problems will provide a measure of difficulty for the new problem. Also, reporting the performance of experimental algorithms on standard challenge problems will provide a comparison point with existing algorithms. Providing performance on both the new problem and standard challenge problems will allow for robust comparisons with other methods.

Scientific fields advance by conducting research on hard and interesting problems. The FERET evaluations and FRVT 2000 have identified three hard problems that have real-world applications: temporal changes between gallery and probe images, pose variations, and recognition from outdoor imagery. By following our recommendations, it will enable the face recognition community to make quantifiable progress while simultaneously working on problems that are relevant to the real-world applications.

7.4 Conclusion

Measuring progress in face recognition, as with any biometric, is a multidimensional process. More than one technique is required for measuring progress. In this chapter we have discussed two methods: independent technology evaluations and meta-analysis.

The conclusions from the FERET and FRVT 2000 evaluations and the meta-analysis appear to be contradictory. The independent evaluations show progress, whereas the meta-analysis suggests that progress is not being made. This contradiction can be explained by the manner in which expectations are established and progress is measured in evaluations and in scientific papers. In scientific papers, it is not possible to directly compare performance among algorithms, whereas in evaluations, the participants know that their performance will be measured against others. This encourages participants to develop new techniques and algorithms with demonstrably better performance than those of rival research groups. This was nicely demonstrated in the September 1996 FERET evaluations. Two groups, Massachusetts Institute of Technology (MIT) and the University of Maryland (UMD), each submitted two algorithms. The MIT group submitted the algorithm from the March 1995 evaluation and a new algorithm designed for the September 1996 evaluation. UMD took the September 1996 evaluation twice: in September 1996 and March 1997. For both groups, there was a substantial increase in performance between the first and second algorithms. The improvement was stimulated by their participation in the evaluations.

In the academic literature, improvement is generally measured by showing an increase in performance over a baseline algorithm. Comparing the performance of an experimental algorithm with a baseline algorithm allows for an indirect comparison of algorithms. What is needed is a mechanism to directly compare performance across experimental algorithms. This can be accomplished by following our recommendations of establishing standard challenge problems and standard baseline algorithms for calibrating the difficulty of a challenge problem. If the face recognition community accepted the standard challenge problems and baseline algorithm, this would change the expectations for performance needed to publish a paper and provide a common yardstick for measuring progress.

Methods for measuring and assessing performance are critical elements in advancing automatic face recognition technology. The FERET evaluations and FRVT 2000 have a proven track record of advancing face recognition technology. Helping to guide and nurture automatic face recognition technology from its infancy to mature commercially available systems. The next step in face recognition technology evaluations is the FRVT 2002, which will evaluate the performance of systems on a data set of 121,000 images. Complementing the maturing of technology evaluations is the development of techniques and methods for performing scenario and operational evaluations [52, 53]. The methods discussed in this chapter are also applicable to other areas of biometrics, and have the potential to assist in advancing all areas of biometrics.

Acknowledgements

DARPA and the DoD Counterdrug Technology Development Program Office sponsored the FERET program. DARPA, the DoD Counterdrug Technology Development Program Office and the National Institute of Justice sponsored the FRVT 2000. The authors thank Tom Gandy for reviewing the chapter.

References

[1] P. J. Phillips, A. Martin, C. L. Wilson and M. Przybocki, An introduction to evaluating biometric systems. *Computer*, **33**, 56–63, 2000.

[2] P. J. Phillips, H. Moon, S. Rizvi and P. Rauss, The FERET evaluation methodology for face-recognition algorithms. *IEEE Trans. Pattern Analysis and Machine Intelligence*, **22**(10), 2000.

[3] P. J. Phillips, H. Wechsler, J. Huang and P. Rauss, The FERET database and evaluation procedure for face-recognition algorithms. *Image and Vision Computing*, **16**(5), 295–306, 1998.

[4] M. S. Bartlett, H. M. Lades and T. J. Sejnowski, Independent component representations for face recognition. *Proc. SPIE Symposium on Electronic Imaging: Science and Technology; Conference on Human Vision and Electronic Imaging III*, San Jose, California, January 1998.

[5] I. J. Cox, J. Ghosn and P.N. Yianilos. Feature-based face recognition using mixture-distance. *Proc. IEEE Conference on Computer Vision and Pattern Recognition*, 1996, pp. 209–216.

[6] K. Etemad and R. Chellappa, Discriminant analysis for recognition of human face images. *Proc. International Conference on Audio- and Video-Based Biometric Person Authentication*, 1999, pp. 127–142.

[7] G. G. Gordon, Face recognition from frontal and profile views. *International Workshop on Automatic Face- and Gesture-Recognition*, Zurich, pp. 47–52, 1995.

[8] S. Lawrence, C. L. Giles, A. C. Tsoi and A. D. Back, Face recognition: a convolutional neural-network approach. *IEEE Transactions on Neural Networks*, **8**(1), 98–113, 1997.

[9] S. Z. Li and J. Lu, Face recognition using the nearest feature line method. *IEEE Trans. Neural Networks*, **10**(2), 439–443, 1999.

[10] S. Lin, S. Kung and L. Lin, Face recognition/detection by probabilistic decision-based neural network. *IEEE Trans. Neural Networks*, **8**(1), 114–132, 1997.

[11] C. Liu and H. Wechsler, Comparative assessment of independent component analysis (ICA) for face recognition. *Proc. International Conference on Audio- and Video-Based Biometric Person Authentication*, 1999, pp. 211–216.

[12] C. Liu and H. Wechsler, Evolutionary pursuit and its application to face recognition. *IEEE Trans. Pattern Analysis and Machine Intelligence*, **22**(6), 570–582, 2000.

[13] C. Liu and H. Wechsler, Robust coding schemes for indexing and retrieval from large face databases. *IEEE Trans. Image Processing*, **9**(1), 132–137, 2000.

[14] S. M. Lucas, Continuous *n*-tuple classifier and its application to face recognition. *Electronics Letters*, **33**(20), 1676–1678, 1997.

[15] B. Moghaddam, C. Nastar and A. Pentland, Bayesian face recognition using deformable intensity surfaces. *Proc. IEEE Conference on Computer Vision and Pattern Recognition*, 1996, pp. 638–645.

[16] B. Moghaddam and A. Pentland, Beyond linear eigenspaces: Bayesian matching for face recognition, in H. Wechsler, P. J. Phillips, V. Bruce, F. F. Soulié and T. S. Huang (eds), *Face Recognition: From Theory to Applications*. Springer-Verlag, 1998, pp. 230–243.

[17] B. Moghaddam and A. Pentland, Probabilistic visual learning for object presentation. *IEEE Trans. Pattern Analysis and Machine Intelligence*, **19**(7), 696–710, 1997.

[18] K. Okada, J. Steffens, T. Maurer, H. Hong, E. Elagin, H. Neven and C. von der Malsberg, The Bochum/USC face recognition system and how it fared in the FERET phase III test, in H. Wechsler, P. J. Phillips, V. Bruce, F. F. Soulié and T. S. Huang (eds), *Face Recognition: From Theory to Applications*. Springer-Verlag, 1998, pp. 186–205.

[19] L. Pessoa and A. P. Leitão, Complex cell prototype representation for face recognition. *IEEE Trans. Neural Networks*, **10**(6), 1528–1531, November 1999.

[20] P. J. Phillips, Matching pursuit filters applied to face identification. *IEEE Trans. Image Processing*, **7**(8), 1150–1164, 1998.

[21] P. J. Phillips, Support vector machines applied to face recognition, in M. J. Kearns, S. A. Solla and D. A. Cohn (eds), *Advances in Neural Information Processing Systems 11*. MIT Press, 1999.

[22] F. S. Samaria, Face recognition using hidden Markov models. *Ph.D. dissertation*, Trinity College, University of Cambridge, 1994.

[23] T. Sim, R. Sukthankar, M. Mullin and S. Baluja, High-performance memory-based face recognition for visitor identification. *Technical Report JPRC-TR-1999-001-1*, Just Research, 1999.

[24] D. L. Swets and J. Weng, Discriminant analysis and eigenspace partition tree for face and object recognition from views. *Proc. Second International Conference on Automatic Face and Gesture Recognition*, 1996, pp. 192–197.
[25] Y. W. Teh and G. E. Hinton, Rate-coded restricted Boltzmann machines for face recognition. *Neural Information Processing Systems*, 2000.
[26] L. Wiskott, J. Fellous, N. Krüger and C. von der Malsburg, Face recognition by elastic bunch graph matching. *IEEE Trans. Pattern Analysis and Machine Intelligence*, **19**(7), 775–779, 1997.
[27] W. Zhao, R. Chellappa and P. J. Phillips, Subspace linear discriminant analysis for face recognition. *Technical Report CAR-TR-914*, Center for Automation Research, University of Maryland, 1999.
[28] L. V. Hedges and I. Olkin, *Statistical methods for meta-analysis*. Academic Press, New York, 1985.
[29] R. Rosenthal, *Meta-Analytic Procedures for Social Research* (revised edn). Sage, 1991.
[30] J. P. Greene, A meta-analysis of the Rossell and Baker review of bilingual education research. *Bilingual Research Journal*, **21**(2,3), 1997.
[31] P. N. Shapario and S. D. Penrod, Meta-analysis of face identification studies. *Psychological Bulletin*, **100**, 139–156, 1986.
[32] L. V. Heggs, Modeling publication selection effects in meta-analysis. *Statistical Science*, 7(2), 246–255, 1992.
[33] G. A. Colditz, A. Miller and F. Mosteller, How study design affects outcomes in comparisons of therapy, I: medical. *Statistics in Medicine*, **8**, 441–454, 1989.
[34] M. Turk and A. Pentland, Eigenfaces for recognition. *J. Cognitive Neuroscience*, **3**(1), 71–86, 1991.
[35] D. M. Blackburn, J. M. Bone and P. J. Phillips, *FRVT 2000 Report*, 2001. Available online at: `http://www.frvt.org/`.
[36] J. Wilder, Face recognition using transform coding of gray scale projection projections and the neural tree network, in R. J. Mammone (ed.), *Artificial Neural Networks with Applications in Speech and Vision*. Chapman & Hall, 1994, pp. 510–536.
[37] H. Moon and P. J. Phillips, Computational and performance aspects of PCA-based face recognition algorithms. *Perception*, **30**, 301–321, 2001.
[38] B. Fischhoff and R. Beyth-Marom, Hypothesis evaluation from a Bayesian perspective. *Psychological Review*, **90**, 239–260, 1983.
[39] P. J. Phillips and E. Newton, Meta-analysis of face recognition algorithms. *Proc. Fifth International Conference on Automatic Face and Gesture Recognition*, 2002, pp. 235–241.
[40] S. Rizvi, P. J. Phillips and H. Moon, A verification protocol and statistical performance analysis for face recognition algorithms. *Proc. IEEE Conference on Computer Vision and Pattern Recognition*, 1998, pp. 833–838.
[41] P. J. Phillips, Human identification technical challenges. *International Conference on Image Processing*, **1**, 49–52, 2002.
[42] P. J. Rauss, P. J. Phillips, H. Moon, S. Rizvi, H. Hamilton and A. T. DePersia, FERET (Face Recognition Technology) program, in A. T. DePersia, S. Yeager and S. Ortiz (eds.), *Proc. Surveillance and Assessment Technologies for Law Enforcement*, 1997, SPIE Vol. 2935, pp. 2–11.
[43] F. Samaria and A. Harter, Parameterisation of a stochastic model for human face identification. *Proc. Second IEEE Workshop on Applications of Computer Vision*, 1994.

[44] D. Maio, D. Maltoni, R. Cappelli, J. L. Wayman and A. K. Jain, FVC 2000: fingerprint verification competition. *IEEE Trans. Pattern Analysis and Machine Intelligence*, **24**(3), 402–412, 2002.

[45] D. Maio, D. Maltoni, R. Cappelli, J. L. Wayman and A. K. Jain, FVC 2002: second fingerprint verification competition. *Proc. 16th International Conference on Pattern Recognition*, 2002, Vol. 3, pp. 811–814.

[46] Fingerprint Verification Competitions. Available online at `http://bias.csr.unibo.it/fvc2000/` and `http://bias.csr.unibo.it/fvc2002/`.

[47] NIST Speaker Recognition Evaluations. Available online at `http://www.nist.gov/speech/tests/spk/`.

[48] A. Martin and M. Przybocki, The NIST 1999 speaker recognition evaluation – an overview. *Digital Signal Processing*, **10**, 1–18, 2000.

[49] J. R. Beveridge, K. She, B. A. Draper and G. H. Givens, A nonparametric statistical comparison of principal component and linear discriminant subspaces for face recognition. *Proc. IEEE Conference on Computer Vision and Pattern Recognition*, **1**, 535–542, 2001.

[50] R. J. Micheals and T. Boult, Efficient evaluation of classification and recognition systems. *Proc. IEEE Conference on Computer Vision and Pattern Recognition*, **1**, 50–57, 2001.

[51] T. Sim, S. Baker and M. Bsat, The CMU pose, illumination, and expression (PIE) database. *Proc. Fifth International Conference on Automatic Face and Gesture Recognition*, 2002, pp. 53–58.

[52] J. M. Bone and D. M. Blackburn, Face recognition at a chokepoint – scenario evaluation results. *Technical Report*, 2002. Available online at: `http://www.dodcounterdrug.com/facialrecognition/`.

[53] T. Mansfield, G. Kelly, D. Chandler and J. Kane, Biometric product testing final report. *Technical Report*, CESG, 2001. Available online at: `http://www.cesg.gov.uk/site/ast/biometrics/media/BiometricsTestReportpt1.pdf`.

[54] P. Penev and J. Atick, Local feature analysis: a general statistical theory for object representation. *Network: Computation in Neural Systems*, 7(3), 477–500, 1996.

[55] D. Bolme, M. Teixeira, J. R. Beveridge and B. Draper, The CSU face identification evaluation system user's guide: version 4.0. *Technical Report*, Department of Computer Science, Colorado State University, 2002. Available online at: `http://www.cs.colostate.edu/evalfacerec/`.

8 The NIST speaker recognition evaluation program

Alvin Martin, Mark Przybocki and Joseph P. Campbell, Jr[1]

8.1 Introduction

The National Institute of Standards and Technology (NIST) has coordinated annual scientific evaluations of text-independent speaker recognition since 1996. These evaluations aim to provide important contributions to the direction of research efforts and the calibration of technical capabilities. They are intended to be of interest to all researchers working on the general problem of text-independent speaker recognition. To this end, the evaluations are designed to be simple, fully supported, accessible and focused on core technology issues.

The evaluations have focused primarily on speaker detection in the context of conversational telephone speech. More recent evaluations have also included related tasks, such as speaker segmentation, and have used data in addition to conversational telephone speech. The evaluations are designed to foster research progress, with the objectives of:

1. Exploring promising new ideas in speaker recognition
2. Developing advanced technology incorporating these ideas
3. Measuring the performance of this technology

The 2002 evaluation included 25 participating sites, by far the largest number to date. Evaluation participants included commercial, academic and governmental research laboratories. The nations represented included Australia, China, France, India, Israel, South Africa, Greece, Spain, Sweden and the USA. As in the past several evaluations, the ELISA Consortium (a group of European laboratories [8, 11]) participated collaboratively and also with individual submissions from each laboratory.

Information on the more recent NIST speaker recognition evaluations, including their official evaluation plans, is available on the NIST Speaker

1 This work is sponsored by the Department of Defense under Air Force Contract F19628-00-C-0002. Opinions, interpretations, conclusions, and recommendations are those of the author and are not necessarily endorsed by the United States Government.

Recognition website [1]. Evaluation data kits used in past evaluations are available as publications of the Linguistic Data Consortium [2].

A scientific evaluation paradigm is used for each NIST evaluation. A blind evaluation is conducted in which the participants do not know the speaker identities in advance. An evaluation begins with each participant registering with NIST and acquiring development and evaluation data as defined in the NIST evaluation plan for that year. The participants run their systems on the development data, for which speaker identities are given, to develop their algorithms, set thresholds, etc. Then the participants are given a limited time period (typically four weeks) to run their systems on the blind evaluation data and submit scores to NIST. NIST then evaluates the scores of the participants' systems and releases the answer key. Participants are encouraged to perform post-evaluation analysis using the answer key in preparation for the workshop. A workshop is held where NIST presents results of the various participants' systems and results on poolings of the data that are of interest to the community. The participants are required to present the details of their systems at the workshop and to submit system descriptions. Any submissions after the answer key is released are considered late and unofficial. Once the data and its corresponding answer key have been released, it is considered exposed and its use in future evaluations is carefully controlled (e.g. it becomes development data in subsequent evaluations). As defined in the evaluation plan [1], each of the evaluation tasks has its own rules and restrictions, but the following ones are in common:

- Listening to evaluation data is *not* allowed.
- Each decision is to be made independently:
  - based on the specified test segment and speaker model
  - use of other test segments or other models is *not* allowed
- Normalization over multiple test segments is *not* allowed.
- Normalization over multiple target speakers is *not* allowed.
- Use of evaluation data for impostor modeling is *not* allowed.
- Use of manually produced transcripts or other information for training is *not* allowed, except when allowed under the extended data evaluation.
- Knowledge of the target speaker's sex *is* allowed and the segment speaker's sex is known to be that of the target in non-cross-sex trials.

Other biometric evaluations are adopting similar scientific paradigms and guidelines appropriately adapted to the biometric under evaluation, as described elsewhere in this book.

8.2 NIST Speaker Recognition Evaluation Tasks

We describe in this section the four types of task that have been included in some of the NIST annual evaluations. Of these four tasks, it is the one-speaker detection task that has been a part of each evaluation and is the one most central to biometric identification [16] using speech.

The two most recent evaluations have each introduced a variant of the one-speaker detection task. The 2001 and 2002 evaluations included an *extended data* one-speaker detection task, which uses much larger amounts of data for speaker training and much longer speech data segments for testing. Transcripts generated by an automatic speech recognition (ASR) system were provided for all of the data. The 2002 evaluation included a *multimodal* test using a Federal Bureau of Investigation (FBI) corpus of forensic-type data. This FBI corpus included data from two types of microphone as well as telephone data. These tests have had some impressive results and are discussed in Sections 8.7 and 8.8.

8.2.1 One-Speaker Detection

This is the basic speaker recognition task that has been part of all the NIST evaluations. The task is to determine whether a specified speaker is speaking in a given single-channel segment of mu-law encoded telephone speech. The hypothesized speakers are always of the same sex as the segment speaker.

The task each year consists of a sequence of trials; the main one-speaker test in 2002 had about 39,000 trials. A *trial* consisted of a single hypothesized speaker and a specific test segment. The system is required to make an actual (true or false) decision on whether the specified speaker is present in the test segment. Along with each actual decision, systems are also required to provide for each trial a likelihood score indicating the degree of confidence in the decision. Higher scores indicate greater confidence in the presence of the speaker. A trial where the hypothesized speaker is present in the test segment (correct answer *true*) is referred to as a *target trial*. Other trials (correct answer *false*) are referred to as *impostor trials*.

8.2.2 Two-Speaker Detection

This task is the same as the one-speaker detection task, except that the speech segments include both sides of a telephone call with two speakers present and the channels summed. Unlike the one-speaker detection task where each training and test segment is limited to a single speaker, here systems must deal with the confounding presence of a second speaker. Note that the task is to determine whether the one specified speaker is present in the combined signal. The segment speakers may be of the same or opposite sex, but the hypothesized speaker is always of the same sex as at least one of the segment speakers.

8.2.3 Speaker Tracking

This task is to perform speaker detection as a function of time. The tracking task uses summed channel segments, as in the two-speaker detection task. Systems are required to identify the time intervals (if any) when the hypothesized speaker is speaking. This task may be viewed as a generalization of the two-speaker detection task. Note that the single target speaker is known, with training data provided.

8.2.4 Speaker Segmentation

This task requires systems to find the intervals within a speech segment corresponding to different unknown speakers. Thus, this task requires clustering speech according to (unknown) speaker identities; in general the number of speakers present is also unknown. The two-speaker detection task may be approached as having this task as a front end, with the number of unknown speakers being two, followed by the one-speaker detection task applied to each of the two clusters.

8.3 Data

The primary data source for the NIST evaluations has been the Switchboard Corpora of conversational telephone speech collected over the last decade by the Linguistic Data Consortium [2]. These all involve 5- to 10-minute conversations between two speakers. A participant calls into an automated operator that connects him or her to another participant and records their conversation as separate sides. The speakers, who generally do not know each other, are paired and assigned a conversational topic by an automatic system. Speaker pairs are never repeated and the assigned topic is never repeated for either speaker. The speakers are recruited adults, frequently college students, approximately half male and half female, who are generally paid nominal fees for their participation. They sometimes do not stick to the assigned topic. Multiple sessions (up to 25) per speaker at least 1 day (up to weeks) apart are recorded using various telephone handsets over the public telephone network [23]. Table 8.1 describes the multiple parts of the Switchboard (SWBD) Corpora.

Table 8.1 The Switchboard Corpora: all two-channel mu-law data.

	SWBD I	SWBD II Phase 1	SWBD II Phase 2	SWBD II Phase 3	SWBD Cellular 1	SWBD Cellular 2
Number of conversations	4870	3702	4575	2728	1309	2020
Number of speakers	543	661	684	640	254	419
Predominant speakers	Adults	College students	College students	College students	Adults	Adults
Data style	On topic	College chit-chat	College chit-chat	College chit-chat	Topic/ chit-chat	Topic/ chit-chat
Collection dates	1990	1997–1998	1998–1999	1999–2000	1999–2000	2000–2001
Targeted location	USA	Mid-Atlantic	Mid-West	South	East coast	East coast

We discuss below how speaker training data and test segments have been selected from the Switchboard Corpora to be used in the NIST evaluations for the one-speaker detection task.

8.3.1 Speaker Training

Training data is provided for the hypothesized speakers of all trials. Such speakers are referred to as *model speakers*. The source of the training data has varied over the course of the evaluations, but the amount of training data has typically been about two minutes. Early evaluations were designed to look more closely at how the source of training data affected performance. Three types of training data were provided:

1. *One session*: two minutes of data taken from a single conversation.
2. *Two session*: one minute of data taken from each of two conversations where the same telephone handset was used for each.
3. *Two handset*: one minute of data taken from each of two conversations where different telephone handsets were used for each.

These early evaluations revealed that a large performance gain was achieved as the training data became more varied (two handset training was best). A later evaluation showed, not surprisingly, that longer training segments improved system performance, but this was a smaller improvement than two handset training. More recent evaluations have used one-session training. Table 8.2 offers information about the training data provided in each of the annual evaluations.

In all of the evaluations, the training data has consisted of consecutive turns of the speaker with areas of silence removed, generally selected from near the end of the conversation. (The end is perhaps a better, more conversational, choice than the beginning, which may contain more formal introductions.) The actual durations of the training segments are allowed to vary within a 10 second range so as to include only whole turns wherever possible.

8.3.2 Test Segments

In earlier evaluations, test segments were chosen to have fixed durations of approximately 3, 10 or 30 seconds. As might be expected, performance improved with longer duration segments. Table 8.2 summarizes the durations of the test segments over the course of the evaluations.

In the recent evaluations, the one-speaker test segments have had varying durations ranging up to a minute and averaging about 30 seconds. They have been selected by choosing a minute of conversation and concatenating the turns of each speaker within that minute into two separate test segments, one per channel. As with the training segments, areas of silence are removed and whole turns included to the extent possible. No more than one test segment is created from each conversation side, and no test

Table 8.2 Information on the NIST Speaker Recognition evaluations, 1996–2002.

Year	Primary condition	Target speakers/target trials	Evaluation features
1996	Not defined	40/3999	Tests of 3 durations, 3 training conditions Switchboard-1 data
1997	Train/test using different handsets 30 second durations Two handset training	~400/3050	Tests of 3 durations, 3 training conditions Switchboard-2 Phase 1 data
1998	Train/test using same handset 30 second durations Two session training	~500/2687	Tests of 3 durations, 3 training conditions Switchboard-2 Phase 2 data Handset type detector information made available
1999	Train/test using different electret handsets Test durations 15–45 seconds Two session training	233/479	Added multispeaker tasks Variable durations used in main test trials Switchboard-2 Phase 3 data
2000	Train/test using different electret handsets Test durations 15–45 seconds One session training	804/4209	Resegmented 1997, 1998 test data for reuse Extra test on AHUMADA Spanish data
2001	Train/test using different electret handsets Test durations 15–45 seconds One session training	804/4209	Repeated 2000 main test with some additional trials Additional test on Switchboard cellular data Additional test allowing human or machine transcripts with extended training data
2002	Test durations 15–45 seconds One session training		

segments come from conversations where training data is selected from either side.

In general, each test segment is used in 11 trials, one of which is a target trial with the segment speaker being the model speaker. The other 10 model speakers are randomly selected from among all model speakers of the same sex as the segment speaker. As discussed in the next section on measuring performance, this 10 to 1 ratio of impostor to target trials is not intended to reflect what is likely in an actual application environment.

8.4 Performance Measure

Two types of error can occur in a detection task, often denoted as missed detections and false alarms. The miss rate, $P_{Miss|Target,}$ is the percentage of target trials decided incorrectly; the false alarm rate, $P_{FalseAlarm|Nontarget}$, is the percentage of impostor trials decided incorrectly. These error probabilities are determined from a system's actual decisions.

NIST has chosen to make the basic performance measure a cost function defined as a weighted sum of these two error rates. This detection cost, referred to as the C_{Det} cost for reasons discussed below, is defined as

$$C_{Det} = (C_{Miss} \times P_{Miss|Target} \times P_{Target}) + (C_{FalseAlarm} \times P_{FalseAlarm|Impostor} \times P_{Impostor}) \quad (8.1)$$

The required parameters in this function are the cost of a miss (C_{Miss}), the cost of a false alarm ($C_{FalseAlarm}$), and the *a priori* probability of a target speaker (P_{Target}). Note that we must have $P_{Impostor} = 1 - P_{Target}$. NIST has used the following parameter values:

$$C_{Miss} = 10; C_{FalseAlarm} = 1; P_{Target} = 0.01 \quad (8.2)$$

The relatively greater cost of a miss compared to a false alarm is probably realistic for many applications. The *a priori* probability of a target speaker is more arbitrary and application-dependent. Note that this specified *a priori* probability need not, and in fact did not, correspond to the actual percentage of target instances in the evaluation data. An advantage of this type of error metric formulation is that the test data need not resemble intended application data in terms of target richness.

The C_{Det} value determined is generally normalized based on the principle that a system without any speaker knowledge should have an expected cost of one. With the parameters specified, a knowledge-free system should opt to decide *false* for every trial, incurring the cost of a miss for all target trials. This would result by Equation (8.1) in a C_{Det} value of 0.1. Thus 0.1 is used as the normalization factor for all C_{Det} values.

The performance of systems on a given task can be shown by bar charts of the C_{Det} scores. The left side of Figure 8.1 shows two such bar charts. The two parts of each bar show the separate contributions of the false alarm and miss rates to the total C_{Det} scores.

This detection cost metric is based on the actual decisions and provides a single numerical value for comparing system performances. We also, however, want to examine the range of possible operating points of a system and compare these across systems. In analyzing factors affecting system performance, it is curves showing the range of possible operating points that are of greatest interest. For this, we must use the likelihood scores that systems are required to provide for each trial.

Receiver operating characteristic (ROC) curves have long been used to show multiple operating points of systems. In [3], NIST introduced the

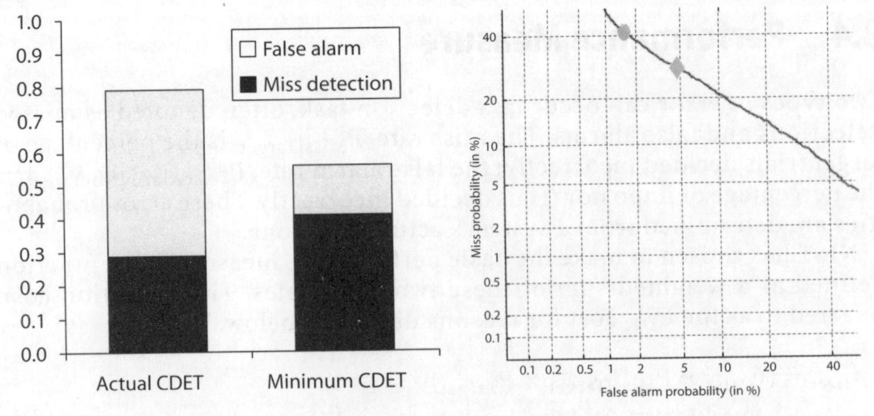

Figure 8.1 Bar charts of actual decision and minimum C_{Det} scores (left) and the corresponding DET curve with actual and minimum C_{Det} points marked (right).

alternative presentation of detection error trade-off (DET) curves for this purpose. DET curves place the possible values of the two types of errors on the horizontal and vertical axes using a normal deviate scale for both. They have the key property that if the underlying distributions of scores for target and impostor trials are normal, then the resulting performance curve plots as a straight line. In the NIST evaluations, the DET plots of system performance curves have almost always been fairly close to linear. Because performance is plotted this way, we have chosen to refer to the error cost function as the C_{Det} value.

There are two special points that we note on each DET curve. One is the actual decision point, denoted by a circle, ●. The other is the point on the curve having minimum C_{Det} value, denoted by a diamond, ◆. These two points are marked in the DET curve of Figure 8.1, and the two C_{Det} bar charts in the figure correspond to these two points. The closeness of these two points is an indication of how well the system chose the likelihood threshold value used for the actual decisions.

8.5 Evaluation Results

The main one-speaker detection task in each evaluation has generally contained hundreds of speakers and thousands of trials. A subset of these trials has generally been specified as constituting the *primary* condition of interest for the evaluation. The primary condition for the 2002 evaluation was specified as trials where the training and test segment were of a cellular transmission type and where the test segment duration was in the 15 to 45 second range. This accounted for the great majority of the trials. In early evaluations with fixed-duration test segments, the 30 second ones were regarded as primary. In some of the earlier evaluations using landline conversations, one handset microphone type, namely electret, was specified

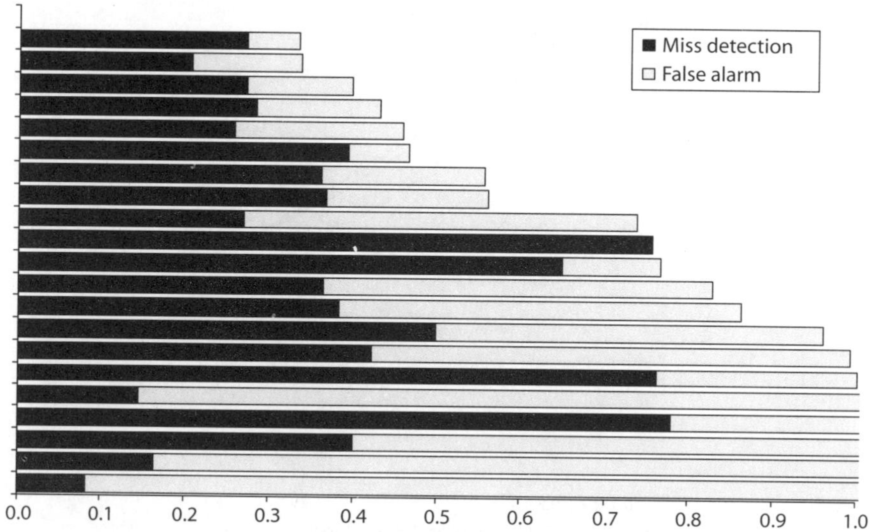

Figure 8.2 Bar chart of primary condition C_{Det} scores in the 2002 evaluation.

for the primary condition. Also in earlier evaluations, the mismatched (and, in one case, the matched) condition between the handsets used in training and in tests of the target trials was defined to be part of the primary condition. Table 8.2 indicates the primary conditions used in each evaluation.

Figure 8.2 shows a bar chart of the C_{Det} scores for the actual decisions by each system in the 2002 one-speaker detection task, using the primary condition trials. Clearly, there is a great deal of variation in system performance for this task. In accord with NIST's understanding with the participants, we are not identifying the various systems in this plot or the DET plots presented hereafter.

Figure 8.3 presents the corresponding DET curves for the 2002 one-speaker detection task. Since the DET plots generally contain more information of interest in terms of performance results and trade-off, we hereafter concentrate on these.

8.6 Factors Affecting Detection Performance

Having access to the raw submission results of all participants gives NIST the opportunity to analyze these results in order to explore various factors that may affect recognition performance. Over the years, NIST has explored the significance of numerous data, speaker and channel attributes on overall performance. Here we discuss several of these using results from recent evaluations.

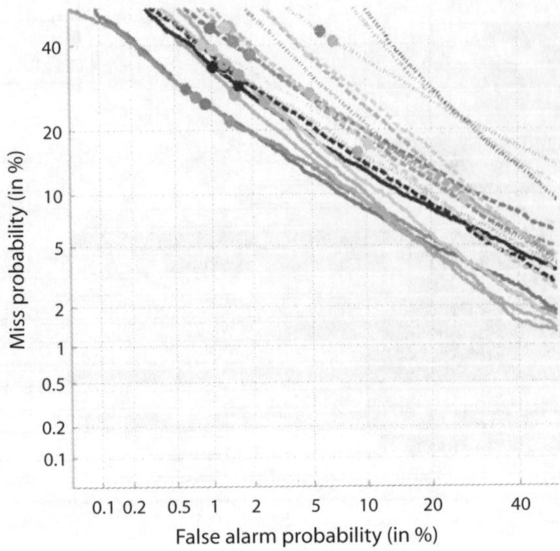

Figure 8.3 DET plot of primary condition systems in the 2002 evaluation.

8.6.1 Duration

Earlier NIST evaluations, as noted, included separate tests in one-speaker detection for segments of 30, 10 and 3 seconds duration. These showed that performance was significantly greater for 30 second segments than for 10 second segments, while performance on 10 second segments significantly exceeded that on 3 second segments. For the past four years, there were no separate tests, and the one-speaker segments averaged around 30 seconds in duration, but varied over a continuous range of up to 1 minute.

Figure 8.4 displays DET curves of one-speaker detection performance in 2002 by ranges of segment duration for one system. The results shown are typical of most of the evaluation systems. They indicate that, for such systems, performance is significantly lower for segments shorter than 15 seconds in duration, but that performance is not greatly affected for segments longer than 15 seconds. This is consistent with the findings of previous years, but indicates that the duration effect seen was limited and that once some minimum duration (apparently in the 10 to 15 second range) is available, the amount of test speech ceases to be a major factor in performance.

8.6.2 Pitch

Pitch would appear to be an important factor in speaker recognition, but attempts to specifically include it in algorithms have had only limited success [4]. NIST investigated ways in which average speaker pitch affects performance in several of the evaluations. Performance was not very consistently affected by limiting consideration within each sex to speakers

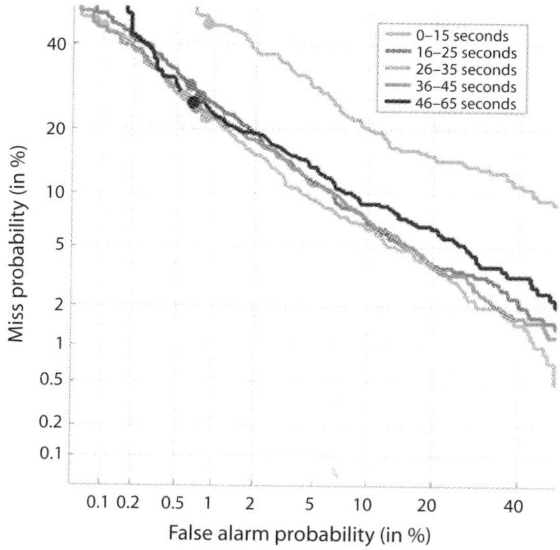

Figure 8.4 One-speaker detection performance as a function of duration for a typical system in the 2002 evaluation.

of particularly high or low pitch [5, 6]. We found, as might be expected, that limiting impostor trials to instances where the impostor's average pitch is close to that of the hypothesized target (in the training data), while including all target trials, degrades performance. But, perhaps surprisingly, a bigger effect was observed when target trials are restricted to those with the largest pitch differences between the training and test segments, while all impostor trials are included.

Figure 8.5 gives an example of a typical system in the 1999 evaluation. For each speaker, the average pitch of the training data and of each test segment was estimated. The plot shows a curve of all primary condition tests and curves limited to target trials where the log pitch difference between the test segment and the target speaker training data are in the high and low 25% of all such differences. Large pitch differences in target trials may correspond to instances where the speaker had a cold or was feeling particularly emotional during either the training or test conversation. Note how large the performance differences are. For example, at a 10% miss rate, the false alarm rate is around 4% when all trials are included. When target trials are limited to the 25% that are closest in pitch, the false alarm rate is less than 1%; when limited to the 25% furthest in pitch, the false alarm rate exceeds 10%.

8.6.3 Handset Differences

The variation in telephone handsets is a major factor affecting the performance of speaker recognition using telephone speech. For the landline Switchboard Corpora, specific handset information is not provided, but

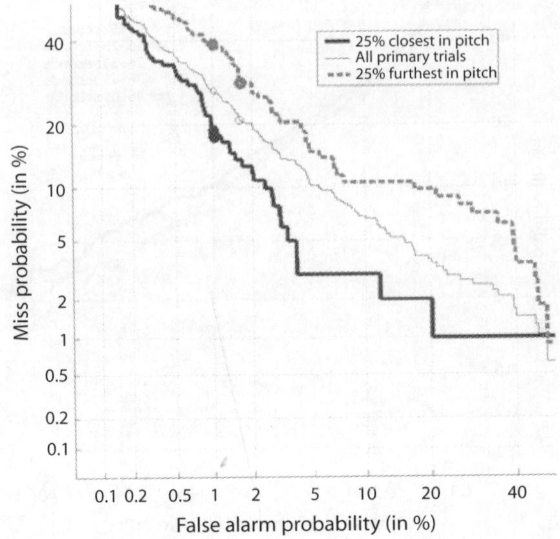

Figure 8.5 Effect of target pitch differences. Shown for one system is the 1999 primary condition performance and performance when target trials are restricted to those with the 25% closest and furthest average pitch differences between the training and test data. The effect of such differences on performance is quite large.

telephone line information (i.e. the phone number) is. We can generally assume that if two calls are from different lines, the handsets are different; if they are from the same line, the handsets are probably the same.

Where possible, we have generally chosen to concentrate in the evaluations on target trials where the training and test segment lines are different. This is clearly the harder problem. Moreover, using same-line calls is, in a way, unfairly easy, since for impostor trials, the training and test segment handsets are always different because, with rare exceptions, speakers do not share handsets. Thus, using same-line target trials could be viewed as handset recognition, rather than speaker recognition.

Figure 8.6 shows the large performance difference in the 1999 evaluation between using same-line and different-line target trials for one system. The impostor trials are the same in both cases.

8.6.4 Handset Type

Most standard landline telephone handset microphones are of either the carbon-button or electret type. We observed in early evaluations that the handset types (i.e. the microphone types) used, in both the training and the test segments, can greatly influence recognition performance.

MIT-Lincoln Laboratory, which has participated in all of the NIST speaker recognition evaluations, developed an automatic handset labeler in a previous evaluation [7]. This handset labeler uses the telephone speech signal from one channel to assign a likelihood that the signal is from a carbon-button handset as opposed to an electret handset. This likelihood is

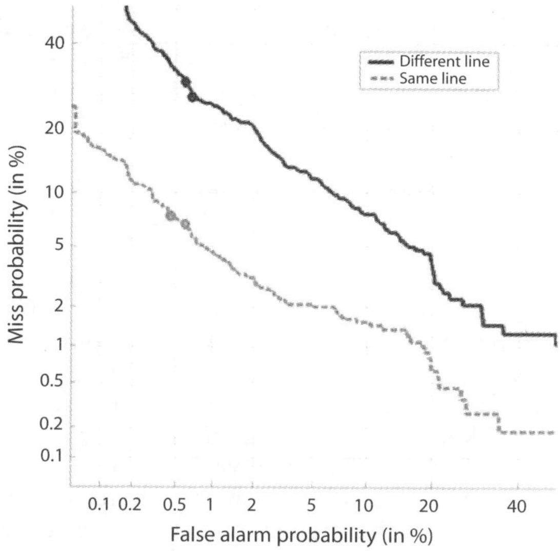

Figure 8.6 Effects of same/different lines for training/test. Performance for a typical system is far higher when target trials are restricted to those involving the same phone line (and presumably handset) in training as in test, as opposed to those involving different lines.

converted into a hard decision (carbon or electret). For subsequent evaluations involving landline data, this hard decision was made available to all of the participating systems for all of the training and test segments. It should be noted that the labeler's decisions were certainly less than perfect, as occasionally different conversations from the same telephone number were assigned the opposite type, but the decisions are believed to be generally quite accurate. Figure 8.7 provides some information on the distribution of

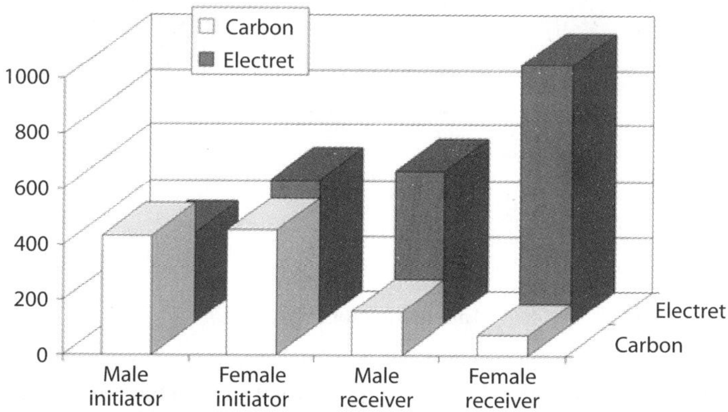

Figure 8.7 Handset type distribution in the 1999 test set. Received calls generally involved electret handsets, but initiated calls are divided between handset types. Presumably this is because the initiators often used pay phones. Females apparently used fewer carbon button handsets than males. This may indicate a bias in the handset-type determination algorithm.

handset types in the conversations used in the 1999 evaluation. The main point to note is that the received calls, generally from home phones, overwhelmingly involve electret-type handsets, while the initiated calls, often made from public phones, are split between electret and carbon-button type handsets. There is also evidence that conversation sides involving female speakers are more likely to be declared to be of electret type. This may indicate a slight bias in the automatic handset-type detection algorithm. It may also help to explain the slightly better overall performance of most systems on male speakers compared to female speakers.

Figure 8.8 shows the variation in performance for different combinations of training and test-segment handset types for one system in the 2001 evaluation. The data plotted here involves a subset of speakers for whom both electret and carbon-button training data was provided, giving two models for each speaker. The trials used are paired by the two models of each target speaker. All target trials involve different telephone numbers. It may be seen that there is a considerable performance advantage to having matching handset types of the higher quality electret type. This was the case for almost all systems. There was some variation in the relative performance of the three other combinations across systems. There appear to be competing advantages to having matched types in training and test data and to having at least some higher quality electret data used, perhaps particularly in the training.

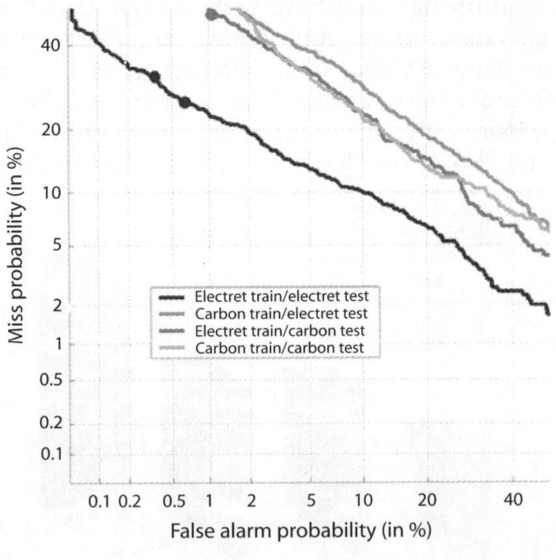

Figure 8.8 Performance as a function of training/test handset. Performance for one system on different number tests for each combination of training and test handset types. All speakers here have both a carbon-button and an electret trained model, and all the trials are paired by such target models. Performance is best when all data is electret.

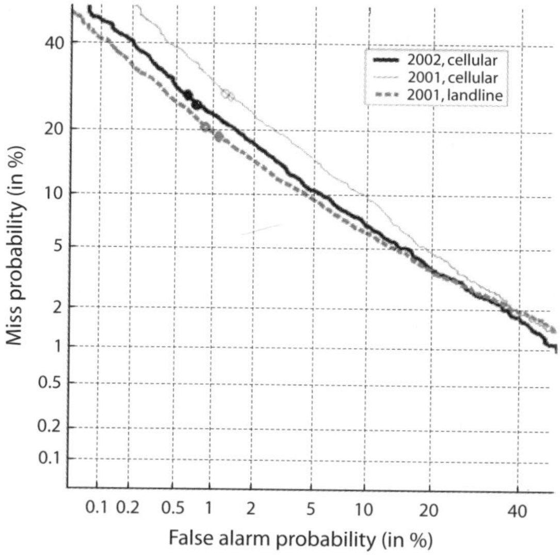

Figure 8.9 DET plots of the best systems on the 2002 cellular, 2001 cellular and 2001 landline test sets.

8.6.5 Landline vs. Cellular

The past few years have brought an increasing interest in the processing of cellular data for both speech and speaker recognition. The two most recent Switchboard collections (shown in Table 8.1) have consisted primarily of cellular conversational data. These collections were used as a subsidiary test in the 2001 evaluation and as the main evaluation data in 2002. Figure 8.9 presents DET plots of the best performing systems on the 2001 landline, 2001 cellular and 2002 cellular evaluation sets.

If the two cellular test sets are of comparable difficulty, and other comparisons suggest that they are, then Figure 8.9 shows some real improvement in the best system performance between 2001 and 2002. The 2001 curves also show that the cellular test sets are measurably more difficult. This comparison, however, rather understates the differences in relative difficulty of landline and cellular data. This is because the landline data is selected so that the target trials always involve different handsets in training and test data, but the collection protocol for cellular does not permit this, and, in most target trials, the training and test handsets are the same.

The greater difficulty of speaker detection in the cellular context will be subject to further investigation in future evaluations. New data collections may make it possible to investigate target trials with different training and test handsets, including mixed cellular and landline combinations.

8.7 Extended Data Evaluation

The accuracy improvements in text-independent speaker detection over the course of the NIST evaluations have been mostly incremental. Each year, evaluation participants would add a few useful ideas to their existing speaker detection systems, often influenced by what had worked for other participants in the previous evaluation. Most systems in recent NIST evaluations have converged on variations of the Gaussian mixture model (GMM) technique [18]. The evaluation paradigm encourages this kind of incrementalism, which can produce markedly, but usually not radically, better systems over the years.

But in 2000, George Doddington suggested a radically different approach to the task, which might, under proper conditions, offer major speaker recognition accuracy improvements [10, 17]. This idea was tested in what became known as the *extended data* portion of the NIST one-speaker detection task in the 2001 and 2002 evaluations. This involved considerably longer training data durations (multiple conversation sides) and test segment durations (a whole conversation side), relative to the conventional one-speaker detection task.

People are generally pretty good and quite robust speaker recognizers [22]. Doddington observed that people do a better job of detecting those with whom they are quite familiar than those they do not know well. They become accustomed to the speaking habits and idiosyncrasies of those they know well. Doddington suggested making use of *idiolectal* characteristics of speakers for whom considerable transcribed speech data was available. He showed that by using the available manual transcripts of the Switchboard-1 conversations, one could make use of the word patterns – specifically the common unigrams, bigrams and trigrams – of individual speakers for detection purposes. In [17], Doddington was the first to show the power of speaker-dependent language models for speaker recognition. As expected, considerable training data is needed to model a speaker's language and realize the benefits of this new technique. The best results used eight or more conversation sides (generally five minutes per conversation) of each speaker for training data.

The 2001 evaluation included a development-type test of the Switchboard-1 conversations, as studied by Doddington. A jackknifing procedure was used to make use of all the conversation sides as test segments, where multiple models are trained for each speaker using 1, 2, 4, 8 or 16 sides as a speaker's training data. Participants were offered uncorrected transcripts of all the conversation sides, exactly as produced by an automatic speech recognition system provided by Dragon Systems [12]. This system had an estimated word error rate on conversational telephone speech of somewhat under 30%. But Dragon Systems had shown in previous work that even partially reliable ASR systems could be quite useful for various tasks, such as topic spotting [13].

Doddington's successful demonstration of idiolect quickly led others to the discovery of recognizing speakers based on their speaking habits and idiosyncrasies at the phonetic level [19, 20]. A speaker's phone patterns – specifically the common unigrams, bigrams and trigrams – can be used for detection purposes. These phone patterns reflect the dynamic realization of phonetic features and are used to model an individual's pronunciation inventory. The phone recognizer needs to be consistent, not necessarily accurate. A strength of the phonetic method is that a speech recognizer, which might not exist, is not required in the language being spoken. Phone recognizers trained in languages other than the spoken language have also been demonstrated to work well, as has the combination of multiple phone recognizers trained in various languages [15, 19, 20].

The 2001 NIST evaluation results of these new speaker recognition methods were quite dramatic. Systems were developed that combined [19, 20] the use of idiolectal information based on the ASR transcripts, phonetic information, text-constrained GMM (TC-GMM) [21], and traditional GMM approaches. Figure 8.10 shows DET curves of the best performing systems on this data. For comparison, the best 2001 DET curve for the main one-speaker detection task, using Switchboard-2 data and much more limited amounts of training and test data, is also shown. These results show dramatic improvements in error rates, approaching an order of magnitude.

The 2002 evaluation also had an extended data test, this time using data from two of the phases of Switchboard-2. The Dragon ASR system was no longer available and, to prepare transcripts in time for the evaluation, a real-time BBN/ASR system was run at NIST on all of the test data. The BBN-provided recognizer was not intended to be of evaluation quality, as it was trained on only 40 hours of speech data and designed for real-time operation. This resulted in a word error rate of approximately 50%, which is considerably higher than that of the non-real-time Dragon recognizer used the year before.

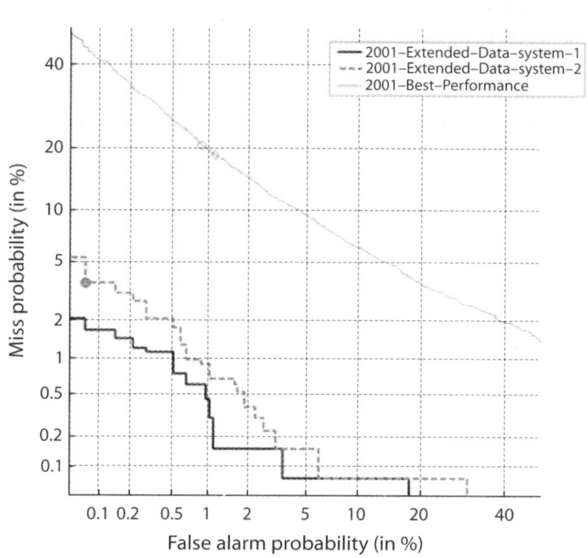

Figure 8.10 DET Plots of the best performing extended data systems of two sites in the 2001 evaluation. For contrast, the DET plot of the best performing system of the main evaluation in 2001 (on a completely different data set) is shown.

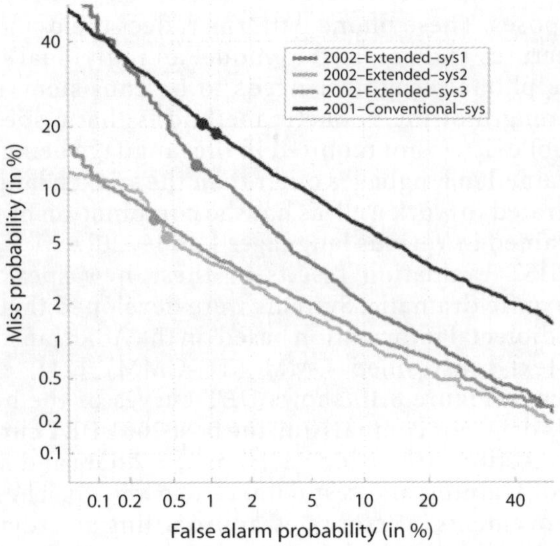

Figure 8.11 DET Plots of the best performing extended data systems of three sites in the 2002 evaluation. For contrast, the DET plot of the best performing system of the main evaluation in 2001 (on a completely different data set) is shown.

Figure 8.11 shows the DET curve performance of the best systems in the 2002 evaluation, along with the best 2001 conventional system DET curve for comparison. The 2002 results, while still impressive, were clearly less impressive than those in 2001. The higher error rates of the ASR transcripts are most likely one of the reasons for this. Another reason might be that the Switchboard-2 conversants, unlike those in Switchboard-1, often did not stick to their assigned topics and preferred casual chit-chat, which could affect idiolectal techniques.

Note, in any case, that these outstanding speaker recognition results currently hold only for situations where extensive training and test data are available, which limits the potential applications. Moreover, the extended task trials were less controlled than those of the main evaluation for differences between training vs. test handsets and male vs. female speakers. To the extent that idiolectal information was used, these differences probably matter little. The best results combined GMM, TC-GMM, idiolectal, and phonetic techniques. It is hoped that future evaluations will be able to include standard and extended data tests that involve the same speakers and conversation sides for a more direct comparison.

8.8 Multimodal Evaluation

An additional type of one-speaker detection test was included in the 2002 evaluation. The FBI has an ongoing interest in forensic applications and was pleased to make available a previously collected corpus for this evaluation. We chose to describe the test as *multimodal* because the corpus contains different speech segments from individual speakers recorded using a telephone handset (internal line), a close microphone and a far microphone. Thus it was possible to evaluate the effects of having different microphone types used as training and test data.

The FBI corpus is described in [14]. The multimodal test made use of 87 male speakers. In addition to its multiple recording modes, it contains training segments of 30 and 120 seconds duration, test segments of 3, 12, 30 and 120 seconds duration, and both spontaneous and read segments. It also is designed for both text-independent and text-dependent testing, though the NIST evaluation was limited to the text-independent parts.

A somewhat different performance measure seemed needed for this type of evaluation. A higher cost for a false alarm than for a miss seemed appropriate in a forensic setting, while it seemed best to view the *a priori* probability of a true speaker (target) trial as equal to that of an impostor trial. Thus we used the values

$$C_{\mathrm{Miss}} = 1;\ C_{\mathrm{FalseAlarm}} = 2;\ P_{\mathrm{Target}} = 0.5 \tag{8.3}$$

We also decided to extend the simple true/false detection model in this case to allow *no-decisions.* We assigned all no-decisions a cost of 0.25:

$$C_{\mathrm{NoDecision|Target}} = C_{\mathrm{NoDecision|FalseAlarm}} = 0.25 \tag{8.4}$$

Thus in this case, the system should decide true, false or no-decision according to which expected cost is lowest:

$$E_{[\mathrm{Cost|FalseAlarm}]} = C_{\mathrm{FalseAlarm}} \times (1 - \mathrm{Pr}(\mathrm{Target|score})) \tag{8.5}$$

$$E_{[\mathrm{Cost|Miss}]} = C_{\mathrm{Miss}} \times \mathrm{Pr}(\mathrm{Target|score}) \tag{8.6}$$

$$\begin{aligned} E_{[\mathrm{Cost|NoDecision}]} = {} & C_{[\mathrm{NoDecision|Target}]} \times \mathrm{Pr}(\mathrm{Target|score}) \\ & + C_{[\mathrm{NoDecision|NonTarget}]} \times (1 - \mathrm{Pr}(\mathrm{Target|score})) \end{aligned} \tag{8.7}$$

Here Pr(Target|score) is the confidence, i.e. the confidence in making a target decision as a function of score. This confidence was a required system output for each trial in the multimodal task. The system should then make a positive detection if this confidence exceeds 87.5% and a negative detection if it is less than 25%. Otherwise, a no-decision is appropriate.

Basically, this was an easier task than the main one-speaker detection task since multiple outside phone lines are not involved and the data is not conversational. The bar charts and DET plot results for one system in Figure 8.12 show the considerably higher level of performance than in previous one-speaker detection plots. The nine curves show all combinations of training and test data from the three microphone source types. For the *matched* cases, the best performance is with the close microphone and the worst with the far microphone. In the mixed cases, those involving the far microphone generally have lower levels of performance.

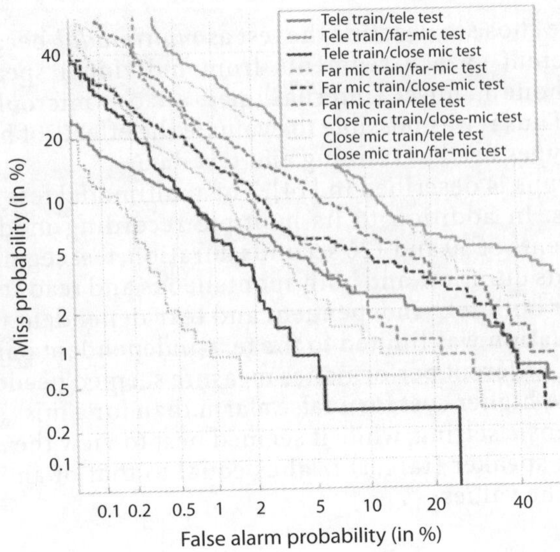

Figure 8.12 DET Plots of the multi-modal system of one site in the 2002 evaluation. Separate plots of all nine combinations of training and test microphone types are presented. The thick lines correspond to the three matched type cases, while the thin lines correspond to the six mismatched cases.

Note in the figure that each actual decision cost bar is divided into four parts corresponding to the four different types of error. Note also the low actual decision costs and that the actual decision points on the DET curves are off the actual curves because no-decisions are allowed. (The percentages of target and non-target trial no-decisions are shown in the DET legend.) One may then ask whether these actual decision error rates are acceptably low for a forensic application.

These results, however, should not be considered typical of actual capabilities in forensic situations. Forensic recordings are sometimes of very low quality, and the emotional stresses that may be involved are not reflected in test recordings in the current evaluation. NIST hopes that data will become available to allow further investigations of these matters in future evaluations.

8.9 Future Plans

Current plans call for the annual NIST evaluations to continue. The Linguistic Data Consortium has plans to collect additional conversational data, both landline and cellular, to support future research, primarily in speech recognition. We hope that suitable portions of this data will also be appropriate for continued research and evaluation in the area of speaker detection.

Speech recognition research has made considerable progress in recent years. It is anticipated that there will soon be available near real-time systems for recognizing conversational telephone speech with word error rates below 20%. The extended data evaluations have shown that such recognition performance can greatly aid speaker recognition systems, particularly when sufficient data is available. Such speech recognizers can be expected to be a part of future text-independent speaker recognizers. Upcoming evaluations may examine the effect of varying training and test segment durations involving long durations that allow systems to get to *know* speakers via idiolectal characteristics.

There is also interest in further multimodal-type tests, to the extent that suitable data can be made available.

It should be emphasized that the NIST evaluations are open to all research sites that wish to participate and share information about their systems at the workshops following each evaluation. Those who may be interested in participating or who know of data sources that could be used in the evaluations are encouraged to contact NIST [9].

References

[1] M. Przybocki, *NIST – Speaker Recognition Evaluations: Benchmark Tests*, 17 December 2001. Available online at: `http://www.nist.gov/speech/tests/spk/`.

[2] Linguistic Data Consortium, University of Pennsylvania, 3615 Market Street, Suite 200, Philadelphia, PA 19104–2608. `http://www.ldc.upenn.edu/`.

[3] A. Martin, G. Doddington, T. Kamm, M. Ordowski and M. Przybocki, The DET curve in assessment of detection task performance. *Proc. Eurospeech*, 1997, Vol. 4, pp. 1895–1898. Available online at: `http://www.nist.gov/speech/publications/`.

[4] G. Doddington, M. Przybocki, A. Martin and D. Reynolds, The NIST speaker recognition evaluation – overview, methodology, systems, results, perspective. *Speech Communication*, **31**(2–3), 225–254, 2000. Available online at: `http://www.nist.gov/speech/publications/`.

[5] M. Przybocki and A. Martin, NIST speaker recognition evaluation – 1997. *RLA2C*, Avignon, April 1998, pp. 120–123.

[6] M. Przybocki and A. Martin, NIST speaker recognition evaluations: review of the 1997 & 1998 evaluations. *RLA2C* presentation, Avignon, April 1998. Available online at: `http://www.nist.gov/speech/tests/spk/1998/rla2c_pres/`.

[7] T. Quatieri, D. Reynolds and G. O'Leary, Magnitude-only estimation of handset nonlinearity with application to speaker recognition. *Proceedings ICASSP*, 1998, pp. 745–748.

[8] The ELISA Consortium, The ELISA systems for the NIST '99 evaluation in speaker detection and tracking. *Digital Signal Processing*, **10**(1–3), 143–153, 2000.

[9] 100 Bureau Drive Stop 8940, Gaithersburg, MD 20899–8940; `alvin.martin@nist.gov`.

[10] G. Doddington, Speaker recognition based on idiolectal differences between speakers. *Proc. Eurospeech*, 2001, Vol. 4, pp. 2521–2524.

[11] The 2002 ELISA Consortium consisted of researchers from the following companies/universities: GET-ENST; University of Avignon; INRIA, CNRS,

l'Université de Rennes 1; University of Lyon II, CNRS; IMAG, University Joseph Fourier; University of Fribourg; and University of Balamand.

[12] B. Peskin, M. Newman, D. McAllaster, V. Nagesha, H. B. Richards, S. Wegmann, M. Hunt and L. Gillick, Improvements in recognition of conversational telephone speech. *International Conference on Acoustics, Speech, and Signal Processing*, IEEE, Phoenix, Arizona, 1999.

[13] B. Peskin, Topic spotting on switchboard data. *Whisper Final Review Meeting*, conference notebook, January 1993.

[14] S. Beck, *User's Manual for Training and Testing – FBI Voice Database For Automated Speaker Recognition Systems*, March 1998.

[15] W. Andrews, M. Kohler, J. Campbell, J. Godfrey and J. Hernández-Cordero, Gender-dependent phonetic refraction for speaker recognition. *Proc. ICASSP 2002*, Orlando, Florida, April 13–17, 2002, Vol. 1, pp. 149–152.

[16] P. J. Phillips, A. Martin, C. I. Wilson and M. Przybocki, An Introduction to evaluating biometric systems. *IEEE Computer,* **33**(2), 56–63, 2000.

[17] G. Doddington, Some experiments on idiolectal differences among speakers, 14 November 2000. Available online at: `http://www.nist.gov/speech/tests/spk/2001/doc/n-gram_experiments-v06.pdf`.

[18] D. Reynolds, T. Quatieri and R. Dunn, Speaker verification using adapted Gaussian mixture models. *Digital Signal Processing*, **10**(1–3), 19–41, 2000.

[19] W., Andrews, M. Kohler and J. Campbell, Phonetic speaker recognition. *Proc. Eurospeech*, 2001, pp. 2517–2520.

[20] W. Andrews, M. Kohler, J. Campbell and J. Godfrey, Phonetic, idiolectal, and acoustic speaker recognition. *Proc. 2001: A Speaker Odyssey, The Speaker Recognition Workshop,* Chania, Crete, Greece, June 18–22, 2001, pp. 55–63.

[21] D. Sturim, D. Reynolds, R. Dunn and T. Quatieri, Speaker verification using text-constrained Gaussian mixture models. *Proc. ICASSP 2002*, Orlando, Florida, April 13–17, 2002, Vol I. pp. 677–680.

[22] A. Schmidt-Nielsen and T. Crystal, Speaker verification by human listeners: experiments comparing human and machine performance using the NIST 1998 speaker evaluation data. *Digital Signal Processing*, **10**(1–3), 249–266, 2000.

[23] J. Campbell and D. Reynolds, Corpora for the evaluation of speaker recognition systems. *Proc. ICASSP 1999,* Phoenix, Arizona, March 15–19, 1999, pp. 829–832.

Large-Scale Identification System Design

9

Herve Jarosz and Jean-Christophe Fondeur

The ability to identify people has been a constant preoccupation of different civilizations and, more recently, of independent and government organizations. The uniqueness and immutable features of fingerprints have been recognized since the dawn of humanity, drawings of fingerprints having been discovered in caves dating from the earliest ages. Early Egyptian potters used to mark their products with their own fingerprints, while Chinese people were using similar methods to authenticate land transfers and criminal confessions. Much later, the advent of modern technology and computers provided a means to efficiently automate biometric identification, and led to the first automated systems in the early 1960s. Early automated applications were primarily by law enforcement agencies, the aim being to identify criminals. In the past 15 years, many civil applications have begun to appear to guarantee the uniqueness of a delivered document – such as ID cards, driver's licenses – or a right – such as welfare, access to a country or corporate area, and on job applications. It is also increasingly common to incorporate biometric data directly into documents, allowing a check that the holder is the rightful owner. It is evident in this scheme that the contribution of the biometric system is essential because it both guarantees the uniqueness of the delivery – thus ensuring the fundamental principle: one person, one right – and checks the holder's identity. This trend has accelerated since the terrorist attacks of September 11, 2001.

Whilst early automated systems dealt with small parts of the population (typically fewer than a million people), they now have to address entire populations – typically tens of millions. It is legitimate to ask what levels of performance can be achieved on such large population databases. The aim of this chapter is to discuss the main difficulties associated with designing such a system and, more particularly, the problems related to feasibility and performance.

9.1 Introduction

9.1.1 Historical Background

The fingerprint image was one of the first biometric elements to attract the attention of scientists at the end of the 19th century. The significance of fingerprints was highlighted by the discovery of two important features: their **uniqueness** – even identical twins have different fingerprints – and **immutability** – fingerprint singularities remain constant throughout a lifetime. By the beginning of the 20th century, police forces such as Scotland Yard (UK) and the Federal Bureau of Investigation (USA), had begun to use fingerprints to solve crimes. The processes of forensic fingerprint card creation and management, and the associated manual search techniques used by police forces, were computerized in the 1960s when computer processing became more widely available. Because fingerprint image processing algorithms required a lot of computational power, the first Automatic Fingerprint Identification Systems (AFIS) were implemented through the use of specific hardware incorporated into the most powerful computers available at that time. With improvements in computer science and electronic engineering – especially the design of microprocessors – AFIS systems became increasingly powerful and cost-effective, thus offering new capabilities to users. The first large-scale identification systems were implemented by federal police forces (BKA in Germany; FBI in the USA), who were responsible for large populations across large geographical areas.

In the 1990s, new applications appeared for such systems in the issuance of documents associated with rights and privileges, such as ID cards, driver's licenses and social welfare documents. The major requirement was to ensure that the right or privilege allocated by the document was not delivered several times to the same person. Biometry was chosen as an efficient means to detect duplicate applicants in a large population. In such applications, biometric measures (e.g. fingerprints) of each applicant are captured at the time of application, and then searched against the biometric measures of the entire enrolled population. If the applicant is already in the database, this biometric measure will be detected and the enrollment will be rejected. Otherwise the right is granted and the applicant's biometric measure is inserted in the database.

More recently, other biometric techniques, such as iris and face recognition, have appeared and have been automated, and whilst they employ less mature technologies than fingerprints, they have a definite role to play in large-scale identification systems. Although a relatively recent technology, iris recognition has proved to be a very efficient and discriminating technology. It is therefore, from a performance point of view, very suitable for large systems, allowing for as high accuracy and throughput as fingerprinting. Another predominant technique, face recognition, has always presented a challenge for the pattern recognition scientific community. Although it has not reached the level of accuracy of fingerprint or iris technology, facial recognition can prove useful in the design of large systems. This is particularly true in applications where face images have already been collected – typically for certain identity documents. In these applications, face recognition can be used with reasonable accuracy without the need to re-enroll the whole population by capturing the fingerprint or iris image, and thus can be easily added to an existing system.

9.1.2 Large-Scale Identification Systems: Requirements and Basic Features

Large-scale biometric systems confront two different types of challenge: those common to all large, general-purpose information systems and those specific to biometric measurement and comparison. The severity of these challenges is such that very few companies have succeeded in successfully designing such systems. We can classify the main difficulties to be confronted in large-scale system design into the following topics:

- *Architecture issues*: distributed CPU, site availability, scalability, redundancy.
- *Administration issues*: logs, reports, user management, backup/restore.
- *Security issues*: integrity, confidentiality of data, virus protection.
- *Process issues*: multiple applications, legal issues, document issuance.
- *Operational issues*: operator training, tuning, maintenance.
- *Performance issues*: accuracy, throughput, response time.

The purpose of this chapter is not to discuss in detail the difficulties related to each of these topics but to focus more precisely on performance issues with respect to:

- Database size
- Desired transaction throughput
- Expected performance

The primary requirement of a large-scale system is the searching of submitted samples against very large databases of enrolled records. For example, the FBI's IAFIS includes more than forty million records of ten fingers each. The National Registration Department's AFIS in Malaysia includes eighteen million records of two fingers each. The Federal Ministry of Internal Affairs' AFIS in Nigeria will include sixty million records of six fingers each. The notion of a large system implies at least ten million registered persons. For this reason 1:1 comparison (verification) does not fall within the problem areas of large-scale identification systems.

Large systems also require high throughput. For example, the FBI's AFIS performs more than 50,000 searches daily, and the Nigerian AFIS is designed for 200,000 searches each day. A large system usually caters for a daily throughput of at least 10,000 searches. A definition of accuracy is provided in the lexicon at the end of this chapter. For large-scale systems employing human checks on potentially matching search–enrollment record pairs, accuracy measures the ability of the system to retrieve the correct record within the top k candidates from a database. The accuracy required is usually not lower than 95% for identification and more than 98% for verification. Addressing the feasibility of performance issues consists of determining the highest accuracy to be expected given average database size and average throughput.

9.2 Extrapolation of Accuracy

Several research studies have been conducted on performance evaluation and large-scale identification system design in the past few years and significant advances have been made on evaluation methods for biometric systems. This chapter focuses more specifically on development of very large-scale biometric systems. Such systems combine the difficulties inherent in large information systems and those associated with the use of biometrics. This chapter also tries to answer several questions that naturally come to mind when addressing large-scale identification systems. Firstly, is the system feasible (i.e. is it technically realistic to ask a system to perform daily hundreds of thousands of fingerprint comparisons against a database of tens or hundreds of millions of persons)? Secondly, what levels of performance can be expected, and most importantly, how can we measure or predict them? We do not aim here to propose a new theory or methodology but rather to present the issues and existing complementary methodologies, in order to see what can be achieved with these methods and how to avoid misinterpretation of the results.

9.2.1 Introduction

In order to estimate accuracy, specific tests (or benchmarks) must be conducted. They consist of creating a test scenario representing the final system, but on a smaller scale. Such a test uses two sets of data: the background database, simulating the system database, and the search database, simulating applicants. Large-scale system performance is generally measured on relatively small databases because of the expense of collecting large amounts of data. This is the case even for fingerprint technology, which has the largest existing biometric databases. Very few companies in the world can conduct benchmark tests on large databases. A typical evaluation consists of searching a few hundred fingerprints against a background database of several hundred thousand records. Even if this appears to be a large data set, this is still several orders of magnitude smaller than that of an operational large-scale system. An effective methodology must address two problems: choosing a data set (background and search) that is both statistically representative of the population and large enough to obtain reliable results. Then these results must be extrapolated to the size of proposed system.

In addressing the first issue, it is sometimes proposed to add synthetic data to the test set or to artificially increase the diversity of the test set by adding artificial noise, thus simulating more difficult acquisition environments, lower levels of user training and cooperation, or other adverse collection conditions. It is generally advised to avoid such methods for *scenario testing* (see [1]), as well as any modifications to the real data, because the resulting bias is almost impossible to predict or remove and makes the results very difficult to interpret. In this discussion we will assume that the data set used to evaluate the performance is statistically representative of the real environment being modeled, and we will therefore focus only on extrapolation of the performance.

Speed/throughput extrapolation will not be studied here (although this is a very critical element for feasibility), as it follows the same rules as any large system and is not specific to biometric systems. People knowledgeable in the design and tuning of such large systems know that the speed/accuracy issue must be addressed with extreme care. It is in particular very tempting and sometimes quite useful to do some quick and simple calculations to extrapolate speed and throughput. However, this can lead to false conclusions for several reasons: system overheads, nonlinearity of the behavior on small databases [17], or simply because a different tuning can be selected based on the size of the database. For these reasons, background test databases smaller that a few hundred thousand people are not suitable for reliable speed/throughput extrapolation.

We will now focus on the accuracy extrapolation problem, since it is specific to biometric systems. In other words, how do we predict accuracy on a very large database from measurements obtained on a small database?

The purpose of the extrapolation is to forecast the accuracy of a system with a large database from results obtained with a smaller database. We will in this section study in some detail the two ratios "False Match Rate" (FMR) and "False Non-Match Rate" (FNMR) (see Section A.3). Operationally, for an identification system, FMR corresponds to "false alarms" and needs to be checked by a human operator. It is a measure of the selectivity of the system: low FMR means high selectivity. FNMR corresponds to "successful fraud". It is a measure of the accuracy of the system. Low FNMR means high accuracy. Interdependence between these two rates is often illustrated by the ROC (Receiver Operating Characteristic; see Figure 9.1), which correlates accuracy with one of these error rates. The ROC directly pinpoints accuracy given the error rate. Each point corresponds to a different tuning (threshold) of the system.

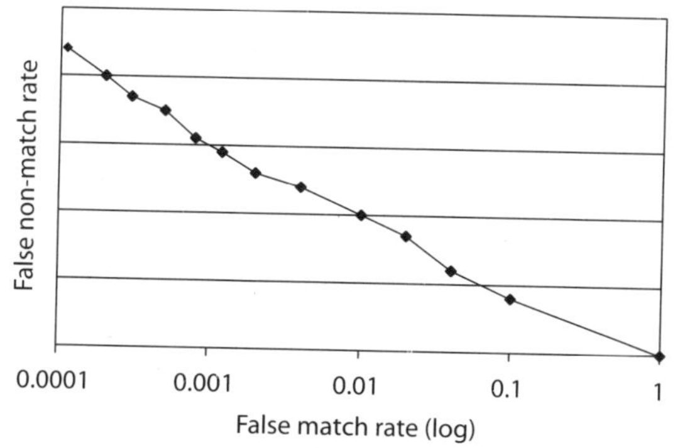

Figure 9.1 Example of ROC curve (Receiver Operating Characteristic). Each point of this curve is obtained for different values of the decision threshold for the same database.

9.2.2 Key Concepts

The extrapolation methods presented in this section are related to the identification process; i.e. searching for a record in a database of several million records. The search result is achieved by measuring the matching distance between the search record and each record in the database. A small distance between two records indicates that they are likely to have come from the same individual. The extrapolation methods are based on the study and analysis of the statistical distribution of this distance.

At this stage it is useful to introduce some mathematical definitions and concepts, and we will take here the formalism developed in [2]. If $d(s,t)$ is the distance between a sample s and a reference t, we will call $f_{\mathrm{gen}}(d)$ the genuine density function of d when both sample and reference come from the same person and $f_{\mathrm{imp}}(d)$ the impostor density function when sample and reference come from a different person. They usually present as in Figure 9.2.

The interesting part of the impostor distribution is its tail on low distances (we can call it the impostor tail), and the interesting part of the genuine distribution is its tail on high distances (we can call it the genuine tail). We will concentrate here on the impostor distribution rather than the genuine distribution for two reasons. First, as we will see later, it is this distribution that determines the behavior of very large systems. The other reason is that the impostor tail (smallest distances for non-matching pairs) follows a clean probability model ("bell curve"-like) and is therefore well analyzed through statistical tools, whereas the genuine tail (highest distances for matching pairs) contains an important component of noise on the data ("bell curve" + some random noise: scars for fingerprints, lenses/eyelids for iris, beards and glasses for faces etc.) and thus does not follow a simple model.

In the case of identification, it is useful to introduce explicitly the database size, and we can define, for a given threshold x, $\mathrm{FMR}_{1:1}(x)$ and $\mathrm{FNMR}_{1:1}(x)$ as:

$$\mathrm{FMR}_{1:1}(x) = \int_0^x f_{\mathrm{imp}}(u)\mathrm{d}u \tag{9.1}$$

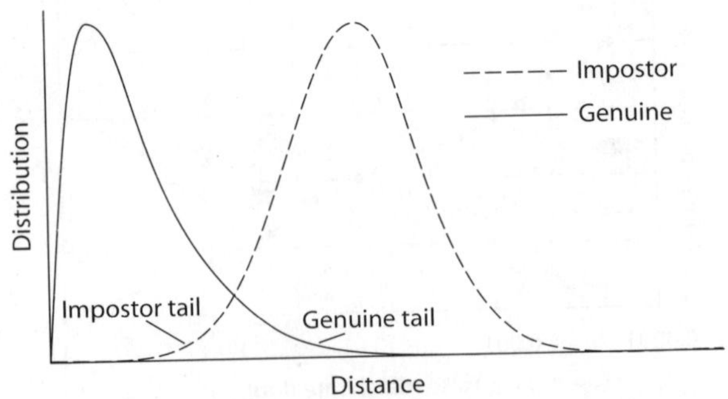

Figure 9.2 Typical genuine and impostor distribution. In an ideal system, these two distributions do not overlap.

and

$$\text{FNMR}_{1:1}(x) = \int_x^\infty f_{\text{gen}}(u)\mathrm{d}u = 1 - \int_0^x f_{\text{gen}}(u)\mathrm{d}u \tag{9.2}$$

$\text{FMR}_{1:1}(x)$ is the probability that one reference and one sample do not come from the same individual, although their matching distance is low (below the threshold x).

$\text{FNMR}_{1:1}(x)$ is the probability that one reference and one sample come from the same individual, although their matching distance is high (above the threshold x).

The index $_{1:1}$ means that one sample is compared to one reference (verification process).

We can now define the probability $\text{FMR}_{1:N}(x)$ that one sample matches one reference out of N but should not have: it is the probability that one of the (non-self) reference templates in the database has a distance below the threshold. We can also define its dual rate, the probability $\text{FNMR}_{1:N}(x)$ that one sample does not match one (self) reference out of a set of N references but should have: it is the probability that the sample and its matching reference have a distance above the threshold.

We will now present four classical methods for accuracy extrapolation, corresponding to different levels of complexity and assumptions:

- Method 1: Extrapolation from experiences
- Method 2: Identification as a succession of N verifications
- Method 3: Extrapolation with extreme value
- Method 4: Extrapolation when the distance can be modeled

9.2.3 Method 1: Extrapolation from Experiences

The first method is very empiric and makes few assumptions about the distance distributions. It just assumes that the tail of the impostor distribution is continuous and regularly variable, building a graph for accuracy against different sized databases as in Figure 9.3. This is done by searching

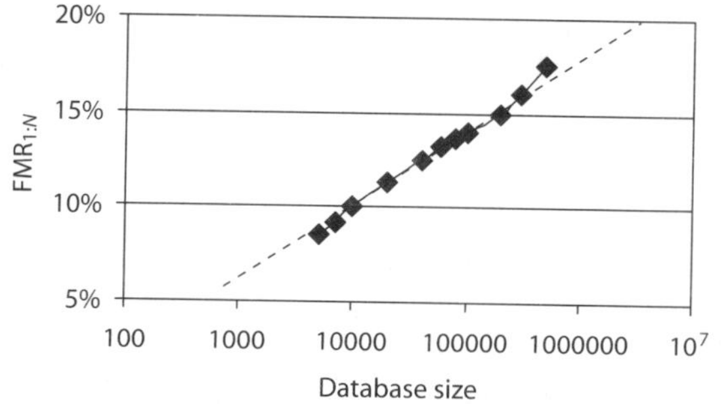

Figure 9.3 This graph shows an measure of accuracy for different database sizes. It can be fit by $\text{FMR}_{1:N} \approx \alpha \log(\mathit{size}) + \beta$.

a data set of fingerprints against different-sized databases of reference prints, and by fitting this graph with a model (linear regression for example). Of course, different biometrics or systems will follow different models in an appropriate representation space. This model can then be used to predict the accuracy (the probability that the correctly matching record will be returned as the top candidate) for larger databases. This generally works locally but may yield false conclusions if it is used for a database that is much larger than the test database. This method is usually reliable if the test set is at least one-tenth the size of the final database. Providing that a large representative set of data is available, this method affords reliable results.

9.2.4 Method 2: Identification as a Succession of *N* Verifications

In this section we make the simplifying hypothesis that an identification against a database of size N consists on N one-to-one independent tests. Under this hypothesis, $\mathrm{FMR}_{1:N}$ is given by:

$$\mathrm{FMR}_{1:N}(x) = 1 - (1 - \mathrm{FMR}_{1:1}(x))^N \tag{9.3}$$

Under this hypothesis, $\mathrm{FNMR}_{1:N}(x)$ does not depend on the database size:

$$\mathrm{FNMR}_{1:N}(x) = \mathrm{FNMR}_{1:1}(x) \tag{9.4}$$

It is then most interesting to use these equations to get a "pseudo-mathematical" feeling of the influence of the database size:

- Equation (9.3) shows that the database size N has a direct impact on $\mathrm{FMR}_{1:N}$: it correspond to the intuitive idea that in a large database the probability of finding a reference close to a given sample is higher than in a small database.
- Equation (9.4) shows no direct impact of the database size N on the FNMR. However, there is an indirect impact. It is most important to clearly understand this as our key point.

Let us take two systems, A and B. Their populations have the same genuine and impostor distributions. They use the same technology, characterized by $\mathrm{FMR}_{1:1}(x)$ and $\mathrm{FNMR}_{1:1}(x)$, but A has a larger database than B ($N_\mathrm{A} > N_\mathrm{B}$). Equation (9.3) implies that if they have the same threshold, $\mathrm{FMR}_{1:N}(x)_\mathrm{A} > \mathrm{FMR}_{1:N}(x)_\mathrm{B}$, which means that A would have more false alarms than B. If we want both systems to have the same selectivity (number of false alarms), the threshold of the system A must be set at a lower value than the threshold of system B ($x_\mathrm{A} < x_\mathrm{B}$), given by:

$$\begin{aligned}\mathrm{FMR}_{1:N}(x_\mathrm{A})_\mathrm{A} &= 1 - (1 - \mathrm{FMR}_{1:1}(x_\mathrm{A}))^{N_\mathrm{A}} \\ &= 1 - (1 - \mathrm{FMR}_{1:1}(x_\mathrm{B}))^{N_\mathrm{B}} = \mathrm{FMR}_{1:N}(x_\mathrm{B})_\mathrm{B}\end{aligned} \tag{9.5}$$

or

$$(1-\mathrm{FMR}_{1:1}(x_{\mathrm{A}}))^{N_{\mathrm{A}}} = (1-\mathrm{FMR}_{1:1}(x_{\mathrm{B}}))^{N_{\mathrm{B}}} \tag{9.6}$$

and thus

$$\begin{aligned}\mathrm{FNMR}_{1:1}(x_{\mathrm{A}})_{\mathrm{A}} &= \int_{x_{\mathrm{A}}}^{\infty} f_{\mathrm{gen}}(u)\mathrm{d}u \\ &= \int_{x_{\mathrm{A}}}^{x_{\mathrm{B}}} f_{\mathrm{gen}}(u)\mathrm{d}u + \int_{x_{\mathrm{B}}}^{\infty} f_{\mathrm{gen}}(u)\mathrm{d}u \\ &= \int_{x_{\mathrm{A}}}^{x_{\mathrm{B}}} f_{\mathrm{gen}}(u)\mathrm{d}u + \mathrm{FNMR}_{1:1}(x_{\mathrm{B}})_{\mathrm{B}}\end{aligned} \tag{9.7}$$

Through Equation (9.4),

$$\mathrm{FNMR}_{1:N}(x_{\mathrm{A}})_{\mathrm{A}} > \mathrm{FNMR}_{1:N}(x_{\mathrm{B}})_{\mathrm{B}} \tag{9.8}$$

This brief analysis illustrates the mechanism of performance extrapolation: first, understand the effect of database size N on FMR, then find a rule of variation of the threshold $x(N)$ to compensate for this effect, and finally measure the effect of this threshold $x(N)$ change on the FNMR.

Another short analysis of these equations leads to the conclusion that the key is the dependency of the threshold x on the database size N: we have seen (Equation 9.3) that $\mathrm{FMR}_{1:N}(x) = 1-(1-\mathrm{FMR}_{1:1}(x))^N$. If x is not a function of N, then

$$\lim_{N\to\infty} \mathrm{FMR}_{1:N}(x) = 1$$

which means that when N becomes very large, the selectivity decreases towards zero. It is therefore necessary to have the threshold x be a function of N to be able to maintain the selectivity.

We can now use Equations (9.3) and (9.4) to get the first extrapolation method for system accuracy ($\mathrm{FMR}_{1:N}$ and $\mathrm{FNMR}_{1:N}$) on a large database of N references, knowing the performance obtained in verification ($\mathrm{FMR}_{1:1}(x)$ and $\mathrm{FNMR}_{1:1}(x)$)

Table 9.1 shows that the $\mathrm{FMR}_{1:1}$ which leads to a constant $\mathrm{FMR}_{1:N}$ depends on the database size.

Such small rates need a large enough database to be estimated with small confidence intervals. Some guidelines exist to help determine the number of required trials to get a relevant estimation of the probability of error – e.g. the "rule of 30" (see [3]), which says that accurate estimation is obtained if the test set is large enough to measure at least 30 errors (false matches). With N samples, estimation of FMR can be done on $N(N-1)/2$ tests (or fewer to avoid correlation between tests). The application of this

Table 9.1 This table shows the value of $FMR_{1:1}$ necessary to obtain a fixed $FMR_{1:N}$ for different sizes of database. $FMR_{1:1}$ must decrease in an exponential way.

N	$FMR_{1:1} \rightarrow$	$FMR_{1:N}$
1	10^{-2}	1.0%
10^2	10^{-4}	1.0%
10^3	10^{-5}	1.0%
10^4	10^{-6}	1.0%
10^5	10^{-7}	1.0%
10^6	10^{-8}	1.0%
10^7	10^{-9}	1.0%
10^8	10^{-10}	1.0%

rule shows that the size of the test required to estimate precisely $FMR_{1:1}$ is of the order of $(60 / FMR_{1:1})^{1/2}$. If $N = 10^8$, $FMR_{1:1} = 10^{-10}$ and 8×10^5 samples are required.

In practice, accuracy and throughput are both improved by implementing a system workflow more sophisticated than just N one-to-one comparisons. This extrapolation method then becomes more difficult to apply. This is the case when the database is partitioned based on "level one" fingerprint classification (such as loop, arch or whorl), or when more than one biometric is used (two fingers etc.) or when the identification process consists of a pipeline of different algorithms, or when the decision algorithm is more sophisticated than a simple threshold. In these cases, the previous model appears to be too simple and does not provide accurate estimation. Some studies [18] interestingly tried to go further in the modeling of such systems – binning, multi-fingers etc. – and provide refined equations taking into account these parameters. These equations improve the accuracy of the estimate and are most useful in understanding the influence of all parameters. However, it is difficult to produce a model that takes into account all of the parameters and the complexity of real systems – especially the correlations between parameters (see [19]).

9.2.5 Method 3: Extrapolation with Extreme Value

The problem of estimation of tails of distributions from only a few observed values has been addressed by mathematicians outside the context of biometrics. For example, in the design of structures such as bridges or ships, natural phenomena such as wind velocity or wave impact pressure must be taken into account. Physical models do not always exist to help forecast the maximum potential of such phenomena. A statistical approach may offer a simple way to identify such extremes. For example, when engineers design a bridge whose life expectancy must exceed 100 years, they need to know the maximum wind velocity which might be expected over this period, even though they may have available only 25 or 50 years of records (see [4]). Extreme value theory proposes a method for estimating the probability of such maximum wind velocity – which could certainly be greater than all speeds observed in the available data. Such values are called Peaks Over Threshold (POT). If the provision of 25 years of wind velocity records is to be sufficient in forecasting the maximum velocity for the next hundred years, it is advisable to take into account all highest observed values – meaning the maximum speed and the values closest to this maximum, i.e. the extreme values. These values only occur in a few instances or in a small part of the data used to estimate the density function. Consequently, they only concern its tail.

The extreme value theory models the tail of the impostor distribution function:

$$F(x) = \mathrm{FMR}_{1:1}(x) = \int_0^x f_{\mathrm{imp}}(u)\mathrm{d}u \tag{9.9}$$

The function $F(x)$ is approximated by the empirical distribution function $F_N(x)$ estimated for N independent distances $X_1, \ldots, X_N$:

$$F_N(x) = \frac{1}{N}\sum_{i=1}^{N} 1_{X_i \le x} \underset{N\to\infty}{\to} F(x) \tag{9.10}$$

$F_N(x)$ is the measured percentage of impostor distances which are smaller than x.

In order to determine the approximation of the curve $F(x)$, let us consider N independent variables $X_1, \ldots, X_N$ measuring distances obtained for sample–reference impostor pairs. Then, recent results show that (see [6]):

$$\text{for } x > u, \frac{1 - F_N(x+u)}{1 - F_N(u)} \underset{N\to\infty}{\to} \mathrm{GPD}(x) = \begin{cases} \left(1 + \dfrac{cx}{a}\right)^{-1/c} & \text{if } 1 + \dfrac{cx}{a} > 0 \\ \exp(-x/a) & \text{otherwise} \end{cases} \tag{9.11}$$

GPD means the Generalized Pareto Distribution, F is defined in Equation (9.9), N is still the database size, a and c are parameters, and u corresponds to the beginning of the tail. Expressions for determining the parameters are given in Section A.4.2.

Figure 9.4 shows the link between the approach presented in the previous section and this one. The step curve $F_N(x)$ (Figure 9.4(a)) represents the best approximation of $\mathrm{FMR}_{1:1}(x)$ that the previous approach allows, whereas the continuous curve of Figure 9.4(b) represents the smooth approximation from extreme value theory. Since we are interested in the inverse of the function $\mathrm{FMR}_{1:1}(x)$ in order to determine the best decision threshold x, the continuous curve in Figure 9.4(b) gives better results (i.e more accurate and more stable) than the empiric distribution function $F_N(x)$ (Figure 9.4(a)).

Table 9.2 compares the estimation of the decision threshold using method 2 and method 3 (extreme value). Method 3 returns a precise value,

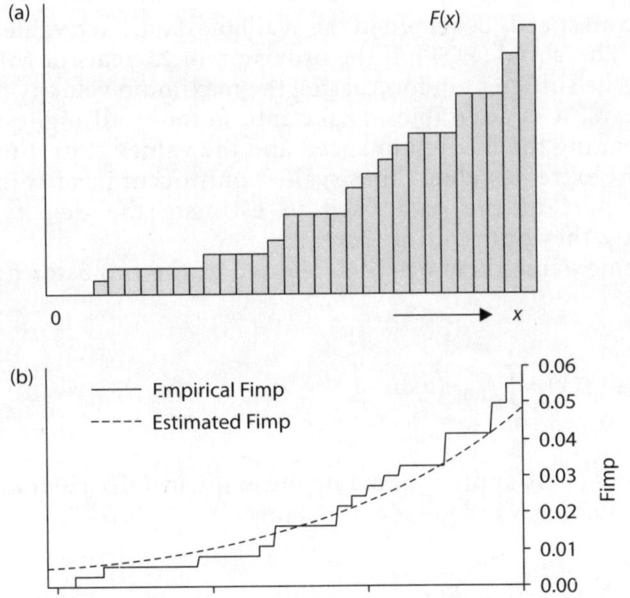

Figure 9.4 Graph (a) gives $\mathrm{FMR}_{1:1}(x)$ for small values of x computed on a database with a high but finite number of references ($F = F_N$). Graph (b) gives the estimation of the tail of $F(x) = \mathrm{FMR}_{1:1}(x)$ using the theory of extreme values.

whereas method 2 cannot do better than an interval whose size grows when N increases. Figure 9.5 shows how the values in Table 9.2 are obtained.

Furthermore, the interval given by the inverse empirical distribution F_{imp} is delimited by two distances that result from the match of two sample–reference pairs. The estimation of F_{imp} by the extreme value method incorporates several values, which makes it much more reliable. Many discussions still continue about the samples that should be used for the parameters estimation, the convergence rate, and the intervals of confidence (see [7]). The field of extreme value still attracts much research attention and has its own scientific journal named *Extremes*. There also

Table 9.2 Comparison of the estimation of the decision threshold with a simple extrapolation (Section 9.2.3) and the extreme value (Section 9.2.4).

	Simple extrapolation		Extreme value	
N	**Threshold x**	**$\mathrm{FMR}_{1:N}$**	**Threshold x**	**$\mathrm{FMR}_{1:N}$**
10	0.50	1.0%	0.50	1.0%
1,000	0.48	1.0%	0.48	1.0%
10,000	in [0.43; 0.45]	1.0%	0.44	1.0%
100,000	in [0.36; 0.40]	1.0%	0.38	1.0%
1,000,000	in [0.26; 0.33]	1.0%	0.30	1.0%
10,000,000	in [0.12; 0.25]	1.0%	0.20	1.0%
100,000,000	in [0; 0.12]	1.0%	0.08	1.0%

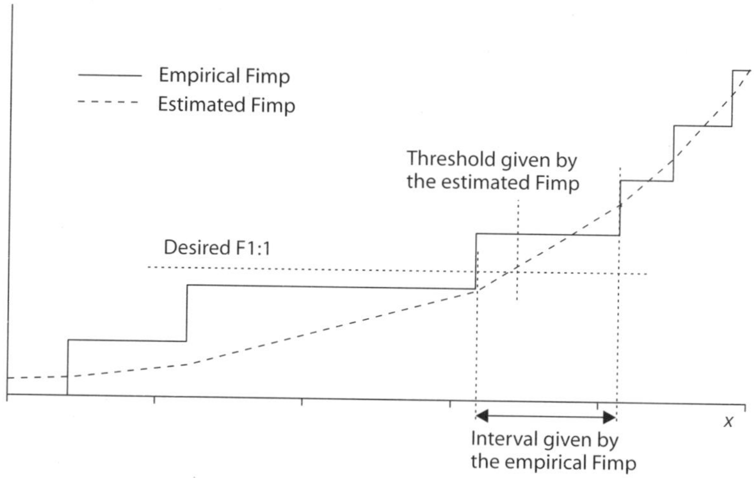

Figure 9.5 The decision threshold is obtained by computing the inverse function of F_{imp}. This operation is possible for the smoothed curve obtained with the extreme value theory, but only leads to an interval with the empirical F_{imp}, which is a step curve.

exist methods to directly compute the quantiles of $FMR_{1:N}$ (see [10]) or provide some intervals of confidence (see [11]).

This method gives the best estimate of the accuracy possible without postulating an analytic mathematical expression for the impostor density function.

9.2.6 Method 4: Extrapolation when the Distance Can Be Modeled

Until now, we have supposed that genuine and impostor distributions were known only by their empirical estimations. However, in some cases, the distribution of the matching distance between samples and references can be modeled. In such cases, the mathematical expression for the impostor density function is known and can be used for extrapolation purposes.

This is the case for iris recognition as developed by J. Daugman [12]. The first step in this process consists of coding an iris by a sequence of 2048 bits. The matching distance between two irises then becomes the number of non-matching bits in corresponding places in the two sequences. The matching distance is the sum of 2048 Bernoulli trials with the parameter p, the probability that any pair of corresponding bits in the two irises are dissimilar. If these random variables were independent, the matching distance would follow a binomial distribution with parameters $(2048, p)$. In reality, the trials are correlated, but correlated Bernoulli trials can still be approximated by a binomial distribution of parameters (n, p) (see [14]) where n is the number of independent variables or the number of degrees of freedom (see Table 9.3):

Table 9.3 Results extracted from [12] with $p = 0.5$ and $n = 244$.

N	$FMR_{1:1}$	$FMR_{1:N}$	Decision threshold
1	10^{-2}	1.0%	0.426
10^2	10^{-4}	1.0%	0.381
10^3	10^{-5}	1.0%	0.365
10^4	10^{-6}	1.0%	0.348
10^5	10^{-7}	1.0%	0.336
10^6	10^{-8}	1.0%	0.324

$$\text{for all } x \in [0,1],\ \text{FMR}_{1:1}(x) = \int_0^x f_{\text{imp}}(u)\mathrm{d}u = \sum_{k=0}^{[nx]} C_n^k p^k (1-p)^{n-k} \tag{9.12}$$

The second example concerns fingerprint identification (see [13]). In this paper a very simple model of fingerprint comparison is provided. A sample and a reference are considered as equal when they share at least 12 minutiae of the 36 possible. Two minutiae match when their positions and directions are similar. The assumption that minutiae are uniformly distributed inside a fingerprint leads to an expression for the probability that there are exactly p corresponding minutiae between the sample and reference. This case is analogous to the previous example for iris recognition, except that the matching score is the number of common minutiae. This gives an exact expression for the impostor distribution under the assumptions taken, corresponding to a very simple matching algorithm.

9.2.7 Influence of Classification

In a large-scale system, optimization of resources and throughput are two key issues for especially for the matching subsystem. In such an architecture, when the number of transactions per day becomes important and when the size database is very large, it is often necessary to speed up the matching process. This issue was also relevant before search automation, when comparison was performed manually and the efficiency of the system was limited by the number of samples that could be compared daily by an operator. The impact of the use of "level one" classification (loop, arch, whorl or the like) on large-scale systems has been widely discussed by several authors: does classification simplify the problem of large systems by reducing the apparent size of the database compared? Does classification reduce the overall accuracy of the system? Can we expect some accuracy improvement at the matching stage because an identification no longer requires as many comparisons as references in the database? We will discuss these issues pragmatically by considering the system as a "black box" and looking at the impact of classification on both the error rates and throughput.

Generally the goal of classification is not to improve the accuracy of the system by restricting the search space but to augment the throughput. In fact, the overall identification process can be modeled as a succession of filters, each reducing the size of the input list in order to arrive ultimately at a candidate list or even a single candidate. Classification is one these filters and is characterized by a filtering rate. The accuracy of the associated filtering process (i.e. the probability of correct print classifications) and the filtering (or "penetration") rate (i.e. the part of the database actually searched) are the two most important characteristics of any classification method and they are linked: increasing the number of database records returned after the filtering process (thus increasing the filtering rate) reduces the risk of dropping the right record (if contained in the database) from the list. At the extremes, if the filtering process returns the entire database, the Miss Rate due to the classification process will be zero, but the filtering rate will be 100%, i.e. there will be no throughput gains from the filtering process. Obviously, this would be a particularly inefficient classification method. A good classification method must provide a high degree of accuracy and a low filtering rate.

Today, among the different biometric technologies, only fingerprinting generally uses classification. This usage was inherited from past experts who observed and modeled fingerprints based on visual pattern analysis. Different pattern classification algorithms or techniques were developed historically, such as the Henry classification or the Vucetich classification schemes. The main advantage of these techniques is that they can be used either manually or automatically. However, their drawbacks are the relative complexity of the rules followed by the experts to reference the fingerprint image and the non-homogeneity of the class distribution (cf. Table 9.4).

The total number of classes may be increased and the efficiency refined when the classifications of several fingers are combined together. Table 9.5 shows that a 10-finger classification based on only five classes per finger appears to be powerful and efficient since the partition space now contains around 1 million classes.

However, because of the non-uniformity of the class distribution and the class correlation between fingers, the efficiency of this approach is reduced. For instance the 10-finger class "RRRRRLLLL" (where R and L stand for Right and Left loops and the position refers to the finger number starting from right thumb up to left little finger) contains around 6% of the

Table 9.4 Classification of fingerprints into seven pattern classes (computed on the FBI's databases see [15]).

Pattern type	Left thumb (%)	Left index (%)	Right thumb (%)	Right index (%)
Arch	3.01	6.09	5.19	6.29
Tented arch	0.40	7.72	0.58	7.96
Right loop	51.26	36.41	0.63	16.48
Left loop	0.46	16.96	58.44	39.00
Whorl	44.77	32.45	35.04	29.93
Scar	0.03	0.17	0.04	0.14
Amp	0.07	0.20	0.09	0.20

Table 9.5 Number of possible classes in relation with the number of fingers.

One finger	Two fingers	Ten fingers
5	$5^2 = 25$	$5^{10} = 9{,}765{,}625$

population, whereas the class "LLLLLRRRRR" is almost empty. In order to measure the efficiency of a classification, we introduce the definition of filtering rate or penetration rate.

Let us assume that the database is divided into C separated [2] classes and the ith class contains n_i references, the penetration rate p or filtering rate is estimated by:

$$p = \sum_{i=1}^{C} \left(\frac{n_i}{n}\right)^2 \quad \text{where } n = \sum_{i=1}^{C} n_i \tag{9.13}$$

If we assume that the class distributions of the samples is the same as the class distributions for the references stored in the database, then the penetration rate gives the average proportion of references having the same classification as the sample. Thus the system throughput is proportional to $1/p$. Table 9.6 shows that the filtering rate of the 10-finger pattern-based classification is significantly higher (and therefore less efficient) than the filtering rate of an ideal classification that would equally divide the space in homogenous classes.

Some techniques exist to refine the pattern-based approach, such as the sub-level pattern classification and the use of ridge count between cores and deltas. However, the classifications presented here are not the only way to achieve the objective. Other approaches continue to be the subject of research such as that presented by Lumini *et al.* [16] or the one developed by SAGEM. These offer as an alternative to the previous exclusive classifications an approach that is called continuous. Their use also depends on the type of the system (forensic or civil) and the nature of the data: fingerprint image or latent print. The many features used to describe a fingerprint (1680 in [16]) are reduced to five using the Karhunen–Loève transform based on a principal component analysis. In this small vector space, it is easier to extract from the database all references that are close to the fingerprint being identified. In fact, for a given threshold ρ, we compute a subset $S_\rho(F)$ of fingerprints in the database for which the matching distance to the

Table 9.6 Effect of the correlation of pattern distribution between finger on the filtering rate (see also [19]).

	Typical filtering rate Ten fingers (pattern-based, five classes, manual classification)	Theoretical filtering rate Ten fingers (five classes, Assuming homogeneous distribution and no correlation between fingers)
P	0.02	0.1024×10^{-7}

search print F is below ρ. $S_\rho(F)$ can be considered as the continuous class of F. This system is parametrized by ρ whose value is a compromise between speed and low error rate. Its value should be small to obtain small subsets $S_\rho(F)$, leading to a high matcher throughput. It should be high to have a low error rate to avoid having the reference corresponding to F fall outside $S_\rho(F)$. As for the previous static classification approach, the continuous approach is characterized by its filtering rate.. However, this continuous processing is more complex and Equation (9.13) cannot be applied.

In conclusion, classification is the primary way to reduce the number of required comparisons in a database search for the identification of a fingerprint. We include here all means to achieve this task, and whatever the algorithm, the common feature is a compromise between increased throughput and decreased accuracy. Tuning a large-scale system is fundamentally a compromise between throughput and accuracy, and classification is one way to achieve this compromise.

Classification usually does not improve the accuracy of the system because the comparisons avoided by the screening process would also have been rejected by the matcher. Use of classification concentrates the matcher effort on the reference fingerprints that have a shape close to the search fingerprint. Finding an optimal classification approach not only depends upon the desired filtering rate, but also on the available data. For example, classification of a sample can involve the images of multiple fingers (as many as 10) from the same individual record, even though the final match comparison is done on fewer fingers. The impact on the accuracy of the system by the filtering process is no longer so obvious. At this level of analysis, classification is just another parameter of the matching subsystem – a trick used inside the system to accelerate the search. Classification allows the introduction of the extra information contained in the fingerprints available in an individual multiple-finger record that are not used directly in the matching process, thus improving the final accuracy of the search.

Table 9.7 summarizes the different cases and the possible impact of these classifications on the system.

9.3 Conclusion

We have seen that extrapolation of performance is a difficult problem and must be considered very carefully. In particular, it is easy to misuse equations, models and mathematics to arrive at wrong conclusions. We have tried to present here the dangers of simple extrapolation and have given four methods to overcome them. No method is 100% reliable and we strongly recommend the use – and consolidation – of at least two of them to arrive at a robust outcome.

Table 9.8 lists the advantages and disadvantages of the four methods.

If the available databases are large enough, the first method should be employed. The second method is often used for a first approximation or to

Table 9.7 The possible impact of classifications on different systems.

Classification method	Impact on performance	Advantages	Drawbacks	Note
Pattern-based approach	Increase system throughput Does not decrease the overall accuracy of the system Accuracy improvement is possible in some specific cases: Latent images, more fingers used during the classification than during the matching etc.	Compatible with human rules (an operator can manually classify low quality fingerprint images or latent images)	Filtering rate limited by the *a priori* definition of fingerprint classes Difficult to implement and follow human rules	Mostly used in forensic systems Usually required when latent images are managed in the system
Continuous approach		Small filtering rate	Manual classification is not possible	Mostly used in civil systems

Table 9.8 Advantages and disadvantages of the four methods.

Method	Advantages	Disadvantages
1. Extrapolation from experiences	Results are very accurate if test conditions are close to real ones.	It requires large databases and extrapolation of the results becomes uncertain when carried out far away from the test conditions.
2. Simple extrapolation	Very simple to compute, it quickly provides an indication of the performance that, which can be reached by the system.	This method is not very precise when the number of samples is not sufficient to estimate the tail of distribution functions (principally impostor).
3. Extrapolation with extreme value	More precise than the previous method when the number of samples is not sufficient to estimate the tail of distribution functions.	The distribution functions must verify some theoretical hypothesis (see [6]).
4. Extrapolation when the distance can be modeled	This method provides the most accurate estimations.	The distribution function of the matching distance must be known. This only happens in particular cases. Some approximations are generally required in order to model the matching distance.

give a general trend. Extreme value is generally used to get a more precise estimation, and method 4 is used when the distribution is explicitly known.

We have also seen that classification is a useful tool to address large-scale system design. It can improve both the system throughput (when used as a tuning parameter of the matching process) and the overall system accuracy (when adding extra information from prints not used directly in the matching process).

The techniques presented here have been used to design and tune large-scale systems which are operational today. These systems efficiently manage several tens of millions people and throughput rates as high as several tens of thousands of searches per day. Such experiences show that even larger biometric systems (hundreds of millions) are feasible. In the future, the combination of several biometric techniques will further improve the accuracy and performance of such large-scale systems.

Appendix

A.1 Estimation of $FMR_{1:N}(x)$ Using Extreme Values

Let $X_1, \ldots, X_N$ be N independent and identically distributed (i.i.d) random variables whose distribution function is F. We denote $Y^N = \max\{X_1, \ldots, X_N\}$. If $F(x)$ is not a function of N, we have:

$$\lim_{N\to\infty} [F(x)]^N = \begin{cases} 0 & \text{if } F(x)=0 \\ 1 & \text{otherwise} \end{cases} \tag{9.14}$$

This result is not interesting, but it could be possible to find a series (a_N, b_N) such as:

$$\text{for all } x, \lim_{N\to\infty} [F(a_N x + b_N)]^N = G(x) > 0 \tag{9.15}$$

The function G is still unknown, so to find it let us denote $X_1, \ldots, X_{Nn}$ as a set of random i.i.d. variables and $Y_i^N = \max\{X_{in}, \ldots, X_{in+n-1}\}$; then:

$$Y^N = \max\{Y_1^N, \ldots, Y_N^N\} = \max\{X_1, \ldots, X_{Nn}\} \tag{9.16}$$

This implies that when N goes to the infinity the function G verifies:

$$G(a_N x + b_N) = [G(x)]^N \tag{9.17}$$

The form of the functions able to follow this equality is:

$$G(x) = \exp(-\exp(-x)) \text{ or } G(x) = \exp(-x^{\alpha}) \tag{9.18}$$

This leads to the approximation that the distribution function of the variable $\max\{X_1,\ldots, X_N\}$ asymptotically tends to:

$$\text{for } x > u, G(x) = \begin{cases} \exp(-(1-kx/\sigma)^{1/k}) & \text{if } 1-kx/\sigma > 0 \\ \exp(-\exp(-x/\sigma)) & \text{otherwise} \end{cases} \tag{9.19}$$

A.2 Use of Extreme Value in a Concrete Case

We still assume that $\{X_1,\ldots, X_n\}$ is a set of i.i.d. variables distributed according to the distribution function F. The previous section considered the maximum of these variables. In this section, we focus not only on this particular value, but also on all values which are close to the maximum, or more precisely, the peaks over threshold (POT). The theory of extreme value shows [6] that if F lies in the domain of GPD (Generalized Pareto Distribution; see Equation (9.10), then:

$$\text{for } x > u, \frac{1-F_N(x+u)}{1-F_N(u)} \underset{N\to\infty}{\to} \text{GPD}(x) = \begin{cases} \left(1+\dfrac{cx}{a}\right)^{-1/c} & \text{if } 1+\dfrac{cx}{a} > 0 \\ \exp(-x/a) & \text{otherwise} \end{cases} \tag{9.20}$$

where

$$F_N(x) = \sum_{i=1}^{N} 1_{X_i \le x}, F_N(x)$$

is the empirical distribution function of the impostor distance.

If $c < 0$, the tail is bounded, and unbounded in the contrary case. Parameters a and c are estimated with the peaks over the threshold u. The threshold u must be chosen low enough to keep enough peaks to make the estimation relevant, but high enough to keep only peaks. The estimation of both parameters a and c needs to sort the set $\{X_1,\ldots, X_n\}$ by increasing order. This ordered set is denoted as $\{X_{1,n},\ldots, X_{n,n}\}$. First, we choose k such that

(9.21)

The following expression [4] gives the estimated values for a and c:

$$c = c_1 + c_2 \text{ and } a = u\frac{M_n^1}{\rho} \text{ with } \rho = \begin{cases} 1 & \text{if } c \ge 0 \\ 1/(1-c) & \text{otherwise} \end{cases} \tag{9.22}$$

$$\text{where } c_1 = M_n^1 \text{ and } c_2 = 1 - \frac{1}{2}\left(1 - \frac{(M_n^1)^2}{M_n^2}\right)^{-1} \tag{9.23}$$

Table 9.9 The parameters of Equation (9.11) for different values of u.

u	12	13	14
a	1.500	1.622	1.823
c	0.126	−0.017	−0.332
Points used to make the estimation	22	12	6

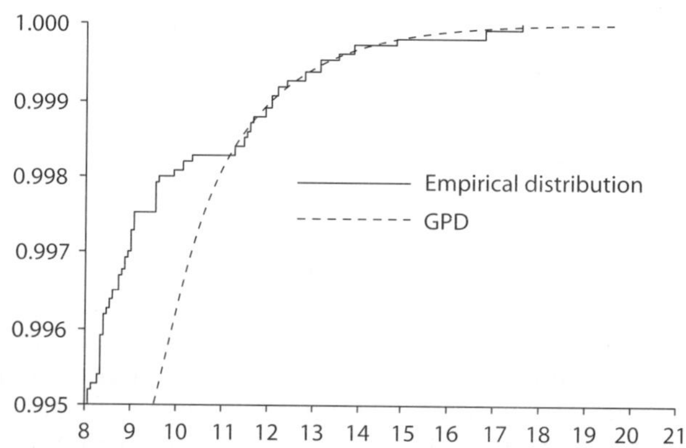

Figure 9.6 The graph shows the tail of the impostor distribution of the distance and its approximation (dotted line) by the method of extreme value for $u = 13, a = 1.622, c = -0.017$.

$$\text{and } M_n^r = \frac{1}{k}\sum_{i=1}^{k}(\log(X_{n-i+1,n}) - \log(X_{n-k,n}))^r \tag{9.24}$$

The parameters a and c are generally estimated for different values of the threshold u and chosen as the average of all the obtained results.

These results are applied here to a particular case: 10,000 matching distances are computing between 10,000 different sample-reference pairs. The matching distance lies in the interval $[0, 17.6]$ (see Figure 9.6). The parameters of Equation (9.11) follow for different values of u: see Table 9.9.

With the estimation of the parameters a and c, Equation (9.11) gives a continuous relation between $FMR_{1:1}(x)$ which is used to find two thresholds:

- The threshold for which the error rate $\text{FMR}_{1:N}$ is below 1% with $N = 10^6$.
- The threshold for which the error rate $\text{FMR}_{10:N}$ is below 1% with $N = 10^6$.
- Both thresholds are estimated on the 10,000 random variables and compared to the true thresholds (Table 9.10).

These values could not have been found without the extreme value theory because they are superior to 17.6, which is the maximum reached by the 10,000 variables used to make the extrapolation.

Table 9.10 Comparison of estimated and true thresholds.

$N = 10^6$	First *x* verifying $FMR_{1:N}(x) < 1\%$	First *x* verifying $FMR_{10:N}(x) < 1\%$
True thresholds	32.9	21.2
Thresholds estimated by extreme value	29.6	20.9
True rate FNMRk:N(x) for the estimated threshold	5.5%	2.8%

A.3 Glossary

Accuracy

Accuracy is related to the quality of service provided by the system, as the user sees it. For a 1:1 comparison(verification), accuracy is measured by FMR or FNMR. For a 1:*N* comparison (identification), accuracy depends upon the type of application. For access control, accuracy is provided by both FMR and FNMR, as for 1:1 comparison. For forensic applications, however, the aim is to retrieve the relevant candidate when the applicant is already in the database, even if a drawback is to often retrieve false candidates. In this case the visual verification step efficiently filters the false candidates and, consequently, the FMR is not very important. Then accuracy concentrates on the "miss" cases and is computed as (1 – FNMR).

Biometric Data

Biometric information acquired from an individual. It is typically a set of one or several fingerprints, irises, face images etc. Most systems today use several data items from the same technology, typically two or more fingerprints, or two irises. To further increase performance, multi-biometric systems are being researched (e.g. combining fingerprint and iris or face).

Classification

Classification is a mathematical technique aimed at organizing *a priori* a set of data in consistent subsets or categories – named classes. A search record is then compared only to the reference records belonging to the same "class", and not to the whole database.

Extrapolation

Since extensive testing is often not possible, or too expensive, before completion of the system design, the issue of estimating system behavior from limited, off-line tests is important. Inference of full system behavior based on such limited test results is referred to as extrapolation.

Failure to Enroll (FTE) Rate

The failure to enroll (FTE) rate is the ratio of persons for which enrollment has not been possible. This typically happens when the quality of the biometric data is too poor.

False Match Rate (FMR)

For a 1:1 comparison, the False Match Rate (in this case identical to the False Acceptance Rate) is the ratio of searches that result in the system accepting an applicant who should have been rejected. For the 1:*N* comparison, the FMR is a measure of the "selectivity" of the system, i.e. the ratio of searches where at least one candidate is found by the matching subsystem whereas the person was not in the database.

False Non-Match Rate (FNMR)

For a 1:1 comparison, the False Non-Match Rate (in this case identical to the False Rejection Rate) is the ratio of searches that result in the system rejecting an applicant who should have been accepted. For the 1:*N* comparison, the FNMR is the rate of "miss" cases, i.e. the ratio of searches where the applicant was not found and retrieved as a candidate by the matching subsystem, whereas the applicant was already in the database.

Filtering Rate (FR)

This notion relates to limiting the portion of the database to be searched based on some criteria. This is the first step for performing a 1:*N* search. The FR is the percentage of the reference database actually compared to the search record. The lower the FR, the more effective is the filtering approach (provided that the accuracy is constant).

Identification and Verification

The 1:1 comparison (Verification) is based on an absolute threshold, the value of which is tuned according to the type of application and to accuracy expectations. The 1:*N* comparison (identification) is based on filtering, sorting and dynamically adjusted thresholds. It should be noted that the larger the reference database, the more effective are the dynamic thresholds.

Matching Subsystem and Matching Distance

A matching subsystem performs a comparison of a search record against a database of reference records. For each pair of search–reference records, it computes a matching distance correlated to the probability that these two records come from the same person. Reference records are then sorted according to that distance and the matching subsystem returns the list of the "closest" records.

Reference Record

A reference record is an individual record – including biometric data – that forms part of the data already registered in the database.

Search Record

A search record is an individual record – including biometric data – that has been acquired for comparison to data already registered.

Throughput

Throughput is directly related to the system's capacity for processing searches in a given amount of time. Usually, throughput is defined as the number of queries processed per day, a query being a 1:*N* identification search based on two fingerprints or a latent search for crime-solving purposes etc.

References

[1] A. J. Mansfield and J. L. Wayman, *Best Practices in Testing and Reporting Performance of Biometric Devices*. Available online at: `http://www.cesg.gov.uk/site/ast/biometrics/media/BestPractice.pdf`.

[2] J. L. Wayman, Error rate equations for the general biometric system. *IEEE Automation and Robotics Magazine*, **6**(1), 35–48, 1999.

[3] G. R. Doddington, M. A. Prybocki, A. F. Martin and D. A. Reynolds, The NIST speaker recognition evaluation: overview methodology, systems, results, perspective. *Speech Communication*, **31**(2–3), 225–254, 2000.

[4] M. D. Pandey, P. H. A. J. M. Van Gelder and J. K. Vrijling, The estimation of extreme quantiles of wind velocity using L-moments in the peaks-over-threshold approach. *Structural Safety*, **23**, 179–192, 2001.

[5] J. E. Hefferman and J. A. Tawn, Extreme value analysis of a large designed experiment: a case study in bulk carrier safety. *Extremes*, **4**(4), 359–378, 2001.

[6] J. Pickand III, Statistical inference using extreme order statistics. *Ann. Statistics*, **3**(1), 119–131, 1975.

[7] P. Hall and I. Weissman, On the estimation of extreme tail probabilities. *Ann. Statistics*, **25**(3), 1311–1326, 1997.

[8] E. J. Gumbel, *Statistics of Extremes*. Columbia University Press, 1960.

[9] J. L. Wayman, Confidence interval and test size estimation for biometric data. *Proc. IEEE AutoID Conference*, 1999.

[10] H. Joe, Estimation of quantile of the maximum of *N* observations. *Biometrika*, **74**, 357–354, 1987.

[11] L. Peng and A. H. Welsh, Robust estimation of the generalized Pareto distribution, *Extremes*, **4**(1), 53–65, 2001.

[12] J. Daugman, Biometric decision landscapes. *Technical Report No. TR482*, University of Cambridge Computer Laboratory, 2000.

[13] S. Pankanti, S. Prabhakar and A. K. Jain, On the individuality of fingerprints. *IEEE Trans. Pattern Analysis and Machine Intelligence*, **24**(8), 1010–1025, 2002. A shorter version also appears in *Fingerprint Whorld*, pp. 150–159, July 2002.

[14] R. Viveros, K. Balasubramanian and N. Balakrisnan, Binomial and negative binomial analogues under correlated Bernoulli trials. *Am. Stat.*, **48**(3), 243–247, 1984.
[15] F. Torpey, *Unpublished Report* using data extracted from the FBI's Identification Division Automated Services database of 22,000,000 human-classified fingerprint record. Mitre Corporation, 1995.
[16] A. Lumini, D. Maio and D. Maltoni, Continuous versus exclusive classification for fingerprints retrieval. *Pattern Recognition Letters*, **18**, 1027–1034, 1997.
[17] J. L. Wayman, Continuing controversy over the technical feasibility of large-scale systems. *National Biometric Test Center, Collected Work*, 2000. San Jose State University.
[18] J. L. Wayman, Error rate equations for the general biometric system. *National Biometric Test Center, Collected Work*, 2000. San Jose State University.
[19] J. L. Wayman, Multi-finger penetration rate and ROC variability for automatic fingerprint identification systems. *National Biometric Test Center, Collected Work*, 2000. San Jose State University.

Biometric System Integration 10

Julian Ashbourn

Much of the literature on biometric *system* design (as opposed to algorithmic, hardware or test design) has focused on system error rates and scaling equations [1–5]. In this chapter, I would like to step back a bit and ask "What are the operational goals and requirements of the system?" It sounds obvious, but before one can design and implement a successful biometric system, one must have a crystal-clear view of the requirements and how the operational processes will fit within the existing business processes and structures [6,7]. It is tempting to get carried away with the technology and think of all the things we *could* do with biometrics and related technologies, when really we should concentrate on what we *need* to do in order to satisfy the requirements at hand, identifying how and where biometrics will be used to support the overall objective.

10.1 Understanding, Describing and Documenting the Requirements

The first step then, is to identify and clearly describe the problem. If we are considering the use of biometrics, then part of the problem presumably lies in personal identity verification. It would help also to check our understanding of the current process for identity verification (if one exists) and to document this accordingly. We can then examine this process in some detail and identify precisely where and how we can integrate biometrics, or indeed, redesign the process from the ground up if required. We shall assume for the purposes of this chapter that a sound business case has already been made for the adoption of biometrics and any supporting technologies required for the application. I stress again how important it is to start off with a clearly identified and documented requirement. It is surprisingly easy to lose sight of the original objective when evaluating interesting technologies, and end up with a cleverly designed system which fails to deliver the fundamental requirements.

Another item to consider in this context is exactly how we describe and document those requirements. Naturally this is important if we intend to issue an RFI (Request for Information) or RFP (Request for Proposal) to systems integrators and/or biometric vendors to help us with our application.

One perennial problem in this respect is that end users are obviously close to their own operation and will describe the situation in a manner which makes sense to them, but which may not be quite so clear to the person developing the application. Add a systems analyst or consultant into the mix as an intermediary, and you have a wonderful opportunity for misinterpretation, which fate usually exploits to the fullest. This situation may be further compounded by each entity speaking their own special dialect and associated jargon according to where they sit. What is needed here is a clearly understood methodology which can be used by all concerned to describe both operational process and fundamental system design, in an unambiguous manner which can be understood and used throughout every stage of the project. Such methodologies are usually referred to as modeling languages, and there are existing techniques available within the broader IT sphere which can be used to good effect in order to describe your biometric requirement and subsequent system design [8, 9]. Indeed, you may already be using such a methodology within your organization. This would allow you to clearly describe and document every aspect of the application, with minimum fuss and in a manner which is easily understood by all concerned. In addition, the information you produce at this stage can be used throughout the project and beyond for training purposes and subsequent support and maintenance.

In understanding and documenting the requirements, it is also important that you capture all relevant views without making assumptions around either existing processes or the envisaged acceptance of the proposed application. Too often, biometric system integrators have proceeded without a thorough understanding of complex and well-developed existing processes. For example, if you are considering a system for prisons, then ensure that you closely involve the prison officers themselves, and understand how they work, how existing systems work and what their concerns are. If you are considering a system for airline check-in or immigration control, then work with the check-in clerks, immigration officers and local systems support personnel and understand the relevant issues in their operational environment. If you are considering a system for banks, speak with some local branch staff and understand the sort of problems they encounter within their day-to-day activities. If you are considering a system for network access control, work with the system administrators and understand how they maintain and use the systems architecture. And most important of all, speak to the prospective users. When you have spoken to the users, speak to some more users. Time spent gathering and understanding all the relative views at this stage will be time well spent, as it may help you to avoid blind alleys as you design and configure your processes for the new application.

We have labored the point a little, but it is absolutely essential that we start off on the right foot with a crystal-clear understanding of the requirements which can be equally clearly articulated and documented, forming a solid foundation for future progress as we design the processes and systems architecture for our new biometric application. It is equally important that we understand the views of those who will be operating and using the system. In all probability, we will learn much from discussion with these individuals, which might usefully be incorporated in our subsequent design.

When we have attended to all these requirements, we shall be ready to consider the operational processes for our biometric application. It is as well to think through these processes, even at a high level, before we decide upon the most suitable biometric methodology. Naturally, we shall have to revisit them later in order to flesh out the detail, but understanding the fundamental processes at this stage, and being able to collate them into the bigger picture, will prove invaluable. This is an area where the use of an appropriate modeling language will prove extremely useful, as it will help us to understand the logical flow of activities throughout the system. Once we have captured the top-level operational processes and thought through conceptually how our system is going to function, we can turn our attention to thinking about the most suitable biometric methodology for the task.

10.2 Choosing the Technology

Whenever people gather together to talk about biometrics, the discussion invariably turns towards the question of performance and to what extent one methodology has an advantage over another. Indeed, I would be quite happy to have a dollar for every time someone has asked me "Which biometric is best?" The answer of course is simple. There is no absolute "best" biometric. Some methodologies have an advantage in accuracy. Some methodologies have an advantage in usability. Some methodologies have an advantage in ease of integration. Similarly, different vendor offerings within the same methodology will often exhibit quite different operational characteristics. This is especially true with fingerprint readers, where even the same OEM sensor may be partnered with different software to deliver different functionality and sometimes different performance characteristics.

There is of course an optimum choice of methodology and device for a given application, but how on earth is one supposed to choose the right combination? Or indeed, should our application be device-independent, or even methodology independent, allowing for transparency or multiple choice biometrics? There are various views on the latter, but unless we perceive a distinct advantage in using dual or multiple biometrics, then it is probably best to stick to one methodology initially and get the system properly designed and working optimally with the chosen technology. The lessons learned from such an exercise will stand you in good stead for any future development.

In terms of where to start, the best place is probably to undertake a little research of your own into the characteristics of different available systems and speak with others in your industry sector who have implemented biometric applications in order to learn from their experience. The choice of methodology may almost be made for you by the very nature of the application you are considering. For example, in a law enforcement-related application, you may be predisposed towards fingerprints in order to interface with related systems. For call center-related applications, you will naturally veer towards voice verification, and so on. If there is no strikingly obvious methodology for your particular application, then your choice will be primarily around usability, resilience, performance and technology "track record" in similar applications. Performance in this context is about providing a suitable level of performance for the application at hand, rather than necessarily looking for the highest available performance. Usability should in most cases be the primary driver, coupled with a suitable level of resilience and robustness according to the environment in which the system will be deployed. This may seem like the opposite prioritization from that employed with certain initiatives, but the recommendation is based upon sound precepts and experience. If the final deployed system is unintuitive in use, or temperamental in its reliability, then absolute performance considerations will become academic as users struggle with the system, in turn causing real performance to plummet, regardless of any impressive-sounding specifications quoted by the device vendor. There is a distinction here between theoretical device performance and total end-to-end realized system performance. The former may be demonstrable in isolation, but the latter will, in reality, be influenced by a host of variables associated with the user, the environment and other considerations.

In understanding and outlining our operational processes, we would have looked at items such as transaction flow rate, user populations and the consequences of incorrect matching (i.e. false matches and false non-matches). From this we can derive an acceptable performance level, in terms of both accuracy and transaction speed. Add this to our expectations of reliability and resilience, and we have the basis for a specification to compare against available products. However, just having a disconnected specification is not enough. We must place it into context by clearly defining the process into which the technology must fit and ultimately support. In this respect, it would be beneficial to use a modeling language which will not only help to describe this situation but will also bring clarity and consistency throughout the whole project life cycle. With a defined process and qualified performance requirements, choosing the technology becomes a logical exercise rather than a game of chance, and this point is stressed accordingly.

Of course, the above needs to be undertaken with due regard to the underlying systems infrastructure and with a clear view of the biometric authentication architecture. At this stage, one should ensure that the products and systems under evaluation offer a flexible approach to architecture, especially with regard to template storage and management. If a given device can only store templates internally, then its application will be limited to scenarios where this is acceptable. If a device and accompanying software allows central template management and storage, but only in a proprietary fashion, then it may not fit with your organization's directory aspirations. We might also consider the security of data passed between device and host node/PC as well as that passed across the broader network, and consider what encryption options are offered. These are the sort of considerations which must be borne in mind while we get down to the fine detail of individual device performance.

Having defined the requirement, thought through the processes, decided upon a suitable systems architecture and settled on a particular biometric methodology, we can now evaluate the available device options in line with our requirements. With certain methodologies, these options will be few, whilst with others (such as fingerprints) the choice may at first seem rather bewildering. However, we can quickly whittle this down to a short list by eliminating those devices which cannot meet our architectural requirements, or which have obvious shortcomings in other areas. When we have our short list, we shall naturally want to receive as much pertinent information as possible from the vendors in order to understand the different approaches taken. We should also ask for details of reference sites and not be shy to make contact with the organizations concerned in order to appreciate their experiences of using the devices and associated systems. This activity may whittle our short list down even further, until we are left with a selection of available devices, any of which might fulfill our requirements. Subject to additional considerations around company stability, device cost etc., we might usefully ask for sample devices to evaluate in-house as a next step.

Evaluating devices is a time-consuming activity which needs careful administration if we are to arrive at meaningful conclusions. Some guidance is given in [10]. The first step is to clearly define our objectives and to set up a test environment accordingly. This test environment should mirror as closely as possible the real-world operating conditions of our planned implementation. We should then identify and confirm our test user population, again ensuring that the user profile distribution matches as closely as possible to the anticipated real-world situation. It will be important to ensure that users are enrolled consistently into the test systems and that they understand the operation of the various devices under test. In this respect, it may be useful to construct an enrollment script (possibly for each device) which may be followed in each case, irrespective of the individual administering the enrollment. The test population will then need to be advised how many times to use the devices over a defined test period, and what to do in the case of failed transactions and other errors. If the devices being tested have an adjustable matching threshold, then this will need to be set at an equivalent level for each device. It would be useful to involve the manufacturers in such an evaluation, ensuring that their respective devices are configured properly and that the results are representative. We can then proceed with the test and log the results accordingly.

Having undertaken the evaluation of our short listed devices, we then need to analyze the results and understand how they might relate to real-world operating conditions (system evaluation and testing is covered in greater detail elsewhere in this book and the reader is advised to digest that information thoroughly before planning a test project). Finally, we shall arrive at conclusions which suggest the most suitable device for our planned system. However, this is not quite the end of the story as we also need to ensure that the associated software is going to integrate seamlessly into our own infrastructure. It may be necessary to write additional interfaces or software modules and we need to clarify exactly who will do this and how they will be tested. Manufacturer support will be important in this context, and may indeed be a contributing factor in your final choice of device.

10.3 Application Development

Application development may be thought of in two broad and related categories: firstly, the core application, which is based upon the overall operational objective, such as a benefits payment system, corporate network access system and so on; and secondly, the integration of the biometric authentication methodology. The two are irrevocably entwined and should be considered together, although the detailed requirements from the developer's position are somewhat different. Depending on the precise nature of the application, it may be that you need two different sets of core skills within the developers – one developer (or team) which really understands the core requirement and all of the associated business logic, and another developer (or team) who really understands authentication, directory structures and networks. Together, the two developers (or teams) will be able to create an efficient, scalable application which will serve well into the foreseeable future. If a suitable application modeling language has been used to define the original requirements and associated business processes, then it will pay dividends at this point, as the development team can work closely to these documented requirements and further produce their own more detailed configuration maps with which to document the development project for archive and standards compatibility purposes.

Taking the first of these tasks, it is important to ensure that the core application is based upon practical and realizable operational processes. It is recommended that an iterative approach be taken to application development in order to regularly check progress against the original requirement with the "customer" within the end-user organization. Such progress checks have benefits in both directions, as the development team ensures that it is on track to deliver the requirement and the customer team ensures that the originally conceived processes are realistic and deliverable. Again, if a modeling language has been used, such discussions will be streamlined and efficient and may be captured within additions to the original documentation where appropriate.

Part of developing the core application will be understanding at which points biometric identity authentication is required and how best to incorporate this functionality. This brings us to the second application development area, which we might usefully refer to as the authentication service. This service will be based upon the necessary architecture, directory, biometric matching engine, data protocols and device operability required to deliver a biometric authentication "result" to the core application and log the transaction accordingly. When this result is received, rules within the application will be followed according to the result, which may entail the provision of certain functionality or the display of related messages to the user and system operator.

Understanding the required functionality and the implications of delivering it will of course be different for different types of application. For example, within a corporate network access system, we may be interested in supplying single sign-on functionality. Within a border control system, we may be interested in supplying an interactive user environment with high levels of automation. Within a benefits payment system, we may be interested in supplying different operator consoles according to role. Within a time and attendance system, we may be interested in automated feeds to accounts systems – and so on. It is important that we focus on this core functionality and ensure that the application is designed to efficiently deliver the core requirements. This in turn requires consideration of the underlying infrastructure and whether enhancements are required in order to support the new application. For example, is our network bandwidth adequate to support the anticipated data traffic at peak periods? Do we have adequate transaction storage capacity? What is our contingency for network or server failure? Do we need to configure periodic batch operations for transaction archiving? These fundamental areas need to be addressed and satisfied before we overlay the biometric authentication functionality. If the network is slow, for example, this might be viewed by the user as poor performance of the biometric functionality, when in reality it has nothing to do with the biometric device or the matching engine, but is a consequence of poor systems performance elsewhere. As is often remarked, the system is only as good as its weakest link, and we need to ensure a consistent level of performance throughout, from the underlying infrastructure, through the core application, to the user interface.

Overall, we should not be thinking in terms of a "biometric application" but rather an application which happens to incorporate biometric authentication functionality. The biometric element is simply a common service in the same manner as directory services. If the application is poorly conceived and inefficient, it will remain poorly conceived and inefficient, whether or not biometric functionality is incorporated. We therefore need to think in terms of a high-performance, efficient, elegantly coded and properly tested core application as our fundamental baseline. Into this, we shall incorporate a well-conceived biometric authentication model which adds minimal operational overhead and interfaces cleanly with synergistic infrastructural components. The whole system should naturally be vigorously tested under real world conditions and properly documented before being rolled out as an operational system. There will of course be attendant user training issues, and it is anticipated that an associated training (and where applicable, marketing) program will be undertaken as part of the overall project.

With regard to application development tools and environments, there currently exists a wide choice of methodologies, from traditional languages such as C and C++, to more recently popular tools such as Delphi,

Java, the recently announced Microsoft C# language and recently established Borland Kylix. There are a couple of points to understand here, especially around the deployment architecture and host machine platforms. If it is not necessary to have cross-platform compatibility within your overall application (for example use on different hosts such as Windows, Unix or Linux) then you may or may not wish to use a "portable" language such as Java. Similarly, your application may or may not be designed as a web-enabled application based upon front-end browser technology, in which case the choice of tools you use for the "presentation layer" will be affected. Lastly, you may or may not need to interface to back-end data sources which rely upon conventional legacy technology and which will therefore require the necessary interfaces. Why mention this? There is a tendency to assume that all current development should be web-based and use associated tools accordingly. However, depending upon the existing architecture, there may be some compatibility and related performance issues involved here. Whilst it is true that many enterprises currently base their networks around TCP/IP technology, there is no unwritten law which states that you must use a web browser as the user interface, linked to a web server, which in turn links to an application server and so on. In some instances, it may make sense to simply design the application for the best performance around the back-end data sources and servers, using native drivers and interfaces wherever possible and designing the user interface around the most sensible and intuitive user processes, rather than trying to make it fit a contemporary browser model. Naturally, it is a question of "horses for courses" and incorporating the longer term view, but these are the sorts of issue which need to be carefully considered.

10.4 Integration with Existing System Architecture

One element of a biometric project which is often not attended to at the right time (and sometimes overlooked altogether!) is how the biometric functionality will integrate into the underlying systems architecture which supports the enterprise in general. This is an important consideration, even if the biometric application is conceived as a totally self-contained entity, with its own access control lists and integral matching engine. In such a case (which would be unusual) we would still need to understand where the application is going to sit (on which server), what the capacity and performance requirements are, and how this deployment will impact the network. It is far more likely that we shall wish to interface to existing components such as a central directory and separate the template matching process from the main application. In this case, we shall need to consider what sits where, what the data exchange protocols are and how best to deploy the application elements for best overall performance. The key to this is to involve the network administrators at an early stage in the project and ensure that the system is designed in accordance with the overall network policy and will sit well within the infrastructure.

An example of where this is important is in the potential interface with an organizational directory. For example, if the directory is based upon Lightweight Directory Access Protocol (LDAP) [11], can the biometric authentication components interface with this? Can the biometric templates produced by the chosen methodology be stored within the directory? What is the directory schema, and do you need to add other fields for the application to use? Who in the organization manages the directory, and what links exist to other applications which may be affected? Naturally, we need to consider these details carefully and understand how all this works in order to ensure that our application will be designed and developed in the optimum manner.

We must also decide where the biometric matching engine will sit, how it uses directory information and how it integrates with the core application. Similarly, as we pass biometric data across the network, is this data encrypted? If so, where do the encryption and decryption take place and what impact does this have in terms of transactional performance overhead? When we understand how all this works, we shall be well placed to design our system correctly. These issues should be addressed in association with our internal network administrators and the biometric vendor in order to ensure a complete understanding.

In conclusion, there is a tendency for biometric vendors to simply state "this is how our system works" and expect that the customer organization will blindly follow this model. Such a model may or may not be sensible for a given organization, depending upon the existing infrastructure and how other organizational systems are configured. The more flexible the biometric product is, the easier it will be to integrate into a given application and existing infrastructure. This factor should be one of the first to be addressed in our consideration of biometric methodologies and products. Clearly, if we do not fully understand the architectural and integration issues right up front, then we are opening the door for potential confusion and misunderstanding further down the line.

10.5 Templates and Enrollment Management

The biometric authentication function depends upon individual biometric templates. The better quality the reference template, the better chance of correctly matching against a live sample. It is therefore important to capture the best possible reference templates from the user population and ensure that users provide live samples with high levels of consistency. What has this got to do with systems design issues, you may ask? Well, it is important to provide an intuitive and flexible enrollment process and to decide upon what is acceptable as an enrollment. We have probably already decided upon the logistics of storing and handling templates from an overall systems perspective. Now we need to consider how to obtain the best possible reference templates. In the majority of cases, user enrollment will be a facilitated process whereby an administrator will explain the procedure to the user and oversee the actual enrollment. But how does this administrator know what a good enrollment is? Clearly, we need to guide them. Part of this can be accomplished by a good quality training package for administrators, resulting in them becoming "expert" users and being able to impart this knowledge to others at the time of enrollment. But we also need to help them by providing the best possible system tools.

The first point here is the user interface itself. This needs to be attractive, uncluttered and intuitive. It also needs to provide a logical sequence of events and clear feedback messages for both the administrator and user where applicable. There are also some desirable functionality requirements such as:

- The ability to roll back a transaction and start again.
- The provision of a reference metric as to the quality of the captured sample.
- The ability to immediately undertake a match against the reference template and provide a "score" accordingly.
- The ability to search the existing database for similar templates which would indicate prior enrollment.
- The ability to create a log of enrollments, both within a given session and overall.
- The ability to capture associated information, such as a photograph.
- The inclusion of comprehensive online help for the administrator.

Having provided such an interface and associated tools, the administrator needs to know how to interpret what he or she is seeing at the time of enrollment. For example, if the reference metric for a captured sample is below a certain quality threshold, then the administrator should cancel the enrollment and start again, perhaps after providing further explanation and guidance to the user. The system can of course help in this context by providing suitable messages according to the result, advising the administrator on required actions.

Similarly, once the user has enrolled and is undertaking the first live match, the administrator may, depending upon the matching score received, advise the user to undertake further transactions before leaving the enrollment station. Attention to detail paid at this stage will pay dividends later, under real operating conditions, where the quality of results will be directly pertinent to the quality of reference template and the users' ability to give consistent live samples. The enrolling systems administrator thus plays a key part in the success of your project overall and should be properly trained and supported accordingly.

But what if you are not implementing facilitated enrollment and are requiring users to enrol themselves, perhaps remotely across the network? The importance of the user interface and the enrollment process now becomes critical. We must provide crystal-clear feedback at each stage which leaves the user in no doubt as to their enrolled status. If we are undertaking background processes, such as checking for the existence of a similar template, then we must decide upon which messages we return to the user under which circumstances. We shall also need to supply comprehensive online help for the user as well as a "primer" tutorial to ensure that they understand the proper usage of the system prior to enrolling. Indeed, we may decide to actually inhibit the enrollment wizard until the tutorial has been successfully undertaken. In such a scenario, we shall also have to consider the credentials of the remote enrollee. How do we know that the identity being claimed is correct and that the person sitting at the remote PC or terminal is who we thought it was? This may be less of an issue within a closed loop corporate system where we can exercise an element of control, but will be difficult when dealing with outsiders. For example, if we were considering the idea of individuals enrolling into a voice verification system remotely via the telephone, then what sort of checks should we impose to confirm identity? If an identity is stolen in relation to our system, then what is the revocation procedure and who is responsible for any fraudulent transactions? There are some significant issues to consider with regard to remote enrollment into biometric authentication systems. There is also the question of covert enrollment and subsequent authentication. The author does not condone such a methodology and recognizes that there are some huge systems-related issues associated with it.

Another area for consideration is the security of templates in association with the application. Templates are directly related to individuals and in most people's view are thus regarded as personal data. Whilst the letter of the law in certain countries is unclear on this point, it might be hard to argue that a biometric template is not personal data. The argument that you cannot derive a person's name from a template may be viewed as irrelevant in some circumstances. The upshot of this is that, to an organization, there is a moral, if not legal, responsibility to take measures to safeguard biometric data with regard to its storage and transmission between system components on the network. Thus the overall system design must reflect this.

Of course, you may decide not to store templates centrally, but instead to store the template on a portable token such as a chip card, which the user may then carry with them and therefore be responsible for. In such a scenario, the user would insert their card into a reader, initiate the comparison of the presented biometric sample to the template stored on the card, and then complete the desired transaction. One immediate question is, "Where exactly is the matching process being undertaken?". If it is being undertaken right on the chip card, with the template never leaving the card, then we can simply focus on the security aspects of the card itself. If, however, the template is being passed from the card to the host system for matching, then how are we protecting the template in transit? The issue is exacerbated if we are using a centrally located matching engine. These are issues which we must not only understand, but for which we need to define a clear policy. A user might understandably ask what is happening to their personal template information within the system, and we should be able to provide an unambiguous, clearly worded answer which leaves no doubt as to the use and storage of this data. Being open and forthright on this issue is very important and will become more so as the use of biometric authentication increases.

Hopefully, this section has impressed upon the reader the need to carefully consider the creation and management of biometric templates. It is a topic which cannot be over-emphasized. Problems in this area have the potential to not only severely affect overall systems performance, but may also lead to complex user relationship issues. It is suggested that organizations contemplating the use of biometric authentication within their processes and systems, pay particular attention to templates and develop a clear policy around their storage and use.

10.6 Understanding User Psychology

What is the key component within a biometric authentication model? Is it the reader? The user interface? The matching algorithm? The host application? Or is it perhaps the *user*? Considering the importance of users and how they interact with the system, they seem to receive relatively little attention from some biometric systems designers. Without users, the whole thing after all becomes somewhat academic. Users are more than mere entities which interact with the system. They are thinking, breathing, individuals who have subjective views and react to events in an individual manner. They also come in all shapes and sizes and have differing levels of dexterity when using devices. Furthermore, they have differing levels of interest in technology and correspondingly different perceptions around the use of emerging technologies such as biometrics. Accommodating such diversities within the user population requires some thought.

Let us consider for a moment the scenario wherein a user has used the system successfully a number of times, but now encounters a problem. Let us assume that this is a bank ATM machine from which the user legitimately wishes to withdraw some money, and that there is a queue of three people behind the user, all waiting to use the machine. How will the user react? There is no single answer. In fact, there may be almost as many answers as there are users. A rational user who has an understanding of technology and a pragmatic disposition may react very calmly, take the time to read and digest any feedback messages and perhaps try again or seek manual assistance according to the situation. An irrational user who has no sympathy towards technology and a quick temper may become quite agitated and perhaps even violent towards the machine and generally abusive. There are a host of variables between these extremes and we must accommodate them all in one way or another.

One of the important considerations in this respect is the provision of an elegant fallback procedure should an error occur, ensuring that the user is not unnecessarily inconvenienced or disenfranchised. Naturally this should be accompanied by a clear explanation and step-by-step instructions for the user to follow in order to deliver the service. Of course, it may be that the user in this case is an impostor, attempting to claim someone else's identity and withdraw money. If this were the case, then the individual, on being refused, would probably wish to exit discreetly from the situation as quickly as possible. It is the genuine user who, for whatever reason, is having difficulty with the biometric authentication, that we need to think carefully about. Some users, for example, will be inconsistent in the way they offer their live biometric. Even with a matching threshold set very low so as not to reject valid users, these individuals may still have difficulty. How we handle such a situation is very important. It is no good saying "Well, it works for everybody else" if it does not seem to be working for this particular individual. It would also be inappropriate to jump to the conclusion that the user is not using the system properly. There may be a perfectly valid, but unforeseen, reason why the template is not being matched. Even if the problem is one of user inconsistency, we need to find a way to gently suggest that they would benefit from a little further instruction. A trained administrator will have no difficulty in handling such a situation, providing of course that the administrator is present. In an unattended situation, we need to rely on system prompts and interactive feedback to guide the user, and this requires careful consideration and design if we are to retain user attention and associated loyalty towards the scheme.

Understanding user psychology is also extremely important when designing the operating environment. First of all, the environment itself must be perceived as an attractive and welcoming one, within which the user may interact with the system in an unhurried and personal manner. There are many related issues here, from color schemes, lighting and signage to available space and relative privacy. If the user is carrying bags, or is perhaps escorting an infant, then they need sufficient space in which to operate the device. Similarly, if they are having difficulty with the procedure, or are perhaps just unfamiliar due to non-habitual usage, then they may not wish to be too conspicuous whilst undertaking their transaction and will require some space and time accordingly. It is important that users do not feel under pressure in this respect and that they are able to concentrate on undertaking the transaction within a comfortable environment. The likelihood for transactional errors is in some ways proportional to the pressure a user feels under whilst making the transaction. Errors lead to more pressure, which leads to a higher propensity towards errors, and so on. This sounds obvious, but it is a factor often overlooked in the design of biometric pilot schemes and even implemented systems.

Another area where understanding user psychology is important is in the initial acceptance of the system. If users feel that their views and wishes have not been taken into account, or if they simply haven't been consulted at all, then there may exist a level of reluctance to use the system in situations where use is optional. Similarly, if communication has been poor and questions around data security are not perceived to have been properly addressed, then this can also have an impact. Obviously, there is also the issue of benefits for the user should they decide to use the system. If such benefits are vague and ambiguous, or if the major benefit is perceived to sit squarely with the system operator rather than the user, then enthusiasm towards the system will be capped. Within the context of a new application initiative, the last thing we need is to start off on the wrong foot with a lukewarm response from prospective users. It is therefore important that we consider the application from a user perspective, identify some real user benefits and ensure these are incorporated in practice. It is then important to market the system enthusiastically, taking care to clearly address any user concerns and to be open and honest about the system architecture and data privacy issues. Even in an application where users are absolutely required to use the system, such as a prison or benefits payment application for example, it will pay dividends to spend some time understanding the user psychology of the target user population and designing the system to be as intuitive and friendly as possible. It will also pay dividends to configure and implement a proper education and training program to ensure that users understand the system and how to use it effectively.

The last point to make about user psychology is the affect it can have on realized system performance. When people are considering biometric devices, they tend to draw attention to quoted performance figures which may differ within fractions of a per cent between different vendor specifications (how these specifications were arrived at is another matter entirely). This is ironic in a way, as the biometric device performance is simply one factor within many which will affect overall system performance. As already noted, the back-end database, network performance and reliance upon associated components will all contribute to the overall perceived transactional performance. Undoubtedly major contributors towards transactional performance are the users themselves and how they personally interact with the system. If you are concerned about published error rates of around one per cent or less, allow me to place this in perspective with the suggestion that errors arising from varying user psychology and interactive response may be measured in several tens of per cent in some cases. In overall throughput terms, this may have a significant affect on the system and its perceived success. It therefore makes sense to fully consider user psychology and all the user-related issues as an integral part of the biometric project. Allow me to stress once more that this is non-trivial. Sufficient time and resource should be dedicated to this area if we are to realize a successful implementation.

10.7 Fine Tuning the System

As with any system, the original implementation may require a little subsequent fine tuning in order to realize the best performance. With respect to biometric systems, there are a number of ways to effect such a tune-up as the system is introduced and we start to analyze transactions and overall performance. Some of these may be simple hardware-related configuration adjustments, some may revolve around software code tweaking and some may be related to users and ongoing user training.

Perhaps the most obvious tuning is setting the threshold for the matching algorithm. Many systems have a range of tolerance in this respect, within which we may bias the system towards the likelihood of increased false matches or false non-matches accordingly. Why should we be concerned with this? It may initially be considered that we simply want the highest accuracy available from the system (rejecting impostors with a very high probability), but adjusting the matching threshold for this characteristic may result in unacceptable levels of false non-matches, with valid users being rejected by the system. Under the assumption that genuine users greatly outnumber the occasional impostor, we should generally wish to lower the probability of false non-matches so as to reduce the total number of instances of rejected customers, especially in the early days of operation. This will allow users to operate the system successfully and build a level of confidence accordingly. Later on, we may wish to tighten the threshold a little in order to achieve the best compromise between the likelihood of false matches and absolute number of false non-matches. In most systems, we will never be able to know how many impostors are ultimately successful in deceiving the system. Estimation of that number will require both knowledge of the probability of a false match and a good guess at the number of impostors attempting access.

The point about rejecting valid users (false rejection) is an important one. Naturally, in a commercial situation, such as an application for bank ATM usage, it will be vitally important not to reject a valid and valued user unnecessarily, as repeated instances could lead to customer dissatisfaction and perhaps even brand switching. It is perhaps no less important in a non-commercial situation, where users might become confused as to the correct operation of the system if they are continually rejected by it, necessitating a higher than necessary degree of support and administration, not to mention the effect on user psychology and individual perception of the application. It is acknowledged that there are some who would argue the opposite case: that they do not mind how often valid users are rejected, as long as impostors are not accepted. Naturally, there are certain high-security applications where this might seem like a reasonable approach, although one is inclined to suggest that it is all relative to the implications of operating at such levels and what procedures are in place to support such an operation, in terms of enhanced training, administration support and so on. Even in such extreme cases, there is probably an optimum threshold setting which strikes an acceptable balance between usability and security, whilst acknowledging the risks associated with opportunist impostor attempts as opposed to more determined attempts to compromise the system. Such a risk assessment is outside the scope of this chapter and would in any case be particular to the application under consideration, but one might suggest that this is another area where understanding user psychology plays an important part.

Adjusting the biometric matching threshold represents an opportunity for fine tuning the overall system, providing that this facility is available with your chosen biometric device and systems infrastructure. Other systems-related areas exist for fine tuning the overall network performance, which can be monitored in conventional ways in order to understand peak load windows, available bandwidth and so on, as well as identifying any bottlenecks due to specific components or the use of common services. This analysis is best left to network administrators who will understand the specific architecture involved and the inherent opportunities for enhancement. Perhaps a less obvious network-related area in certain applications is the physical deployment of biometric points of service and the associated throughput in terms of transactions. If bottlenecks are found to exist at certain times simply due to the number of persons requiring the service, then we may wish to consider the addition of further points of service or biometric "stations" in order to distribute the load. Load balancing at the back end of the system, around directories, databases and matching engines, is unlikely to be a problem with all but the largest of operational systems, providing of course that the systems infrastructure has been properly designed.

The opportunities for software fine tuning will depend upon the application in question, its specific architecture and who provided it. If the application has been developed in-house, then naturally we shall have local expertise to manage and optimize the code. However, it is not just a question of optimized code. There may be alterations required to the user interface, for example, in order to make operation more intuitive. Or perhaps the functional process might be altered in order to effect an operational improvement, or updated drivers or interfaces installed to facilitate better integration with other components. The key point here is to be fully aware of the systems architecture, the functional design of the software and exactly what is happening "under the hood" inside the equipment. Without this information, we shall be at a loss to know whether subsequent alterations are likely to effect an improvement or otherwise. If the application is brought in from an external vendor, then we may or may not have such detailed information, depending upon how we negotiated with them during the procurement phase. In this respect, the use of a modeling language and associated methodology would ensure at least some understanding of systems architecture and functionality, which could be further exploited as the system is fine tuned and bedded in. In most cases, scope will exist for incremental system software tuning in order to obtain the best performance within a given infrastructure, but it is an area which should be approached with caution if the software was developed externally.

Assuming that the physical implementation of the biometric system has been constructed and configured for optimum performance and stability, there may still be much we can achieve in the area of ongoing user training. Typically, users may have received an introduction to the system and its operation at the time of enrollment and may have undertaken their first few transactions under supervision in order to ensure consistency in their presentation of the biometric sample. As time moves on, individual users may or may not retain this understanding and may have a greater or lesser opportunity to practice their operational skills. Infrequent users may be more or less consistent in their approach to transactions, with instances of errors being equally varied. If this situation prevails, there is much we can achieve by having periodic 'refresher' training sessions and workshops where users can air any concerns they might have and be reinstructed as to the correct operation of the system. Such an initiative can have a significant impact on the overall system performance realized within a typical application. Naturally, good communication plays a role in this also and we should ensure that an ongoing communications program is included as part of the ongoing biometrics project.

In conclusion, there is much we can do to fine tune our system after it has been installed and when we have the benefit of hindsight as to actual transactions and usage. There may be some obvious problem areas that need to be addressed, or there may be an opportunity for more subtle tuning as we move forward, both with the system and its users. In any event, we should be open minded as to the potential of improvement through tuning, and not be shy to look into this area. It just might make the difference between overall success or failure for our biometric project.

10.8 Ongoing Management

Having designed, implemented and fine tuned our system to the point at which it is operating well and everybody is happy with it, we still have an ongoing requirement for overall management and support. The system will need to be maintained, individual biometric devices may need to be replaced, users will come and go, organizational requirements may change, and so on. The first principle to establish in this respect is that of "ownership". The system must have a clear owner or sponsor who is responsible for it on an ongoing basis and who makes any decisions necessary for its support and maintenance. This authority may indeed choose to delegate some of the day-to-day operational issues, but must be ever-present as the final arbiter when decisions need to be made with respect to the application. Furthermore, this authority must be visible in the form of an individual who is an enthusiastic and effective champion for the application overall. They must have a passionate interest in the system and its users and live and breathe the application day by day. They must have an equally passionate interest in its success and be ready to vigorously defend the cause in the face of competing organizational resources if need be (whilst obviously understanding the bigger organizational picture). Without this level of commitment, the ongoing success of the system will be open to question.

Part of this commitment will of course manifest itself in the day-to-day operation and management of the system. In turn, part of this will be concerned with actual systems support and fallback procedures in the event of failure, and part will be concerned with producing the required management information. Yet another part will be concerned with managing users and user-related issues. The overall manager must therefore have not only a strategy, but a process methodology for handling all situations that may arise within the real operating scenario. They must also interface with other elements of the organization as appropriate to ensure that the application is meeting the broader objectives and that all potential efficiencies and opportunities are realized. In practice, this authority might consist of a team or project office which has a diverse and specialized skill set in order to meet the varying requirements, led by the enthusiastic champion referred to earlier. Together they will manage the system on an ongoing basis and handle any related situations as they arise. Naturally, all this is rather "top level" and the actual tasks and reports will depend upon the precise nature of the application in question, although there are perhaps some generic elements for inclusion, such as the following:

- A user and application help desk function.
- An ongoing communications function.
- A systems maintenance and support function.
- An application futures board or steering group.
- A general administration and reporting function.

Depending on the nature of the application there may be additional elements around ongoing supplier relations and external liaison as well as variations on the points noted above, but suffice it to say that this is an important area which needs to be adequately supported. It is also worth bearing in mind that users of biometric systems are likely to have more questions than would typically be the case for familiar office-type applications and that these will all need to be dealt with in a knowledgeable, friendly and authoritative manner by suitably qualified individuals. There is obviously a cost associated with such levels of support and this factor should be taken into consideration within the original business case for the application.

One element of ongoing management which is perhaps worth singling out is that of performance tracking and analysis. As part of our original design, we will have no doubt incorporated a transaction database which captures each transaction event. It would be most useful if this database, in addition to capturing the fundamental elements, such as user name, point of service, time and date, and result, also captured a matching score for each transaction and, where applicable, the relative threshold setting. This information, together with the more usual system performance metrics, would enable us to extract a wealth of practical information as to how well our application was really working and what, if any, trends are emerging around errors. For a given individual, we would thus have not only a usage pattern including successful transactions and errors, but a means of analyzing the matching scores over time and understanding how the user is adapting to the application or otherwise. If certain users are experiencing higher than average error rates, then we shall at least have some pertinent information to work with and should be able to identify the cause. For example, erratic and wildly varying scores would indicate inconsistency in use. Consistently poor scores might indicate an inappropriate threshold setting or a poor quality reference template. The culmination of this information over time will help us become expert in understanding and administering the biometric system.

10.9 Related Issues

Several times in this chapter we have referred to the use of a modeling language within the process and systems definition context. It is perhaps worth expanding on this idea a little and explaining some of the associated benefits.

One of the potential problems with any application development lies with the potential for misinterpretation of ideas and requirements. This is hardly surprising when one considers the typical diversity of the individuals involved and their particular spheres of operation. For example, within the business or operational area there will exist individuals who have a good grasp of the current processes, but may not have an understanding of how technology may be applied in order to improve them. They may call in a business analyst to help them shape their thinking, who in turn may call in a systems analyst in order to understand the current infrastructure. Both of them may be liaising with an external consultant who is defining an overall vision. The consultant will be liaising with one or more technology vendors with regard to specific components, and somewhere in the mix will be the application developers, who in turn will be liaising with the infrastructure and network architects. The trouble is, all these individuals tend to speak a different language, complete with a particular jargon which is typically incomprehensible to those outside that particular sphere. The chances of the original requirements surviving intact down through all these layers and being translated into an efficient application which fulfills all of the original objectives whilst remaining universally understood and accepted can be pretty slim with any complex application development. If you add to this mix the relative lack of detailed awareness around biometric systems, their precise operation and the issues involved, then you have a golden opportunity for misunderstanding which the law of Murphy will vigorously exploit. One way around this is to utilize a clear cut methodology throughout the project in order to ensure a common understanding at each key point and be able to articulate and document processes and systems functionality accordingly. If the same methodology can be used by all those involved with the project, then we can all start speaking the same language and ensure that the understanding does indeed cross the various boundaries. This is the general idea of using a modeling language.

Such a language may typically be used for:

- Defining the current processes.
- Defining and specifying the requirement.
- Conceptually designing and describing the new application.
- Defining the detailed functionality of elements within the new application.
- Defining the existing infrastructure.
- Designing the new application systems architecture.
- Defining relationships and dependencies between components.
- Defining logical operational flows.
- Scoping the software development.

In addition to the above, it may be useful to incorporate the modeling language methodology and diagrams into the RFI (request for information) and RFP (request for proposal) documents for procurement purposes. This will leave vendors and consultants in no doubt as to the requirements whilst introducing them to a way of working which is understood within the organization, thus facilitating ongoing discussion and program management in a structured manner. Much time can be saved in this respect, helping to keep overall projects costs down whilst promoting efficiency and clarity throughout. A key benefit in this respect is the portability of diagrams and associated documentation from the front line operational area right through to application development and the ongoing support function. Indeed, depending upon the application, such documentation may also be useful for marketing and training departments in understanding the system functionality. Lastly, such an approach also ensures a well-documented archive of the original idea and subsequent development of that idea through to application development and implementation.

Organizations may already have a modeling language policy and preferred methodology with which they are familiar, in which case it makes sense to utilize it accordingly. If this is not the case, then there exists a lightweight, easy-to-use modeling language specifically designed for biometric and related applications named Bantam [12], which would be well worth evaluating in order to gauge its usefulness within the proposed biometric project. The Bantam methodology includes the graphical notation and standard document templates along with RFI and RFP templates and other information and utilities to help bring clarity and consistency to this important area. Unlike some methodologies, it covers both hardware and software and yet remains simple and intuitive in use, with a minimal learning curve.

In conclusion, this chapter has taken a brief look at some of the systems-related issues and tried to place them within an overall program context. Naturally, there are a host of other related issues, especially around marketing and user training, which will have an impact on the overall success of the project. Designing, building and implementing a significant biometric system may be compared with building a ship. You know roughly what a ship looks like and what it is supposed to do (float on water), but this is not enough. You also need to know the environment in which it will be used, the purpose for which it will be used, how many passengers it will carry and of what type, who the crew are and what their capabilities need to be, what the performance requirements are, and a host of other parameters. You will also need a specification for engine type, fixtures and fittings, construction methodology and so forth. You would probably not start building the ship without a detailed blueprint, and you should not start on your biometric project without a blueprint either. In this respect, clearly understanding and documenting the requirements, defining and documenting the operational processes, defining and documenting functional elements of the system, describing the existing architecture, designing the required infrastructure, specifying the required components and defining the software requirements are all part of the biometric program blueprint. This all needs to be undertaken in a structured and coherent manner and documented in such a way that any project team could pick up the documentation, understand the requirements and build the system, just as any shipyard could pick up the blueprint and build the ship.

References

[1] R. Germain, Large-scale systems, in A. Jain *et al.* (eds), *Biometrics: Information Security in a Networked Society.* Kluwer Academic, 1999, pp. 311–326.

[2] A. Jain and S. Pankanti, Automated fingerprint identification and imaging systems, in H. C. Lee and R. E. Gaensslen (eds), *Advances in Fingerprint Technology* (2nd edn). CRC Press, Boca Raton, 2001, pp. 275–326.

[3] J. L. Wayman, Error rate equations for the general biometric system. *IEEE Robotics and Automation*, **6**(1), 35–48, 1999.

[4] J. L. Wayman, Large-scale civilian biometric systems – issues and feasibility. *Proc. CardTech/SecurTech Government*, Washington, DC, 1997. Available online at: `http://www.engr.sjsu.edu/biometrics/nbtccw.pdf`.

[5] J. Phillips, A. Martin, C. Wilson and M. Przybocki, An introduction to evaluating biometric systems. *Computer*, **33**(2), 56–63, 2000.

[6]] J. Ashbourn, *Biometrics – Advanced Identity Verification.* Springer-Verlag, London, 2000.

[7] UK Biometric Working Group, *Use of Biometrics for Identification and Authentication: Advice on Product Selection*, Issue 2.0, March 22, 2002. Available online at `http://www.cesg.gov.uk/site/ast/biometrics/media/BiometricsAdvice.pdf`.

[8] T. Quantrani, An Introduction to the Unified Modeling Language (2001). Retrieved March 15, 2003 from `http://www.rational.com/media/uml/intro_rdn.pdf`; now available at `http://www-106.ibm.com/developerworks/rational/library/998.html`.

[9] Object Management Group, *An Introduction to OMG's Unified Modeling Language.* Retrieved March 15, 2003 from `http://www.omg.org/gettingstarted/what_is_uml.htm`.

[10] A. J. Mansfield and J. L. Wayman, *Best Practices of Testing and Reporting Biometric Device Performance*, version 2.01. UK Biometrics Working Group, August 2002. Available online at: `http://www.cesg.gov.uk/site/ast/biometrics/media/BestPractive.pdf`.

[11] *Open LDAP 2.1 Administrator's Guide*, January 10, 2003. Retrieved March 15, 2003 from `http://www.openldap.org/doc/admin21/`.

[12] J. Ashbourn, *Bantam User Guide.* Springer-Verlag, London, 2001.

11 Biometrics and the US Constitution

Kenneth P. Nuger and James L. Wayman

11.1 Introduction

11.1.1 Privacy Versus Security; Mankind Versus Machine

Is biometric identification a practical aid to ensuring individual and collective safety in an increasingly crowded and complicated world, or an Orwellian vision with the potential to strip us of our right to be let alone? In these early years of the 21st century, the sciences and technologies of surveillance and monitoring have advanced to the point where we now, collectively, have the capability to threaten the most basic democratic notions of individual autonomy and privacy. Yet at the same time, autonomous acts by unmonitored individuals and groups have cost the lives of thousands and have threatened our fundamental right to feel, and indeed be, safe and secure. This inherent conflict between collective safety and individual privacy rights needs to be addressed carefully by those who create and implement these technologies.

Biometric identification technology in particular has raised concern among some civil libertarian groups for its potential to allow for government, and commercial, monitoring of activities of *named* individuals. We now have the technological means not just to monitor statistically the behavior of groups (which is the historic core of all social sciences), but to directly associate specific behavior with identified individuals. Although this capability, through fingerprinting, dates at least to the 19th century [1], it is the automatic association of actions to individuals through the use of measurement and computational machinery that some find so unsettling, particularly if such capability lies in the hands of government. Yet we fully expect our government to develop and utilize technologies to protect us from the actions of individuals with violent or otherwise criminal intent. What are the reasonable governmental uses of biometric identification technology under the limitations imposed by the US Constitution?

In a democratic society which values government accountability and individual privacy, it is the rule of law which shapes the manner by which the government may interact fairly with its citizenry [2]. One of the basic principles shaping the rule of law in the USA is the right of individuals not to be treated unfairly by government. This broad notion manifests itself in two legal principles in particular: due process and individual privacy. The concept of due process requires the government to acknowledge the possibility of errors, allowing means for their mitigation. The concept of privacy goes beyond simply acknowledging the possibility of errors to set limits on the power of government to meddle in the lives of individuals. These court-protected guarantees require the government to respect the rights of individuals by limiting intrusions to those which directly further recognizable and legitimate societal interests. This historical balance between individual rights and societal interests is at the heart of all democracies, and is placed under a new strain by the advent of biometric technologies. The purpose of this chapter is to explore ways in which this balance might be maintained, reconciling government use of biometric technologies with the US Constitutional requirements of due process and privacy protection.

11.1.2 The Growth of Both Anonymous Public Transactions and the Complexity of Identification

Human identification using automatic (computer-aided) means dates back only somewhat over 40 years [3], but the public identification of individuals has a much older history. In 1850, the population density of the USA was about 8 persons per square mile. In 1990, it was over 70 [4]. In 1850, only one US city, New York, had a population of over 170,000 people. Today, well over 100 cities do. In earlier, less urbanized, times it was the rule, not the exception, that parties in a civil transaction were previously acquainted. Identity verification was done humanly as a matter of course, and was formally required in many public proceedings. The role of the required witnesses to civil proceedings, such as marriage ceremonies and deed transactions, is to verify the identities of the participating parties as well as to attest that the transaction did, indeed, take place.

In many US states, election laws require voters to declare at the polling place, in an "audible tone", both their name and their address [5]. These laws are precisely for the purpose of allowing the public verification of a claimed identity, thus ensuring the sanctity of the "one person, one vote" concept of our democracy. Although we do have a tradition of anonymity with regard to public political activities [6], the idea that more general public transactions can be accomplished anonymously is a very new one to society. The idea of accomplishing financial transactions based on credit with no previous relationship between the creditor and debtor dates only to the growth of the credit card industry over the last 30 years.

It might be that our desire for anonymity has grown because of the pressures of living in an increasingly crowded society. The process of identification, whether by human or machine methods, has certainly become increasingly difficult with increasing population size. Identification requires the recollection of personal characteristics previously associated with an identity and the recognition that these characteristics do not more closely match those of any other individual. The identification of a number, say N, of individuals requires the comparison of each of the presented characteristics to N previously learned patterns, so the process grows in complexity as N^2, whether done by man or machine.

Additionally, if these personal biometric characteristics are represented in a machine as points in a high-dimensional space (as most biometric measures are), the space becomes increasingly crowded as the number of points increases. This increases the possibility that a sample biometric will be incorrectly matched to one stored in the same area of the space. To minimize this "false match" problem, comparison thresholds for matching samples to stored points must be tightened as *N* increases. Tighter thresholds, of course, increase the probability that a sample will incorrectly *not* match a stored vector from the correct individual, the "false non-match" problem. So not only does the complexity of the identification problem increase with increasing *N*, both false positive and false negative recognition errors increase as well.

So given the increasing demands for impersonal transactions, the increasing difficulty of identifying individuals in a crowded society, and the recent availability of automatic means for such identification, it seems inevitable that society would turn to biometric identification as a tool. It is our belief that the fundamental question is not the appropriateness of a general requirement for public identification, but the risks added when that identification becomes handled by automatic means, most particularly in the hands of the government.

The use by government of biometric technologies has increased rapidly in the past decade. Biometrics is currently used on the state level as a condition of qualification for some government programs, such as social service benefits and driver's licenses. This use of biometrics is aimed at preventing people from enrolling under multiple identities in violation of the system policy. It is increasingly common for state and local governments to require welfare applicants to have fingerprint images scanned into a database as a condition for receiving welfare. When this unique personal identifier becomes part of the database, an individual engaging in a fraudulent attempt to enroll twice in the same welfare program will be matched and apprehended when submitting to another fingerprint scan. When a qualified recipient later presents her- or himself to collect benefits, the fingerprint image may again be scanned to verify the presenting individual's identity. Similarly, many states' driver's licensing agencies, like California's Department of Motor Vehicles, require applicants to provide a fingerprint as a condition for obtaining a driver's license. As with welfare agencies, motor vehicle agencies wish to prevent individuals from fraudulently obtaining a driver's license in another name when a previous license has been revoked.

Government use of biometric technology as a tool of public policy, in a society where personal identification has become increasingly difficult, must recognize the balance between personal rights and societal interests. In the USA, this balance is protected by the Constitution, as interpreted by the courts, both state and federal.

11.1.3 Constitutional Concerns

In the USA, the rights of privacy, due process and security from unreasonable search and seizure are grounded in the Fourth, Fifth and Fourteenth US Constitutional Amendments:

Fourth Amendment
The right of the people to be secure in their persons, houses, papers, and effects, against unreasonable searches and seizures, shall not be violated; and no warrants shall issue, but upon probable cause, supported by oath or affirmation, and particularly describing the place to be searched and the persons or things to be seized

Fifth Amendment
Any person shall be held to answer for a capital, or otherwise infamous, crime, unless on a presentment or indictment of a grand jury, except in cases arising in the land or naval forces, or in the militia, when in actual service, in time of war, or public danger; nor shall any person be subject, for the same offense to be twice put in jeopardy of life or limb; **nor shall any person be compelled, in any criminal case, to be a witness against himself, nor be deprived of life, liberty or property, without due process of law**; nor shall private property be taken for public use, without just compensation

Fourteenth Amendment
Section 1. All persons born or naturalized in the United States, and subject to the jurisdiction thereof, are citizens of the United States and of the State wherein they reside. No State shall make or enforce any law which shall abridge the privileges or immunities of citizens of the United States; **nor shall any State deprive any person of life, liberty, or property, without due process of law**, nor deny any person within its jurisdiction the equal protection of the laws.

It has been the role of the courts, both state and federal, to provide a contextual interpretation to the adjectives "unreasonable" and "due" in these three amendments when assessing the appropriateness of government behavior toward its citizens. It is most instructive, therefore, to study both the philosophical history and recent court rulings in these areas as they might apply to biometric technologies. In the next section, we will survey "due process". Privacy, including "self-incrimination" and "search and seizure" issues, will be addressed in Section 11.3.

11.2 Due Process

11.2.1 Entitlements and Rights

The logic of due process is rooted in the notion that personal freedom in a constitutional democracy can only be preserved when there is some consistent way to check arbitrary and capricious actions by government [7]. Simply stated, the Fifth and Fourteenth Amendment concepts of due process require both federal and state governments to carry out their obligations fairly. With regard to biometric technology, our primary due process concern is with the denial of service by a government agency on the basis of a biometric measure. In such a case, distinction is made under the law between the denial of an entitlement or privilege and the denial of a right.

If a government agency denies an individual what is normally considered an entitlement, due process requires the agency to demonstrate the reasonableness of its action in light of the hardship placed on the individual. In this case, the government must establish a rational basis for the denial. If the denial is of what is normally considered a right, the government must demonstrate, to the satisfaction of the court, a compelling government interest. The compelling government interest must be shown to be preferred over the denied human freedom. In other words, if the denial is of an entitlement, the test is "reasonableness" or "rational basis"; if a right, the test is "compelling government interest" or "preferred freedom". These two, the *rational basis* and *preferred freedoms* tests, comprise the primary analytical methodologies the courts utilize when judging the constitutionality of government decisions that affect individuals' rights.

11.2.2 Instrumental and Intrinsic Approaches

Philosophically, two rationales have been used to justify the importance of due process in a democratic society [8]: the intrinsic and the instrumental approaches. The intrinsic approach suggests that society gains an important moral benefit by allowing individuals to participate in the governmental processes affecting them. To be a "person", rather than a "thing", implies the right to be consulted about what is done with you [9]. As Justice Frankfurter opined in *Marshall v. Jerrico*:

> no better instrument has been devised for arriving at truth than to give a person in jeopardy of serious loss notice of the case against him and opportunity to meet it. Nor has a better way been found for generating the feeling, so important to a popular government, that justice has been done. [10]

The second approach to due process, the instrumental, focuses less on the right of people to be part of a decision-making process affecting them and more on the need to ensure that rules for distributing government services are accurately and consistently followed. The instrumental approach creates due process requirements to minimize substantially unfair or mistaken decisions that lead to deprivations of entitlements conferred to people by law. Under the instrumental approach, the purpose of due process is not primarily to ensure participation by affected parties, but rather to ensure that government makes an accurate decision [9].

In both intrinsic and instrumental approaches, due process protections increase as the individual identities of the affected parties becomes clearer [11], and as the effect of the government decision becomes increasingly under the control of an identifiable decision-maker. Therefore, under both approaches an individual has a right to greater due process of law in challenging an adverse government decision than would a group [12]. Similarly, an individual has a right to greater due process of law as the severity and harm of the adverse decision increases [13].

Regardless of the philosophical approach taken, the specific form of due process is not fixed in constitutional law. Some elements of procedural due process appear to be based on notions of natural law [14], and others on notions of basic fairness [15]. The government has the duty to give the affected party: (1) notice [16]; and (2) an opportunity to be heard in an open hearing [17], before a neutral and detached magistrate [18], with no direct, personal, substantial pecuniary interest in reaching a conclusion against him in his case [19].

While both have influenced due process, the instrumental approach has emerged over the last 25 years as the primary engine guiding due process protections, while the intrinsic approach has fallen into disfavor. Under the intrinsic approach, where citizen participation is valued, judges are inevitably required to make judgments about the degree of harm done to an individual by an adverse government decision, whether involving an entitlement or a right. While perhaps compassionate toward the human condition, this "political" power given to the courts under the intrinsic approach is often criticized as unduly subjective and distorting of the concept of a politically neutral judiciary. Under the instrumental approach to due process, the political nature of the judiciary is de-emphasized because the focus is less on the moral right of individual participation and more on ascertaining whether government administrators followed proper procedures. Therefore, under the instrumental approach, the judge is spared from the subjective dilemma of determining how a government decision will adversely affect an individual and, instead, will focus on the more objective task of assessing whether government administrators followed the statutory procedures.

The now-popular instrumental approach guides courts to be less concerned with the effect of the government decision on the individual and more with whether the government decision was made consistent with statutory public policy. Therefore, obtaining due process before the deprivation is less important than obtaining due process at some point in the policy process. This is bad news for anyone receiving a government denial based on a biometric (or other) measure because it allows the deprivation hearing to come after (post-deprivation), rather than before (pre-deprivation), the imposition of the adverse decision.

The intrinsic approach, with its emphasis on individual participation, gives denied citizens a more intimate ability to challenge the fairness of government action in a timely manner. By minimizing the requirement for individual participation, the trend away from the intrinsic and toward the instrumental approach might make society more skeptical of biometric technology as a tool of public policy. If government utilizes biometric technologies to deny individuals what are commonly considered constitutionally protected rights, with the justification that all statutory procedures were properly followed, both the legitimacy of government and the usefulness of biometric technology as a public policy tool could be undermined.

11.2.3 Constitutional Development: From the Intrinsic to the Instrumental Approach of Procedural Due Process

In 1970, with the decision of *Goldberg v. Kelly* [20], the Supreme Court began a procedural due process revolution. The main focus of procedural due process is on whether, and to what degree, individuals can participate in a government decision adversely affecting them and whether this participation may be exercised before the adverse decision takes affect. Generally, if the adverse government decision jeopardizes a fundamental right, as opposed to an entitlement, the government must meet the preferred freedoms test, demonstrating a compelling justification. When rights are in jeopardy, government agencies can more effectively assuage judicial concerns over due process by allowing individuals the ability to plead their case to the agency in a pre-deprivation hearing. If however, entitlements are in jeopardy, an agency may be able to satisfy the judiciary's concerns by merely providing a post-deprivation appeal.

In *Goldberg*, the State of New York terminated welfare benefits without first allowing the individual an evidentiary hearing. The Supreme Court ruled against New York, arguing that because both the federal Aid to Families With Dependent Children and New York's Home Relief programs were statutorily created entitlement programs, these forms of assistance should be considered more a form of property than a gratuity. Since the Court viewed welfare as an entitled property right created by statute, and because welfare benefits give a recipient the means to survive, the Court ruled that benefit denial without a pre-deprivation evidentiary hearing would violate due process. The Court reasoned that the nature of the entitlement was so fundamentally linked to the basic survival of the individual, that termination without a *prior* hearing would gravely impair a basic right to life and liberty [21].

During the early 1970s, the Court expanded the concept of a pre-termination hearing in several areas: revocation of parole [22], probation [23], early release for of imates for good behavior [24], suspension of a driver's license [25], high school suspension [26], public posting of people unfit to consume alcohol [27], housing evictions from public projects [28], repossession of property [29], garnishment of wages [30] and denial to students of state residency eligibility while attending college [31]. Cumulatively, these decisions constituted a mini due process rights revolution. The Supreme Court, utilizing the intrinsic approach, appeared to emphasize, as an indispensable component to due process, the pre-deprivation participation by those adversely effected.

If left unabated, this use of the intrinsic approach would have strengthened the ability of individuals to challenge adverse decisions contemplated by government, including those based on a biometric identifier. The costs of this approach on government, in terms of financial burdens, time delays, and jeopardized policy implementation, could have become staggering. The Supreme Court had previously appeared much less inclined to force government to provide pre-deprivation hearings. For example, the Court had held earlier in the 20th century that government could, without a pre-deprivation hearing, seize mislabeled vitamins [32] and spoiled food inventory [33], impose rent controls [34], disqualify a contractor from doing business with the federal government [35] and deny to an individual employment as a cook for a concession in a defense contractor's plant [36]. In these various settings, government was able to successfully demonstrate that the action taken comported with due process because of the obvious and immediate threat to the public safety.

However, in 1976, when the Court decided *Mathews v. Eldridge* [37], it shifted its procedural due process focus away from the intrinsic approach and toward the instrumental. Although biometric identification was relatively unknown in the 1970s, this decision would ultimately have a tremendous impact on the ability of government to incorporate biometric technologies into public policy areas some twenty years later. In the *Mathews* case, the issue was whether the Social Security Administration could terminate disability payments without first holding a hearing. In ruling for the Social Security Administration, the Court developed a three-part balancing test to assess the necessity of a pre-deprivation hearing. This test applies, of course, to deprivations based on biometric measures as well.

The first part of the *Mathews* test requires the Court to assess the nature of the private interest at stake. In *Mathews*, the private interest at stake was the property interest Mr Eldridge had in continuing to receive his disability payments. Here, the Court distinguished disability payments from the welfare payments at issue in *Goldberg*. To receive welfare payments, an individual had to demonstrate poverty, a condition with grave consequences. However, in *Mathews*, the Court ruled that the criterion for disability payments was merely the demonstration of the inability to work, which does not necessarily imply poverty. While some might find the distinction somewhat shrill, the Court reasoned that Mr Eldridge would not necessarily suffer gravely adverse affects from termination of his disability payments, as he may have had other means of support.

The second part of the balancing test requires the Court to assess the chance that government administrators had made an erroneous decision. In *Mathews*, the Court reasoned that the probability of the Social Security Administration being in error was very slim as its decision was based on a battery of medical examinations showing that Eldridge was no longer disabled. This second part of the balancing test gives the courts wide latitude in evaluating the error rates compatible with due process, and supports the use by government of scientifically based data to demonstrate the accuracy and, therefore, reasonableness of technologically based service delivery systems, including biometrics.

The third part of the balancing test requires a court to balance the first two parts, the nature of the affected private interest and the error probability, against other government interests that might be adversely affected by a pre-deprivation hearing. These factors include administrative efficiency, fiscal responsibility, and the potential impact of the pre-deprivation hearing on the government decision. In *Mathews*, the Court ruled that disability payments were not fundamental to the liberty of Eldridge, that there was little risk of agency error, and that the government had a substantial interest in preserving both its administrative and fiscal efficiency. Therefore, all the process due Mr Eldridge was a post-termination hearing, ruled the Court.

The *Goldberg* and *Mathews* rulings form the basis for procedural due process requirements in the use of fingerprint imaging in current social service programs. With these cases serving as a guide, agencies can be instructed on the proper treatment of a welfare applicant whose fingerprint image appears to match another already in the database. From *Goldberg*, agencies recognize that courts will logically treat welfare as a statutorily created entitlement, thereby conferring a property right on those who qualify. Further, its denial would cause grave hardship. Thus the first part of the *Mathews* balance test sides clearly with the denied recipient.

The second and third parts of the *Mathews* test, however, can be employed by social service agencies using fingerprinting to strengthen their legal position. The current generation of two-print [38] fingerprint systems is extremely accurate with regard to false matches. This accuracy [39], coupled with the usual protocol of human visual comparison of computer-matched images when fraud is suspected, decreases the chance, under the second part of the *Mathews* test, that the agency decision was based on error. Under the third part of the *Mathews* test, the fingerprint imaging requirement serves the important, even compelling, purpose of reducing welfare fraud. Thus, it is justifiable on the grounds of both administrative and fiscal efficiency. Finally, to ensure the integrity of the decision and the dignity of the recipient, the agency should provide a timely process to denied parties to allow challenge of the decision.

An example of the reconciliation of constitutional due process and biometric measurement to satisfy both individual and government interests is demonstrated by New York's enactment of fingerprint imaging as part of its state social service program. In 1992, New York amended section 139-a of the New York State Social Services Law to require fingerprint imaging as a condition for obtaining welfare in both Rockland and Onondaga counties. Since initial assessments of this experimental program were favorable [40], New York expanded the fingerprint imaging requirement to several other counties in an attempt to reduce both intra- and inter-county welfare fraud. However, if the agency believes, as a result of a matched fingerprint, that an applicant is guilty of fraud, section 139-a provides that benefits will not automatically be denied. Before the denial of benefits, the individual must be given notification and then may request a hearing to be conducted within 45 days [41].

At the hearing, the individual may present evidence contesting the accuracy of the biometric comparison and will have the opportunity to examine the state's evidence. This hearing preserves both the intrinsic nature of due process, by granting the individual the right to participate in the decision-making process, and the instrumental nature of due process, by allowing the careful assessment of the accuracy of the finding.

While engineering and algorithmic advances may be making biometric devices extraordinarily accurate in benchmark tests, numerous conditions exist in practice that increase the false match rate. Therefore, as government agencies begin using biometric data to deny recognized rights, whether welfare checks to an indigent, or a driver's license to a commercial driver, courts will be forced to consider two distinct problems. Firstly, they will have to determine whether the device's accuracy in the considered application is sufficient to meet the rigors of due process. Secondly, they will have to determine whether agencies have appropriate and timely remedies for aggrieved individuals challenging an adverse decision. If these two conditions are met, the chance that reviewing courts will find due process violations against agencies that use biometric technologies will be greatly reduced.

11.2.4 The Enigma of Substantive Due Process

Even if a government agency acts under color of statute and its decision-making processes are accurate, its policies may be challenged as being arbitrary, capricious and wholly unrelated to legitimate government interests, and thus lacking "substantive" due process. As discussed earlier, if the policy adversely affects a fundamental right, government must demonstrate that it has a compelling interest in implementing the policy. If the policy adversely affects an entitlement or a privilege, the individual must demonstrate that there is no rational relationship between the policy and a legitimate government interest. In either case, aggrieved parties can challenge government policy as arbitrary or capricious, lacking substantive due process. Courts, then, are required to determine the nexus between the policy and legitimate government interest, as well as to assess whether the benefit denied is a right or a privilege, with greater protection afforded to a finding of right [42].

The California case of *Christopher Ann Perkey v. Department of Motor Vehicles* [43] applies directly to biometric technology and serves as an excellent example to illustrate the problem of substantive due process [44]. California instituted a requirement that each applicant for a driver's license submit a fingerprint to the Department of Motor Vehicles [45]. Ms Perkey refused to be fingerprinted and was denied a license solely on the basis of this refusal. She took legal action, claiming that the fingerprint requirement violated substantive due process because there was no relationship linking it with the state's stated interest in promoting highway safety. In other words, she claimed the fingerprint requirement to be arbitrary and capricious, unrelated to any legitimate government interest.

In response to her challenge, the California Supreme Court first determined the right to drive as not fundamental and not protected by the Constitution. Therefore, the appropriate level of analysis was determined to be the rational basis test. This rendered the state fingerprinting requirement constitutional as long as it remains procedurally fair and reasonably related to a proper legislative goal [46]. To substantiate the state's fingerprint policy, the Department of Motor Vehicles offered the following argument. The state has an obligation to ensure that individuals who are issued a driver's license meet the qualification for safe driving. These qualifications are, at least, a demonstrated knowledge of California's vehicle laws and a demonstrated competence in observing these laws while driving. When an individual fails to demonstrate competence in either of these areas, California can deny or revoke the driver's license. Studies in California during the 1980s showed that the incidence of driver's license fraud was increasing [47], leading to the suspicion that people whose licenses had been revoked or suspended for reckless or impaired driving were illegally applying for new licenses under assumed names. Therefore, the Department of Motor Vehicles contended that the interception of applications from those who pose a serious danger to public safety constituted a proper legislative objective. The remaining question, then, was whether the fingerprint requirement was reasonably related to that objective.

In *Perkey*, the Department of Motor Vehicles asserted that fingerprint technology was the only reliable way to ensure the integrity of its driver's licensing records. Handwriting samples are too variable, and photographs, or more precisely, one's appearance, can too easily be changed. Therefore, the Department of Motor Vehicles argued, and the California Supreme Court agreed, that the fingerprint requirement bore a rational relationship to the legitimate goal of furthering highway safety by giving the state a reliable method of checking the identity of driver's license applicants.

This example demonstrates the basic requirements of substantive due process. If an individual challenges the use of a biometric identifier in this way, the reviewing court must assess no fewer than three different questions. First, is the interest furthered by the government a valid exercise of governmental power? Both the federal government and the states have broad police powers to further the health and safety of individuals within their respective jurisdictions. Therefore, this first question requires the court to determine whether the policy is consistent with governmental police powers. The second question requires the reviewing court to assess whether the challenged policy is arbitrary or capricious. In so doing, the court must determine the potential for error in the challenged program. In the case of biometric technologies, courts will have to determine whether the methods are reliable enough to ensure the absence of a matching error. The third question is whether there is a clear connection between the challenged government program and a legitimate government interest. If the denial is of a fundamental right, the reviewing court must be convinced that there is a *compelling* interest in doing so. If, however, as was the case in Perkey, the right denied is deemed non-fundamental, or a privilege, the reviewing court need only assess whether the challenged government program is rationally related to a *legitimate* government interest.

The difference between a compelling interest and a legitimate government interest is subjective. Under the preferred freedoms test, denial of a fundamental right must be done only in the pursuit of a particularly important governmental responsibility, and then, only in the most narrowly prescribed way . However, under the rational basis test, the one used in *Perkey* based on the finding that a driver's license is not a fundamental right, the government need only demonstrate that its goals are beneficial. In these instances, the challenged program needs only the smallest relationship to the legitimate government interest asserted to not contravene the standards of substantive due process.

11.3 Individual Privacy

11.3.1 The Basis of an Inferred Right to Privacy

The primary problem in discussing privacy is that the US Constitution contains no specific guarantee to privacy [48] and only some state constitutions afford citizens a guarantee to privacy against state and local government activities. Still, the right to privacy does have a recognizable heritage. In 1890, Samuel Warren and Louis Brandeis, in their seminal law review article [49], argued that even without a specific guarantee, the US Constitution gives people "the right to be let alone". Since that time, the legal understanding of the basis for privacy protection has increased dramatically [50]. Professor William Prosser, in his seminal article on privacy, notes that individuals have no less than four areas in which they can expect privacy protection [51], from disturbance of one's seclusion, solitude and private affairs; from embarrassing public disclosures about private facts; from publicity that places one in a false light; and from appropriation of one's name or likeness for someone else's gain.

Perhaps privacy's best definition remains as given by Warren and Brandeis over a century ago: "the right to be let alone". It is the fear of many that the use of biometric technology by government will erode this fundamental human right. These fears are aptly summarized by Justice Douglas in *Osborn v. United States* [52] in the warning of an

> alarming trend whereby the privacy and dignity of our citizens are being whittled away by sometimes imperceptible steps. Taken individually, each step may be of little consequence. But when viewed as a whole, there begins to emerge a society quite unlike any we have seen–a society in which government may intrude into the secret regions of man's life at will.

A few examples serve well to underscore Justice Douglas' concerns. Like a Social Security Number, a biometric identifier can serve as a "key" to sensitive and potentially damaging information about a person. When an individual provides, for example, a fingerprint for use in a legitimate identification system, a degree of individual autonomy is potentially ceded. Without strong protections, government personnel who lift a fingerprint from the scene of a political meeting could feed this data into a computer and match the fingerprint with a name, thereby revealing a personal identity [53].

Not only does a biometric identifier create a key that can aid in surveillance, it also creates a means, via "data mining", for linking an individual identified by the biometric measure to information stored in both private and governmental data banks. As the interchange of information grows, relating data becomes strikingly simple, allowing the creation of "cradle to grave" profiles of individuals. Even more disconcertingly, the strands of one's life can be tied together without one's knowledge or consent, by individuals who have never personally collected any of the information in question. Individuals, and ultimately society, might accept biometric technology if the limits of the use of collected information can be ascertained in advance. Certainly, the indiscriminate dissemination of biometric information will be broadly opposed.

Anonymity is recognized as important in the exercise of some other constitutional rights [54]. For example, the United States Supreme Court, in *Talley v. California* [55], ruled that ordinances that prohibit public distribution of leaflets not identifying the name of the author violated the first amendment. If, by matching a latent fingerprint to one in a database, a government agency could attach a name to an anonymous leaflet, a chilling effect could be had on the right to free expression, especially if the anonymous speech is critical of government policy.

If individuals can claim a constitutional right to personal anonymity in different contexts, the collection and dissemination of biometric identifiers could prove to be a vexing problem. Unlike names, addresses, phone numbers and even physical appearances, which can change or be altered over time, some biometric identifiers, such as iris patterns or finger ridges, are relatively stable. Once a biometric identifier and a person are linked together in a database, that person forever sacrifices a degree of personal anonymity [56]. To the degree that the information is accessible by more people, basic privacy rights are eroded.

Therefore, the most serious privacy dilemma confronting biometric technology is not one of physical intrusiveness, or the kind of probing typically related to search and seizure, but rather one of personal autonomy. Today's social realities lend support for the collection of personal and sensitive information, by both the government and the private sector, for a variety of reasonable purposes. Usually in the case of the public sector a clear case can be made that this information is needed to provide public services efficiently. The impact on personal privacy can only be limited if the collection, storage and use of these information fragments are narrowly specified.

11.3.2 Privacy and the Fourth Amendment

The use of biometric technologies in government applications must be sensitive to an individual's reasonable expectation of privacy under the Fourth Amendment guarantee of the right to be "secure in their persons, houses, papers, and effects, against unreasonable searches and seizures". [57] In *Katz v. United States* [58], decided in 1967, the Supreme Court clarified the meaning of the Fourth Amendment by opining that it protects people, not places, and wherever a person has a reasonable expectation of privacy, he is entitled to be free from unreasonable government intrusion. However, a person's Fourth Amendment expectation of privacy can be offset if the government can demonstrate that doing so advances a legitimate government interest [59]. This particularly applies to administrative, non-criminal searches where courts generally give a broader interpretation to the word "reasonable".

There are two important questions to be addressed. First, if a government agency collects biometric data, especially in non-criminal applications, does that act constitute a search for purposes of Fourth Amendment analysis? Second, if collecting biometric data is considered a search, is the search reasonable?

There is little doubt that courts will quickly answer the first issue in the affirmative; that biometric data collection by government agencies constitutes a search in both administrative and criminal settings, thereby acknowledging a privacy right under the Fourth Amendment. It is well established, from fingerprinting [60] to drug testing [61], that the gathering of physiological information by a government agency is considered a search under the Fourth Amendment.

With regard to the "reasonableness" of a biometric search, we turn again to the California case of *Perkey v. Department of Motor Vehicles* [43]. Here, Ms Perkey contended that the mandatory fingerprint requirement violated her Fourth Amendment guarantee against unreasonable search and seizure. In disposing of this argument, the California Supreme Court examined the existing case law and concluded the fingerprint requirement did not exceed Fourth Amendment thresholds for obtrusiveness. Its inquiry first examined whether fingerprinting involved the type of intrusive invasion of bodily integrity that has, in the past, been found to violate either the due process clauses of the Fifth and Fourteenth amendments or the implicit privacy guarantees of the Fourth Amendment. The court concluded that fingerprinting alone does not infringe upon an individual's right to privacy, holding that fingerprinting involves none of the probing into an individual's private life and thoughts that marks an interrogation or search. Unlike forcible stomach pumping [62] or the forced taking of a semen sample [63], both previously disallowed by the courts as unreasonable, the physical process of taking a fingerprint does not require penetration beyond the body's surface [64]. Therefore, fingerprinting does not readily offend the principles of reasonableness under the Fourth Amendment. Indeed, because fingerprinting is so inherently unobtrusive, courts have routinely permitted it in many non-criminal contexts including, for example, as a requirement for employment.

While it is possible for non-penetrative searches to still offend individual notions of privacy, this particular objection is not likely to prevail against fingerprint scanning technology. Fingerprint scanning does not cause any discomfort, penetrate one's body, or leave any residual traces on the individual. Its similarity to rolled ink fingerprinting also gives it a sense of social familiarity [65, 66]. Therefore, it is not likely that reviewing courts will rule that fingerprint scanning technologies are unreasonably physically intrusive.

Still left unresolved, however, is whether other forms of biometric technologies will be deemed unreasonably intrusive. Technologies like signature and speaker recognition systems, hand and finger geometry devices, infrared face and vein recognition systems, body odor detectors, and retinal [67] and iris scanners, may slowly meet with general social acceptance, as neither unusual behavior nor penetration of the body is required. Courts have held, as in *National Treasury Employees Union v. Von Raab* [61], that urinalysis, used for drug screening as a condition of employment even in the absence of suspicion, is not so intrusive as to constitute an unreasonable search. It is therefore highly unlikely that the less intrusive and humiliating forms of biometric technology will be found unreasonable in court challenges.

In the wake of catastrophes like "9/11" and the Oklahoma City bombing of a federal building, might our society acquiesce to biometric identification as a condition of entry into airports and other public buildings? In light of a perceived decrease in public safety, society may develop the view that there is a decreased expectation of privacy under the Fourth Amendment for people in certain public settings. Much like random drug testing is becoming accepted as reasonable in advancing specific public interests, might not large-scale biometric identification become accepted as an aid to public safety? Certainly, as witnessed by the increasing use of security systems at airports and other public places, our society is reevaluating privacy claims of the individual in favor of more public security. As long as biometric devices are used as an administrative tool to further legitimate, recognizable public interests, courts will likely follow precedent and rule that such use does not constitute unreasonable search and seizure under the Fourth Amendment.

11.3.3 Privacy and the Fifth Amendment

The Fifth Amendment insulates an individual from being compelled to testify against him- or herself in a criminal proceeding. Therefore, from welfare eligibility to driver's license renewals, any use of biometrics that furthers governmental administrative purposes would not generally support a Fifth Amendment self-incrimination objection. The Supreme Court has, however, extended the right against self-incrimination to non-criminal procedures that may lead to criminal prosecution [68].

The use of biometrics in direct relationship to criminal investigations, however, does allow for Fifth Amendment challenges. Courts have consistently taken the position that forced extraction of physical evidence, especially after some suspicion of criminal activity, does not automatically amount to forced self-incrimination, as long as the methods are relatively non-intrusive, do not affront society's sensibilities, and do not provide the state with evidence of a testimonial [69] or communicative nature [70]. However, the Supreme Court drew the line in 1952 in *Rochin v. California* [62], invalidating evidence of drug use obtained from a forced stomach pumping, because such government behavior shocked the conscience of a civilized society.

If *Rochin* begs the question as to what forms of forced extraction do not shock the conscience, the case, *Breithraupt v. Abram* [71], decided in 1957, provides a valuable clue. In *Breithraupt*, police took a small blood sample from an unconscious person involved in a fatal car accident. The Supreme Court ruled this extraction was constitutionally permissible, stressing that clinical blood extraction was not significantly intrusive and had become both commonplace and accepted by society. Nine years later, the Court reiterated this point in *Schmerber v. California* [72], by recognizing that both federal and state courts have held that the right against forced self-incrimination does not extend to forced subjection to fingerprinting [73], photographing, or physical measurements, nor to forced writing [74], speaking, standing, walking or gesturing for identification purposes [75, 76].

The courts' positions on Fifth Amendment issues have generally followed the path of *Breithraupt* [71] and *Schmerber* [72] in allowing non-intrusive, non-verbal data collection in criminal settings. It is therefore likely that courts will rule that biometric data taken in a criminal context, or in a non-criminal context where the data may later be used in a criminal proceeding, is not so obtrusive that it violates elements of privacy inferred within the Fifth Amendment's self incrimination clause. Obviously, this conjecture is based on the premise that biometric applications remain relatively unobtrusive and involve no physical penetration.

The key may be well summed up in a passage from *Miranda v. Arizona* [77], where the Court offered insight on the concept of forced self incrimination when it said:

> to maintain a fair State-individual balance... to require the government to respect the inviolability of the human personality, our accusatory system of criminal justice demands that the government... produce evidence against him by its own independent labors, rather than by the cruel, simple expedient of compelling it.... Compelled submission fails to respect the inviolability of the human personality.

11.3.4 Privacy of Personal Information

Much more troublesome than the privacy issues posed by biometrics in the Fourth and Fifth Amendment context is the threat to the more general privacy right to control the use of one's own personal information in varying contexts. In many ways, this right is similar to the notion of inviolability of the human personality expressed above in Miranda. It is equally well established that both national and State governments have broad powers to regulate privacy rights to further broad public interests. These concepts are rooted in Article 1, Section 8 of the United States Constitution, granting Congress an array of explicit powers, and the tenth amendment in the Bill of Rights, granting to states broad police powers to further the health, safety and morality of people within its jurisdiction. Together, these provisions provide a basis upon which government may pass legislation carefully defining when individual privacy rights must be protected and when other public interests or competing private rights may prevail.

Especially as administrative government has taken root in the 20th century, it is a common and accepted practice for government to collect, keep and use information about individuals to further legitimate government interests. However, a problem arises when biometric data, collected for authorized purposes, is transferred to other government or non-government agencies for purposes not originally intended.

In *Perkey*, a privacy objection was raised over how the fingerprint information was used. Although the Department of Motor Vehicles initially collected fingerprints to develop a driver database with the goal of reducing driver's license fraud and increasing highway safety, it routinely provided or sold information in the database to third parties for non-safety related purposes. Because California's state constitution acknowledged a right to personal privacy and state statute recognized the privacy of one's personal information, the California Supreme Court ruled the Department of Motor Vehicles was prohibited from disseminating the fingerprint data to unauthorized third parties [78]. It is significant to note, however, that individuals in states with specific state privacy protections, like California, will likely be afforded a greater degree of privacy than individuals in states without.

The challenge for society is to create policies ensuring that biometric identification data is used for disclosed, accepted and legitimate purposes, and is not made available for purposes for which the data was not originally intended. Perhaps the most effective manner to accomplish this is through legislation. For example, when New York amended section 139-a of its Social Services Law to allow finger imaging as a qualification for obtaining welfare in selected counties, the law stipulated that the fingerprint data could not be used for any other purpose than the prevention of multiple enrollments for home relief [79]. Therefore, this statute allows fingerprint imaging for the purpose of disbursing welfare, but may not be used to identify the individual for any other purpose. Provisions like this may go far in developing social acceptance for the use of biometrics as well as protecting government due process requirements and individual privacy rights.

Certainly, provisions like those imposed in New York will prove fruitful in permitting the widespread use of biometrics in both the public and private sectors. In an era of increased state autonomy and less national oversight, states may well be saddled with the responsibility of working with each other to coordinate effective privacy provisions that make biometrics a socially acceptable form of data collection. Historically, however, the American experience fully demonstrates that state regulation of nationwide programs is uneven at best and devastating to the national public interest at worst. Last century, during the last heyday of states' rights, problems like child labor, unscrupulous business production methods, environmental degradation, employee and product safety, sanitation, and many other socially important issues, were routinely ignored in favor of free market forces. Given the nation's failed attempt at state controlled regulation a century ago, and the recent re-emphasis on "states' rights", broad social acceptance of biometrics may be stalled if a patchwork of privacy regulations fail to uniformly secure the reasonable privacy expectations of people throughout the nation

Perhaps the most realistic long-term, legitimate and thoughtful balance between individual control of private information and the public and private sectors' need to access that information lies in national legislation, where standards for authorized and unauthorized biometric data collection and transfer can be applied uniformly. In the short term, federal legislative efforts to comprehensively define the acceptable use of biometrics in both the public and private sector appear unlikely. In the wake of devolution of federal authority and evolution of state authority, an uneven patchwork of varying privacy protections of biometric data may emerge, inevitably raising difficult issues in the courts.

In a limited way, the Privacy Act of 1974 [80] may serve as a useful model of existing national legislation to balance the need for biometrics with individual privacy. This act established a framework of rights and remedies for the subjects of records maintained by federal agencies. The law's many provisions generally require federal agencies to keep records with due regard for the privacy interests of the people who are the subjects of those records and gives these people some control over the collection, maintenance, use and dissemination of information about them [81]. If state and local jurisdictions, in collecting biometric data, fail to consider adequately the privacy interests of individuals, Congress could attempt to apply many of the principles inherent in the Privacy Act to state and local agencies. While any such attempt may elicit cries of federal invasion into states' rights, these may fall on deaf ears in the courts if egregious privacy invasions result due to uncoordinated efforts by the states to protect individuals.

11.4 Conclusions

There is little doubt that, in the coming years, more government agencies will begin using biometric technology to increase the perception of security and to assist their service delivery systems. It is also true that our society's technological genius has advanced computer and biometric technology such that it is becoming easier each day to link the identity of an individual to a life history of personal information. With this fascinating but troubling reality must come responsible and thoughtful decision-making. In a democracy, power ultimately rests with the people, and if the people ultimately feel threatened by biometric technologies, they certainly have the collective power to carefully modify, or even stop, its use in even reasonable situations.

Ultimately, before biometric technologies can become successfully integrated in public sector settings, several challenges must be met. First, government agencies must provide timely and substantive procedures for individuals wishing to challenge adverse decisions based on biometric measures. This acknowledges that comparisons based on biometric measures can be in error. Second, government agencies should not use biometrics to gather information about people covertly. For example, prior to required fingerprinting, the California Department of Motor Vehicles would routinely lift latent fingerprints from a driver's license application if the applicant refused to volunteer them during the application process [82]. The use of biometrics by government agencies in public settings should be fully disclosed [83]. Third, government should not collect unnecessary personal information about individuals. Government must not serve as a warehouse of personal information on citizens, especially if the information is unrelated to the implementation of specific public policies. An overzealous data collection strategy will lead to public distrust in government policy-making and the potential for abuse through "data mining". Fourth, government should not improperly use information obtained for a proper purpose nor disclose the information for unauthorized use. An agency, legally required to accommodate particular privacy rights, in giving private information to other government jurisdictions for other, even reasonable, purposes, will find itself facing legitimate privacy challenges. An interesting dilemma, however, is posed by the possibility of a government agency, unrestrained by privacy provisions, giving biometric information to another agency similarly unshackled. The complexity of our federal relationship in the USA, and the uneven patchwork of privacy rights from state to state could pose serious obstacles for the widespread use of biometric technologies and vexing challenges for the courts. Finally, government should be vigilant in ensuring the accuracy of the information collected.

The importance of these precautions cannot be overstated. Each of these can become a significant measure in determining whether government is behaving fairly and within the perimeters of due process of law and privacy protection. Additionally, from a non-constitutional perspective, each of these precautions, if heeded, should significantly assuage societal fears about infusing biometric technology into daily life. The higher the social trust in government, which is currently at record low levels, the greater will be the chance of both the public and legal acceptance of this emerging technology into our society.

References and Notes

[1] H. Faulds, On the skin furrows of the hand. *Nature*, **22**, 605, 1880.

[2] For a discussion of the relationship between the rule of law and fairness or due process, see W. H. Simon, The legacy of *Goldberg v. Kelly*: a twenty year perspective: the rule of law and the two realms of welfare administration. 56 *Brooklyn Law School* 777 (1990).

[3] M. Trauring, On the automatic comparison of finger ridge patterns. *Nature*, **197**, 938–940, 1963.

[4] U.S. Department of Commerce, Bureau of Census, *Population, Housing Units, Area Measurements, and Density: 1790 to 1990*. Available online at: `http://www.census.gov/population/censusdata/table-2.pdf`.

[5] For instance, California State Election Code 14216.

[6] The idea that public political activity can be accomplished anonymously has a long tradition, dating to Revolutionary War days, upheld by the courts in decisions such as *Talley v. California*, 362 U.S. 60 (1960) and *McIntyre v. Ohio Elections Commission*, 514 U.S. 334, (1995).

[7] S. H. Kadish, Methodology and criteria in due process application – a survey and criticism. 66 *Yale Law Journal* 319 (1957).

[8] In *Carey v. Piphus*, 435 U.S. 247 (1978), Justice Marshall wrote that the two concerns of procedural due process are the prevention of unjustified or mistaken deprivations and the promotion of participation and dialogue by affected individuals in the decision-making process, at 259–262 and 266–267.
[9] L. Tribe, *Constitutional Law*. The Foundation Press, 1988.
[10] *Marshall v. Jerrico*, 446 U.S. 238 (1980).
[11] In *O'Bannon v. Town Court Nursing Center*, 447 U.S. 773 (1980), which dealt with the unwilling transfer of elderly patients form a decertified nursing center, Justice Blackmun, in his concurrence, noted that as individuals are more specifically singled out from a larger group, their due process rights grow stronger, at pp. 800–801.
[12] *Id.*, at pp. 799–800. Justice Blackmun noted that as "governmental action affects more than a few individuals, concerns beyond economy, efficiency, and expedition tip the balance against finding that due process attaches. We may expect that as the sweep of governmental action broadens, so to does the power of the effected group to protect its interests outside rigid constitutionally imposed procedures".
[13] *Id.*, at 803. When a government action entails a serious likelihood of added stigma or of harmful treatment that increasing puts an individual at risk of death or grave illness, government action must include input by those affected by the decision.
[14] *Hannah v. Larche*, 363 U.S. 420, at 442 (1960).
[15] *Solesbee v. Balkcom*, 339 U.S. 9, at 16 (1950), Frankfurter, J., dissenting.
[16] See for example, *Mullkane v. Central Hanover Bank and Trust Co.*, 339 U.S. 306, at 314–315 (1950), in which the Court opined that notice must be reasonably calculated, under all circumstances, to apprise interested parties of the pendency of the action and afford them an opportunity to present their objections.
[17] *In re Oliver*, 333 U.S. 257, at 266–273 (1948).
[18] *Coolidge v. New Hampshire*, 403 U.S. 443, at 449 (1971).
[19] *Tumey v. Ohio*, 273 U.S. 510, at 523 (1927).
[20] *Goldberg v. Kelly*, 397 U.S. 254 (1970).
[21] The Court noted that welfare provides the means to obtain essential food, clothing, housing and medical care and that terminating the welfare benefits, rightly or wrongly, will have the inevitable effect of depriving an individual of the very means by which to live. Therefore, considering the grave consequences, terminating welfare benefits for which an individual had statutorily qualified, without first allowing the individual to demonstrate the decision to be erroneous, violated constitutional due process.
[22] *Morrissey v. Brewer*, 408 U.S. 471 (1972).
[23] *Gagnon v. Scarpelli*, 411 U.S. 778 (1973).
[24] *Wolff v. McDonnell*, 418 U.S. 539 (1974).
[25] *Bell v. Burson*, 402 U.S. 535 (1971).
[26] *Goss v. Lopez*, 419 U.S. 565 (1975).
[27] *Wisconsin v. Constantineau*, 400 U.S. 208 (1971).
[28] *Caulker v. Durham*, 433 F.2d. 998 (4th Cir., 1970) cert. denied 401 U.S. 1003 (1971).
[29] *Fuentes v. Shevin*, 407 U.S. 67 (1972).
[30] *Sniadach v. Family Finance Corporation*, 395 U.S. 337 (1969).
[31] *Vlandis. v. Kline*, 412 U.S. 441 (1973)
[32] *Ewing v. Mytinger and Casselberry, Incorporated*, 339 U.S. 594 (1950).
[33] *North American Cold Storage Company v. Chicago*, 211 U.S. 306 (1908).

[34] *Yakus v. United States*, 321 U.S. 414 (1944).
[35] *Gonzales v. Freeman*, 334 F.2d. 570 (D.C. cir., 1964).
[36] *Cafeteria and Restaurant Workers Union v. McElroy*, 367 U.S. 886 (1961).
[37] *Mathews v. Eldridge*, 424 U.S. 319 (1976).
[38] Almost universally in the USA, two forefingers are imaged in social service applications requiring fingerprints.
[39] In this context, we can quantify "accuracy" as the complement of false match rate. That is, accuracy = (1 – false match rate).
[40] See J. J. Killerlane III, Finger imaging: a 21st century solution to welfare fraud at our fingertips. 22 *Fordham Urban Law Journal* 1327, at 1331. The State determined that for every $1.00 spent on the program, the two counties saved $4.50 in welfare fraud.
[41] *N.Y. Soc. Serv. Law* 139-a (3) (f), 1994.
[42] The rights/privilege distinction has been much debated of late. The courts have generally ruled that individuals have no due process protection if government chooses to deny to an individual a privilege. The debate gets murky, however, because increasingly, society is having a harder time determining what is a right and what is a privilege. For example, although there is no constitutional right to a driver's license, driving privileges have become so engrained in our society, and perhaps so necessary for an individual's basic survival, that through usage, the privilege of driving has become, over time, a right.
[43] *Christopher Ann Perkey v. Department of Motor Vehicles*, 42 Cal. 3d. 185; 721 F.2d. 50: 228 Cal. Rptr. 169 (1986).
[44] While the findings of *Perkey* are binding only within California, the principles regarding due process are instructive in a broader sense and may indicate how similar cases could be argued in federal, or other state, courts.
[45] California Vehicle Code, Sect. 12800 (c).
[46] *Hale v. Morgan*, 22 Cal. 3d. 388, at 398, n. 4 (1978).
[47] One study conducted in 1980 indicated that there were approximately 186,000 more driver's licenses issued to males between the ages of 20 and 44 than there were eligible persons in this age group. *Perkey*, at 190.
[48] Even without a specific right to privacy, the United States Supreme Court has interpreted the Constitution to protect the privacy rights of individuals in some types of circumstances. See E. Alderman and C. Kennedy, *The Right to Privacy*. Vintage Press, New York, 1997.
[49] S. D. Warren and L. D. Brandeis, The right of privacy. 4 *Harvard Law Review* 193 (1890).
[50] J. R. Reidenberg, Privacy in the informational economy: a fortress or frontier for individual rights. 44 *Federal Communications Law Journal* 195 (1992).
[51] W. Prosser, Privacy. 48 *California Law Review* 3 (1960).
[52] *Osborne v. United States*, 385 U.S. 323 (1966).
[53] *Perkey, Supra* n. 41, Bird, C.J., concurring.
[54] S. Kent and L. Millett (eds), *Who Goes There?: Authentication Technologies through the Lens of Privacy*. Committee on Authentication Technologies and Their Privacy Implications, National Academies of Science Press, 2003.
[55] *Talley v. California*, 362 U.S. 60 (1960).
[56] This is not withstanding "trusted third party" approaches suggested by some vendors for biometric pattern encryption. See N. Ratha, J. Connell and R. Bolle, Cancelable biometrics. IBM Exploratory Computer Vision Group, *Proc. Biometrics Consortium 2000*, September 2000 and J. Cambier, U. von Seelen, R. Glass, R. Moore, I. Scott, M. Braithwaite and J. Daugman, Application-specific

biometric templates. *Proc. Third Workshop on Automatic Identification and Advanced Technologies*, March 14–15, 2002. Tarrytown, New York, USA.

[57] United States Constitution, Amendment Four of the Bill of Rights.

[58] *Katz v. United States*, 389 U.S. 347 (1967).

[59] A literal reading of the Fourth Amendment requires government to have probable cause to justify a search. However, the Court has eroded probable cause standard and allows searches when doing so furthers the public interest. For a more complete analysis of the erosion of probable cause standard, see K. Nuger, the special needs rationale: creating a chasm in Fourth Amendment analysis. 32 *Santa Clara Law Review* 1 (1992).

[60] In *Smith v. U.S.*, 324 F.2d. 879 (D.C.Cir. 1963), the court ruled in part that, at least as part of the normal identification process, taking a photograph or fingerprint is consistent with Fourth Amendment requirements.

[61] *National Treasury Employees Union v. Von Raab*, 489 U.S. 656 (1989).

[62] In *Rochin v. California*, 342 U.S. 165 (1957), the Supreme Court declared that forcing a drug suspect to have his stomach pumped was police behavior that shocked the conscious of a civilized society and therefore, violated the Fourth Amendment ban against unreasonable searches and seizures.

[63] *People v. Scott*, 21 Cal. 3d. 284 (1978).

[64] *Perkey*, at 191.

[65] In a 1990 study funded by the Federal Highway Administration, less than 8% of 10,000 users of a fingerprint scanner agreed to the statement "It invades your privacy". See "Personal Identifier Project: Final Report", The Orkand Corporation, April 1990, DMV 88–89, page IV-D-9, available from the Biometric Test Center at San Jose State University.

[66] In a 2001 study, when given a choice between the statements, "Some people oppose the use of finger-imaging biometrics to verify a person's identity, because they feel this procedure is like fingerprinting and seems to treat people like presumed criminals" and "Other people approve finger-imaging identification as an easy-to-use process that helps protect individuals and the public against fraud" 20% agreed most strongly with the first (oppose) and 77% agreed more strongly with the second (approve). A. Westin, *Public Attitudes Toward the Uses of Biometric Identification Technologies by Government and the Private Sector: A SEARCH Survey*. Available online at: `http://www.search.org/policy/bio_conf/Biometricsurveyfindings.pdf`.

[67] Less than 13% of the 10,000 users of a retinal scan system agreed with the statement "It invades your privacy" in the 1990 Orkand Corporation study noted previously. At this writing, retinal scanners are no longer commercially available.

[68] *McCarthy v. Arndstein*, 266 U.S. 34 (1924).

[69] For example, in *Amalgamated Transit Union v. Sunline Transit Agency*, 663 F.Supp. 1560 (C.D.Cal. 1987), the District Court iterated that "the taking of a urine sample is no more testimonial in nature that the blood test in *Schmerber*, the handwriting exemplar in *Gilbert v. California*, 388 U.S. 263 (1967) or *United States v. Wade*, 388 U.S. 218 (1967)".

[70] *Malloy v. Hogan*, 378 U.S. 1 (1964).

[71] *Breithraupt v. Abram*, 352 U.S. 432 (1957).

[72] *Schmerber v. California*, 384 U.S. 757 (1966).

[73] In *New York v.Quarles*, 467 U.S. 649, at 671 (1984), Justice O'Connor concurring, noted how an interrogation does not offend the values underlying the Fifth Amendment any more than compulsory taking of blood samples, fingerprints or voice exemplars.

[74] *Gilbert v. California*, 388 U.S. 263 (1967).
[75] See also *U.S. v Chibbaro*, 361 Fed. 2d 365 (Cal. 3d Cir.) (1966) and *People v Graves* 64 C.A. 2d. 208 (1966).
[76] See R.J. Weintraub, Voice identification, writing exemplars and the privilege against self incrimination. 10 *Vanderbilt Law Review* (1957).
[77] *Miranda v. Arizona*, 384 U.S. 436 (1966).
[78] California's Vehicle Code, Section 1808, provides certain exceptions to this privacy requirement, including release of the home addresses of certain public officials (1808.4), the records of individuals convicted of substance abuse laws that relate to their mental or physical health (1808.5), records of certain convictions (1808.6) and dismissals of traffic offenses (1808.7).
[79] N.Y. Soc. Serv. Law 139-a (3)(b). This law stipulates in part that any finger imaging data "may not be used, disclosed or redisclosed for any purpose other than the prevention of multiple enrollments in Home Relief, may not be used or admitted in any criminal or civil investigation, prosecution, or proceeding, other than a civil proceeding pursuant to section 145-c of this article", which addresses individuals attempting to defraud the welfare system.
[80] 5 U.S.C. Sect. 552a.
[81] See M. N. Flics, Employee privacy rights: a proposal. 47 *Fordham Law Review* 155 (1978).
[82] *Perkey*, at 191.
[83] For a discussion on the surreptitious governmental use of facial recognition at the Super Bowl XXXV sporting event, see J. Woodward, *Super Bowl Surveillance: Facing up to Biometrics*. RAND Corporation, IP-209, 2001. Available online at `http://www.rand.org/publications/IP/IP209/`.

12 Privacy Issues in the Application of Biometrics: a European Perspective

Marek Rejman-Greene

12.1 Introduction

Biometric methods of authentication offer a more secure link between a specific individual and a non-human entity. Numerous trials and deployments demonstrate the wide range of possible application: restriction of access to physical spaces and electronic resources to those individuals who have been previously cleared; denying the opportunity for potential fraudsters to assume multiple identities; enforcing accountability for individuals undertaking electronic transactions; and matching facial images from CCTV cameras to databases of criminals. It is not surprising that such breadth of application has prompted concern. In the main, concerns appear to centre on threats to the end user's privacy, but we believe that the issues are more complex and that a clearer appreciation of end user perceptions would enable system integrators, designers and the customer to respond in a more appropriate manner. Should the issues be primarily those of privacy[1], there is a Europe-wide legal framework against which all future deployments should be assessed. As this framework is often inaccessible to those unfamiliar with new political and legal developments in Europe, we introduce it in the context of the historical development of the individual's rights to privacy.

For many Europeans, the widespread use of biometric technologies in films and the perception of these techniques as "perfect" have reawakened the fears of an all-knowing computer system able to track every citizen and consumer from the cradle to the grave. The somewhat more informed citizens express a worry about the security of such a biometric-based identifier and the consequences of reuse of templates in other ways, perhaps placing the reputation of the individual at risk. And beyond these concerns are the future possibilities of the use of DNA data in tracking people, and in the linking of biometrics with parallel developments in other surveillance technologies.

---

1 Note that in this context, "privacy" refers to issues of personal rights as perceived by individuals, and not the alternative definition of confidentiality of communications.

European sensitivities towards privacy issues have developed against fears of a growing power imbalance between government and citizen, and between corporation and consumer with innovations in data mining and data matching techniques contributing to the imbalance. The development of philosophical notions of inalienable rights of the individual was heavily influenced by the experience of many countries both during the Second World War, and under post-war Central and Eastern European governments where (generally) manually operated filing systems tracked dissident citizens and members of minorities. Such notions were first codified in the 1950 European Convention on Human Rights (ECHR) [1]. Two decades later, with the commercialisation of large mainframe computers, the first laws to protect "personal data" about individuals were drafted, based upon an internationally agreed framework but with a local interpretation.

As personal information began to be used in services offered by multinational organizations, and with the expansion of the European Union, the need for harmonisation of these laws required a Europe-wide legal consensus. The 1995 Personal Data Directive [2], and its transposition into national laws, offers the legislative underpinning to any discussion about the use of biometrics in modern systems in Europe. However, there have been many criticisms of its approach. Its approach predated the age of the Internet, and its complexity rendered it opaque to the average person. More specifically it has been criticized for difficulty in application to new technologies, even though it was designed to be a framework directive, allowing more specific legislation to be framed based upon its guiding principles [3]. Discussions are beginning on changes to this underlying framework. However, in this chapter we examine how the application of biometrics could be influenced by the current legislative position, and suggest possible changes in the future. We note that the scarcity of case law on data protection (especially as it might apply to biometric methods) makes it difficult to move significantly beyond a restatement of statutory provisions viewed through the lens of comments from a few noted legal observers. Note also, that in addition to these Europe-wide data protection and privacy laws, specific national laws may restrict the use of certain biometric methods that are considered to be particularly sensitive, such as those applied to fingerprint records or databases of facial images.

Many deployments of biometric-enabled systems will not just take account of what appears to be legally possible. A well-designed system (making use of 'socio-technical' design principles) will also be sensitive to the perceptions and concerns of end users and their social groups. Biometric technologies are almost unique as a security mechanism in the need for cooperation by the end user to ensure their correct operation[2]. To ensure their success, early deployments should be grounded on research into user perceptions, ascertaining their underlying concerns through

2 Use of automatic face recognition on images from databases of photographs and CCTV outputs may be an exception to the requirement for user cooperation.

more advanced techniques than simple questionnaires. Based upon the results of these studies, the agency can launch an educational program to reassure future end users. Such research and the output of Privacy Impact Assessments might also mandate the implementation of additional security procedures and more advanced technical solutions. Early sight of these requirements during the initial stages of the system design cycle will simplify their secure integration.

Some of the user concerns can be addressed directly, for example by impartial studies into any health and safety issues, although it is clear that attitudes may take time to change. Those concerns that are less clearly articulated will require more extended studies. We shall refer to the results of preliminary work in European organizations that indicate a possible route forward.

12.2 Privacy – from Philosophical Concept to a Human Right

The notion of individual privacy appears to be a modern phenomenon – at least for the majority of the population in a society. In less mobile societies with poor roads, few people would venture outside their immediate neighborhood and the arrival of fairs or itinerant travelers was subject to closely circumscribed laws [4]. Fears of the spread of disease, the supply of poor quality goods and the possibility of having to care for any ill or unemployed newcomers, constrained the influx of immigrants. In these societies, the daily lives of the ordinary people were led without much privacy. Indeed, the strong Puritan tradition in 17th century England and the American colonies seemed to encourage a panoptic[3] surveillance by one's neighbors.

Shapiro regards the partitioning of rooms in a household as the first step to a culture of privacy and individuality [5]. The impact of this architectural development was minor compared with the expansion in transport infrastructures in the 18th and 19th centuries, such as improvements in road quality and the creation of a canal and railway network (the latter going in hand with the first electronic communications – the telegraph). The rapid urbanization of much of Western Europe and parts of the USA completed the options for many citizens to move outside of their place of birth and schooling, and to assert an individuality apart from their kinship groups. The accompanying increase in crime rates required a curbing of the anonymity that life in a city could offer. Hence the second half of the 19th century saw the introduction of the census and codification of laws on recording births, marriages and deaths. This was also the time of first use of

3 Jeremy Bentham's *Panopticon* was an architectural design that ensured prisoners were always aware that they could be under observation, whilst not being certain that warders were watching them.

biometric identities for tracking and recording criminals, albeit not automatically identifying individuals in the way we understand biometrics today. Initially this aimed to collect as much information about externally visible features and easily measurable dimensions, Bertillon's anthropometry [6] being the most celebrated scheme. This short-lived approach was superseded a few years later by the discovery of the remarkable individuality of fingerprints. By the turn of the 20th century, Scotland Yard had embarked on the use of the hugely successful Galton–Henry classification system and the fingerprint as a key forensic tool had arrived.

Public authorities could be overzealous in their tracking of criminals; and the middle classes, growing rapidly in influence and numbers as the complex economy demanded more in the service sector and more services from government, articulated a need to limit their powers. In much of Europe, the tradition of the Roman Empire and its successor, the Church of Rome, had already determined a right for state, economic and religious institutions to maintain close surveillance on its peoples. Westin comments that this contrasted with English common law traditions and the Protestant emphasis on individual rights [4]. Although the legal systems based upon these two traditions had still to be reconciled in the long debate on a Europe-wide personal data framework, by the beginning of the 20th century, the plea for a 'right to be left alone' from the civil authorities was already being articulated[4]. With the questioning of the power of a state to affect all facets of the life of the citizen, one part of the personal privacy debate had started. The other aspect, that of giving individuals a right over the way that information about them is collected and used, was to remain less pressing for another half century. Manual records on populations of millions were always going to be inefficient in impacting the lives of individual citizens.

The Second World War demonstrated how the accumulation of personal data in the hands of unscrupulous authorities could be abused. Some countries already had well-established manual systems with lists of people of Jewish origin, while in others no such data had ever been gathered. The difference between these two approaches to the collection of personal records was tragically confirmed once these countries were invaded. The use of mechanical sorting equipment, which was already being imported, could have greatly simplified the organization of information about the citizens of occupied Europe. It is not surprising that the publication of Orwell's *1984* in early post War Europe should have struck such a chord. In 1950, the participating states to the Council of Europe articulated a response in Article 8 [7] of the ECHR guaranteeing a right of privacy: "Everyone has the right to respect for his private and family life, his home and his correspondence". This was linked to freedom from interference by public authority in the exercise of this right except in closely defined circumstances. The Convention offered individual redress against governments

4 The phrase is generally attributed to Warren and Brandeis in 1890: The right to privacy, reprinted in [4]

abusing their authority by an ultimate personal appeal to the European Court of Human Rights in Strasbourg.

The increasing prosperity of the 1950s and early 1960s was accompanied by a belief in the benefits of technological progress and organizational efficiency. In particular, governments in Europe were attracted to the potential of computerization of records, firstly, of the census, and then of other services such as social welfare payments. Just as the first wave of such proposals were about to be implemented, the climate of thought amongst Europeans changed. Although the events of 1968 have been characterized as a rebellion by the youth of Europe, other currents of opinion were questioning the wisdom of concentrating power, and the information on which power is built, without countervailing checks and balances. In Germany, the notion of "informational self-determination" seemed to capture the essence of the second meaning of privacy protection.

The world's first data protection act, passed in the German state of Hessen in 1970, was directed at offering this check on the operations of a regional government, but as more countries recognized the need for such legislation, the scope widened to take in commercial use of personal data as well. Increasingly, the limitations of national laws in a rapidly globalizing world led to calls for an international system for data protection, to protect against states with no laws or inadequate laws from becoming "data havens" with no controls on the processing of data. For example, shortly after Sweden passed a data protection act in 1973, export licenses to the UK were not given in two instances on the basis of a lack of a corresponding act in the UK at that time. In 1980, the Organisation for Economic Cooperation and Development (OECD) adopted guidelines on cross-border data flows [8], while a year later, the Council of Europe Convention [9] set common standards for legislation based upon a human rights approach, aiming to harmonize the differing laws in continental countries.

Although these agreements were influential in determining the course of subsequent laws – such as the first UK Data Protection Act in 1984 – by 1990 it was clear to the European Commission that the lack of a common framework, under which personal information could be gathered, processed, stored, transmitted and disposed of securely, was likely to impede the commercial development of both existing and novel services. In its introduction to the rationale for the harmonization of data privacy laws, the Commission states [10]:

> Developments of a frontier free Internal Market and of the so called "information society" increase the cross-frontier flows of personal data between Member States of the EU. In order to remove potential obstacles to such flows and to ensure a high level of protection within the EU, data protection legislation has been harmonised. The Commission also engages in dialogues with non-EU countries in order to insure a high level of protection when exporting personal data to those countries.

Over the course of the following five years, the Commission (together with its two counterparts in the European law-making process, the European

Parliament and the Council of Ministers from the then 15 member states) agreed the principles for an EU-wide directive of 1995 [11]. This required governments in each of the countries to transpose the directive into national law by 1998. (The current status of this implementation can be checked on the Internet [12].) The Personal Data Directive was designed to be a framework for other, more specific, directives and laws that would apply the principles to specific circumstances. The first additional directive applied to the telecommunications sector, a directive that has recently been updated and is required to be implemented in the national laws of member states by 31 October 2003. In spite of its recent agreement, there have already been calls to make changes in the light of experience in applying the framework directive, 95/46/EC; see, for example, [14]. As such changes are unlikely to impact on the application of biometrics for a number of years, this article will confine its remit to the current directive, and wherever appropriate, to the implementation of the directive in the UK as the Data Protection Act 1998 [15].

A number of practical interpretations of the directive and the national laws have been prepared. Some are available publicly, while others are offered as part of a consultancy package. For example, the European Commission GUIDES project has produced a set of guidelines for those active in the e-Business sector [16]. Another useful resource is the Privacy Audit Framework for compliance with the Dutch Data Protection Act, and for which an English translation is available [17]. This framework enables an external auditor to check that an organization's services operate correctly under this Members State's law, and to issue a certificate against the successful completion of the audit.

12.3 The European Personal Data Directive

In overview, this directive establishes Eight Principles[5] of personal data protection which determine the legality of the processing of such data. Personal data must be:

1. Processed fairly and lawfully.
2. Collected for specified and lawful purposes and not processed further in ways that are incompatible with these (the "finality" principle).
3. Adequate, relevant and not excessive in relation to the purposes for which they are collected or processed.
4. Accurate (and where necessary kept up to date).

5 This "user-friendly" organization into eight principles – as used in the UK Data Protection Act – is not made explicit in the directive. Principles 1–5 are contained in Article 6 of the directive, Principle 7 is mirrored in Article 17 and Principle 8 derives from Article 25. The specific rights referred to in the Sixth Principle are covered in a number of articles in the directive.

5. Not kept for longer than is necessary for the stated purposes (that is in a form that permits identification of the data subjects[6]).
6. Processed in accordance with the data subject's rights (which are explained in further detail).
7. Secure (against accidental or unlawful destruction, accidental loss, alteration, unauthorized disclosure or access, using measures that have regard to the state of the art and costs of implementation, and ensuring that a level of security is maintained that is appropriate to the risks represented by the processing and the nature of the personal data to be protected).
8. May only be transferred to those countries that ensure an adequate level of protection for the personal data.

There are several additional provisions of note, before we examine the application of this law to biometric-enabled systems:

- Processing of data extends over the whole life cycle of data, and covers (or will cover after a transitional period) data held in paper records as well. (Article 2b of the Directive).
- Each country must provide for a supervisory authority (in general a Data Protection Commissioner's Office, although in the UK this is fulfilled through the office of the Information Commissioner, with additional responsibilities for Freedom of Information) to whom controllers of data processing operations must, in general, notify their intentions[7] (Articles 28, 18–21).
- Coordination of the activities of the Commissioners is undertaken by an Article 29 Data Protection Working Party, which publishes regular statements on issues common to all member states [18].
- Certain categories of data are deemed to be specially sensitive and processing of this type of information is allowed only under specific conditions (Article 8).
- Article 15 grants a right to individuals not to be subject to a decision that produces legal effects or significantly affects them, is based solely on automated processing of data intended to evaluate certain personal aspects relating to them, e.g. performance at work, creditworthiness, reliability and conduct.
- The Sixth Principle offers specific rights for data subjects to obtain copies of certain data about themselves, to request corrections, etc. (Articles 10–12, 14).
- There are exemptions for many of the Principles, but, overall, the Directive does not apply to three categories of activities: firstly, people who undertake processing of data for purely personal or household reasons

6 "Data subject" is the identified or identifiable natural person whose personal data is being processed.

7 In Germany, large institutions and organizations may have their own Commissioners.

are exempt; secondly, certain activities of governments or their agencies are exempted, including those relating to public security, defense, state security and the activities of the state in areas of criminal law; and thirdly, exemptions for certain journalistic and literary activities are listed (Articles 3, 13).
- In a few instances, national laws can be framed selecting one of a number of optional clauses. Note, in particular, that individual member states can extend protection to individuals beyond those mandated in the directive (Article 4).

12.4 Applying the Directive and National Laws to Biometric Systems

Although Data Protection Commissioners recognize that biometrics offers a challenge to the legal framework on personal data and privacy, to date only three have explicitly considered the ground rules for operation of biometric-enabled systems. The Dutch Commissioner's office overviewed biometric methods in 1999 and assessed the application of the European Privacy Directive to them. A report was published and is available on the web in English [19].

More recently, CNIL, the French data protection commission, has undertaken a major study into the privacy implications of biometrics. It found that there was a lack of reliable information about how biometric-enabled systems operate in practice and confirmed that, in general, technologists and data controllers were not aware of the rights of end users. In view of the potential harm that could result to end users from systems not designed in accordance with data protection principles, CNIL has proposed a number of measures. In its 2001 annual report [20], CNIL categorized applications using biometrics into two broad groups. It maintained that there was no problem with systems where the template storage is under the end user's control, e.g. stored on a card, a PC or a cellphone in the possession of the user. The second class, where the template is stored in a centralized database, is more complex. Where the biometric record is of a type that leaves no trace or is not easily captured without the cooperation of the end user (such as eye-based systems or those applying hand geometry devices), integrators can use these methods, provided that the usual data protection principles such as finality and proportionality are observed. In contrast, centralized template storage using biometrics that leave a trace or can be easily obtained (such as systems with face, fingerprint or DNA recognition) should only be applied in high security systems. Furthermore, CNIL proposed that, in the near future, systems using biometrics should be specifically authorized, and (in the absence of any independent assessment process) a labeling scheme for IT products (including biometrics) that conform to the data protection principles should be instituted.

In response to the world-wide reaction against recent acts of terrorism, the Article 29 Working Party emphasized the exceptions already offered under both the Framework Directive and the then Telecommunications Directive, warning against "the proliferation of the use of means of identifying, and more generally, gathering of data relating to individuals through the use, for instance, of biometrics" [21]. The German data protection authority has also recently published a more detailed analysis of the role of biometrics in this context.

As we shall see later, the transposition of the Personal Data Directive into national laws, and the probable way in which the application of these laws to biometric systems will be interpreted, differs across the 25 nations of the European Union. As relatively few systems have been implemented in European countries, and fewer still are deployed to the public at large, there has not been an opportunity to test their status. The observations in this paper aim to summarize some of the aspects that have been discussed in the legal community. As usual, advice on deployment of systems in a specific country should be sought from legal counsel in those member states. Their advice should cover individual national laws on the use of specific biometrics such as fingerprints, as these may add further restrictions on the application of these technologies.

The European Commission funded BIOVISION roadmap project [22] has reviewed the biometric context of the directive and national laws, and provide initial materials towards the definition of a code of conduct for applications making use of a biometric in a privacy-compliant manner. A parallel activity is being undertaken by the UK government managed Biometric Working Group [23].

12.4.1 Biometric Data as "Personal Data"

Perhaps the aspect of personal data protection law that has been debated most extensively is the question of application of the law to biometrics. To what extent is biometric data, "personal data" within the meaning of the directive and the national laws?

The directive defines personal data to be "any information relating to an identified or identifiable natural person", making the distinction with legal entities such as companies. Furthermore, it amplifies the definition by stating that an identifiable person is one who can be identified directly or indirectly, in particular by reference to

- an identification number; or
- to one or more factors specific to his physical, physiological, mental, economic, cultural or social identity.

Recital 26 of the directive states that "account should be taken of all the means likely reasonably to be used either by the controller or by any other person to identify the said person" and that the principles should not apply once the data is rendered anonymous in such a way that the data subject is no longer identifiable. The UK implementation of the directive limits the application to living persons (the directive itself is silent on this aspect) and takes a different approach towards defining personal data as that which can identify an individual:

- from that data; or
- from that data and other information which is in the possession of, or is likely to come into the possession of, the data controller... Section 1(1).

Possible personal data that relate to the implementation of a biometric can include:

1. The image or record captured from the sensor at the initial enrollment.
2. Any transmitted form of the image or record between sensor and processing systems.
3. The processed data, whether completely transformed to a template or only partially processed by an algorithm.
4. The stored image or record or template.
5. Any accompanying data collected at the time of enrollment.
6. The image or record captured from the sensor during normal operation of the biometric (verification of identity or identification).
7. Any transmitted form of the image or record at verification or identification.
8. The template obtained from the storage device.
9. Any accompanying data obtained at the time of verification or identification.
10. The result of the matching process.
11. Any updating of the template in response to the identification or verification.

Situations where biometric data is not treatable as personal data are likely to be relatively rare. Indeed, the Dutch Data Protection Authority says that "(biometric) data involved will remain personal data in most, if not all stages of their processing"[8]. One case where the data is unlikely to fall within this definition is for a biometric application where all of the following conditions are met:

- The identity of a previously enrolled individual is only represented by a "one way" template with no possibility of reconstruction of the original record.
- The template could also be generated by a sufficient number of other subjects in the population.
- The template is stored on a card (or token) held by the end user.
- The comparison, at verification, of the output of the sensor with the template, is made on the card (or token) itself.
- All images and records relating to the enrollment are securely disposed of at the time of enrollment.

8 Section 3.1 of the 1999 report [19].

- No other data is available that, combined with the biometric data, could link the user uniquely to a template.
- The backup alternative, in case of failure of the biometric, does not expose the biometric to a process whereby a subsequent verification could reveal the person's identity.

Any divergence from this protocol would need to be examined in the light of the definitions of personal data. Of course, bringing the sensor onto the portable storage medium itself (e.g. a smart card), storing all data on the card and making the comparison on the card, and then transmitting the result of the comparison in a secure way off the card would move considerably towards the required level of non-identifiability. (Such a solution has been developed under the European Commission-funded Finger_card project [24]).

However, applications making use of the two modes of *Identification* and *Exclusion of Already Identified Persons* are likely to fall within the definition of personal data, and therefore within the remit of the Directive, unless one of the many exemptions offered in the Directive are invoked.

12.4.2 Biometrics and Sensitive Data

Article 8 of the personal data directive lists the following special categories of data that demand specific additional attention:

- Racial or ethnic origin
- Political opinions
- Religious or philosophical beliefs
- Trade union membership
- Processing of data concerning health or sex life

In general, the subject should have given *explicit* consent to the processing of such data[9], although there are a number of exemptions from this requirement. Note that data relating to offences, criminal convictions or security measures may only be carried out under the control of an official authority.

Those aspects that might impact on the operation of biometric methods are racial or ethnic origin and data relating to health. It is inevitable that the initial photographic image captured by the camera in a face recognition system will have some indication of race. However, it seems excessive to label it as sensitive data at this early stage of processing (as the Dutch Data Protection Commissioner believes[10]) with all of the attendant requirements for consent, if it is to be discarded following processing into a

9 Contrast explicit consent with consent offered in the form of an "opt out" box in an application form.

10 Section 3.6 of the 1999 report [19], and footnote 3 that refers to an assessment of a parallel study into privacy and CCTV.

template. Indeed, this very aspect is addressed in a current proposal for minor revisions to the Directive, jointly tabled by a number of European countries. The proposed amendment aims to make it clear that these "essentially incidental revelations... do not amount to sensitive data for the purposes of the Article 8.1" [25].

Most biometric systems have been developed, validated and tested by organizations in the USA and Europe. It not inconceivable that the algorithms that are used operate preferentially for ethnic groups that are highly represented in those geographical areas; and that, for example, directed searches for templates of facial images relating to non-Caucasians could be successfully initiated – albeit with results outputted on a probabilistic basis. Another example is the fear that images of eye features could indicate a tendency towards certain illnesses. There is little evidence for such claims, but there is little research so far to prove otherwise. It has been suggested that some systems using speaker verification could be tuned for other functions, such as detection of states of anxiety or even likelihood of lying. Any such applications could contravene the principles of personal data protection, specifically the First Principle of fair and lawful processing. The implication of the sensitive data principle for people with disabilities who require special forms of biometric, or default to a non-biometric alternative, may need to be considered.

12.4.3 Proportionality Principle

A fundamental principle in European law is that of proportionality, which some writers maintain would rule out the use of a biometric method, if the objective could be achieved in some other, less privacy-threatening way. Jan Grijpink describes how a hand geometry device is likely to be acceptable for access to buildings critical for the operation of an organization, whereas access control by means of a fingerprint biometric to a secondary school might be more difficult to justify [26]. Furthermore, many privacy advocates question the need for identification when verification against an anonymized identity will suffice. Even when a biometric identification is required – for example to confirm whether a person has already been enrolled in a given scheme – the link to other data about the individual (name, etc.) may not be necessary for the application.

12.4.4 First Principle Compliance – Fair and Lawful Processing

Processing of personal data needs to be carried out in a fair and lawful manner. This includes the act of obtaining the biometric data in the first place. Covert collection of biometric data is not permitted unless it falls within one of the defined exemptions. Wherever possible, the subject's consent should be sought, since that consent removes many of the problems for an agency deploying a biometric-enabled system. If the system includes additional software that also uses personal data, perhaps of a different form, it would seem prudent to explicitly mention the use of the biometric in the consent statement. A further decision is on whether to adopt an "opt-in" or "opt-out" approach, and again, this will depend on the specifics of the planned deployment.

12.4.5 Fourth Principle Compliance – Accuracy

By their very nature, biometric systems could occasionally return a false accept, and with it the possibility of an inaccurate record of activity against another individual. Whether this is considered as a failing in accuracy or in security (the Seventh Principle), the system designer and implementer should take appropriate steps to ensure that the personal data of the individual whose identity has been assumed is not compromised.

12.4.6 Seventh Principle Compliance – Security

The Seventh Principle (Article 17 of the Directive) requires the controller (the person or agency that determines the purposes and means of processing of the personal data) to implement appropriate technical and organizational measures to protect the personal data. It requires the controller to offer a measure of protection against

- unlawful destruction or accidental loss;
- alteration;
- unauthorised disclosure or access; and
- all other unlawful forms of processing;

in particular where the transmission involves the transmission of data over a network. If controllers do not carry out the processing themselves, then the processors actually undertaking the work must provide guarantees that the specified security measures are carried out. In addition, a legal contract must be in place between the controller and the processor, and the controller must ensure compliance with these measures. The measures should take account of the state of the art and assess the costs and risks involved.

It is clear that the framers of the directive understood the impact of inadequate security on the integrity and confidentiality of personal data. With biometric-enabled applications, this may be an important consideration, and maintaining the security of biometric data is particularly critical in reassuring end users that their identities will not be stolen in this way. It will also hinder the "function creep" that could extend the range of uses beyond those for which the original system design was intended.

Biometric methods themselves may well be required in order to provide the requisite levels of security for other IT systems processing personal data. In such applications, use of traditional user authentication methods based on "what you know" and "what you have" may not be sufficient, and the risk–benefit analysis based on the state of the art will require the use of a stronger authentication. This biometric itself has to be secure and securely integrated for the Seventh Principle to be met.

The UK Data Protection Act reminds controllers of the need to ensure the reliability of employees who have access to personal data and the accompanying detailed advice from the Information Commissioner makes reference to ISO standard 17799 [27] on information security.

12.4.7 Eighth Principle Compliance – Transfer to Third Countries

Transfer of data outside of those countries that have an adequate level of protection is not allowed except under specific conditions (Article 25 of the Directive). The adequacy of the protection is judged against a number of criteria:

- Nature of the data
- Purpose and duration of the processing operation(s)
- Countries of origin and final destination
- Rules of law, both general and sectoral, in force in the third country
- The security measures that are complied with in that country

A number of derogations from the strict requirements of this principle are listed in the directive (Article 26). Transfer of personal data to countries lacking this level of adequacy may take place on a number of grounds. Among these is the unambiguous consent [28] of the data subject, consent being defined as "any freely given specific and informed indication of his wishes... (signifying)... his agreement". (Note the distinction between this type of consent and "explicit consent" required in the case of processing of sensitive data.) Other grounds include necessity for the performance of a contract or protection of the vital interests of the data subject. It is not surprising that ensuring compliance in all states for services offered by multinational companies can result in escalating costs for legal advice.

One approach for organizations is to agree on bilateral transfers of data using approved contractual commitments in accordance with section 26(4). A better solution is for the laws and practices in that target country to be declared to provide this degree of adequacy. The European Commission, after obtaining the best evidence available for the adequacy of the personal data frameworks within specific countries, has issued a number of decisions in respect of the adequacy of protection in certain states. These decisions are available on the Internet [29].

12.4.8 Automatic Decision-Making

Article 15 could be interpreted as limiting the operation of certain biometric-enabled systems that significantly affected the individual. Since all biometric systems are based on the probability of a match, it is likely that there will be many instances of failed authentication. Human intervention at this point would remove the Article 15 restriction, although there will be many situations where individuals could be disadvantaged. They could miss a plane connection while they wait for an immigration officer alerted by a false reject, or be embarrassed on failing to be authenticated when together with friends or business colleagues. This article, originally in the pre-directive French legislation, requires further clarification by the data protection authorities in respect of its application to biometrics[11].

Testing of all types of devices has demonstrated that some individuals have difficulty in even enrolling. If biometric-enabled systems become the norm, will those unable to enrol using a popular from of biometric become a new underclass [30]?

12.4.9 Exemptions

The directive has many exemptions, primarily for government in national security, crime and health applications. Exemptions are also allowed for parties to a contract to agree to waive some of their rights, for the data subject to consent to exemption or where the interests of the data subject are paramount. There are additional exemptions for journalists and researchers. This illustrates one of the main criticisms of the Directive and the national laws that implement it: its complexity. With differences in interpretation already impacting on the application of biometrics, system integrators looking to cross-border deployments may experience a degree of uncertainty as to the legality of their proposals.

12.5 Article 8 of the European Human Rights Convention

Article 8 of the European Convention on Human Rights [31] offers a wide-ranging protection for individual privacy. (The ECHR has formed the basis for challenging unfair decisions of public authorities since coming into force in 1953.) This article states:

1. Everyone has the right to respect for his private and family life, his home and his correspondence.
2. There shall be no interference by a public authority with the exercise of this right except such as is in accordance with the law and is necessary in a democratic society in the interests of national security, public safety or the economic well-being of the country, for the prevention of disorder or crime, for the protection of health or morals, or for the protection of the rights and freedoms of others.

Note the two principal tests in the exemption clause: its necessity for democratic society and any exemption being in accordance with the law. Necessity entails a response to a pressing social need and has to be in proportion

11 In Germany, the automated decision making article does not apply to the biometric process itself.

to the aims of the law. Wadham and Mountfield [32] comment that the second test of accordance with the law requires:

- The need for a specific legal rule to authorise this interference.
- Adequacy of access to the specific law by an individual.
- The law must be sufficiently precisely formulated to allow the individual to foresee the circumstances under which the law could be applied.

Challenges to the legality of biometric schemes based upon this right could arise in government applications, especially where the Personal Data Directive offers exemptions, e.g. in national identity schemes, for security systems in critical infrastructures, in the criminal justice system and in the provision of medical services.

12.6 The Role of Privacy-Enhancing Technologies

Many innovative services will use personal data in order to improve the customer experience as well as providing valuable feedback to the service provider. This may allow the provider to improve the service, to market it more effectively or to sell the data to others who may configure their services accordingly. The European Commission recognized the benefits of such innovation, but was also concerned that consumers and citizens might not appreciate the significance of agreeing to such reuse of personal data. As a result, it has promoted Privacy-Enhancing Technologies (PETs [33]) that would provide a measure of protection. Their studies distinguished two types of PET:

1. Where the design of a biometric-enabled system has been specifically tailored to be privacy-respecting using the best available technologies.
2. Where measures are offered to the end users individually to enable them to protect their privacy.

Simple PETs of the second type, such as P3P, Platform for Privacy Preferences, allowing Internet users to specify at a high level what information should be passed back to a web site and with whom it should be shared, have been criticized for being too simplistic. In any case, the default conditions on purchase or installation of software were unlikely to be changed by the majority of users.

Biometric devices could be designed in accordance with PET principles, although only one system has been commercialized so far. For biometric-enabled systems to be designed for privacy and security, the customer for the deployment and the customer's system designers need to consider such requirements from the inception of a project. One example of a PET-based system of the first type stores the template only on a card held by the user.

The willingness of end users to make a system work, either on account of an immediate direct benefit (such as the use of fingerprint verifiers for the distribution of South African pensions), or to support the safety and security of the community (such as trusted traveler schemes), will only be gained following an understanding of their fears and concerns. PETs can offer some reassurance, and accreditation of systems by external entities, such as the Common Criteria scheme for IT security [34], or conformance with ISO 17799, may reassure the more technically aware.

There is little appreciation of the role that biometric methods could play in protecting the privacy of individuals, i.e. in acting as a PET, or a key component of a PET. It may be that applications that use biometrics in this manner are not afforded sufficiently high visibility. For example, in Europe, awareness of identity theft and the impact of poor authentication practices appears to be limited, mostly restricted to those who have suffered a loss and their immediate circle of colleagues and friends. Repeated warnings of imminent disaster followed by little or no impact on the average citizen (the "crying wolf" effect) leads to disbelief and failure to take the most basic precautions. Marketing the positive value of biometrics in minimizing the opportunities for identity theft will be needed once solutions become commercially attractive.

12.7 Looking to the Future

When the Personal Data Directive was agreed upon, provision was made for a review of its implementation together with the preparation of reports on possible amendments. This review has started and the 2002 conference [14] is one contribution to the debate. However, it is already clear that countries in the EU have chosen to transpose the directive in different ways, thereby adding to the confusion that the harmonization of laws was meant to address[12]. Some member states, such as the UK and the Netherlands [35], have decided, in general, to follow the wording of the directive, applying it to both the public and private sectors, clearly identifying the exemptions for government activities in the areas of national security, criminal justice, health etc. Germany has delineated its national law into two sections that deal with private and public applications separately [36]. The Irish bill (still to be passed at the time of writing) amends the pre-existing legislation on a clause-by-clause basis in order to conform to the 1995 directive [37]. The detail in the Swedish national law makes explicit the right to revoke at any time a previously given consent for processing of personal data if they are of the sensitive class or if they are to be transferred to certain third countries [38].

With such a complex legal infrastructure, it is no surprise that organizations processing personal data have raised concerns about the application of the directive and the national laws that follow from them. For example, in

12 Many of these national laws are available on the Internet together with unofficial translations into English.

the UK, the results of a government consultation exercise are available, together with a response by the Information Commissioner [39]. However, recent comments by the European Commissioner appear to rule out any early radical change to the Directive [40]. In his view, priority should be given to the uniform and consistent application throughout the member states. To this end, the Europe-wide CEN/ISSS activity, IPSE (Initiative for Privacy Standardisation in Europe), launched in 2000, may be one way forward [41]. Its first recommendation is for the collation of Best Practices for compliance with data protection laws, making it widely available either freely or on a low-cost basis. We note that there are alternative viewpoints, including ones that advocate complete abolition of the personal data protection regime on the grounds that it is fundamentally misconstrued and works against the interests of both organizations and individuals (see, for example, [42]).

Compliance with the requirements of a poorly understood data protection law may not be sufficient. A clearly articulated Code of Practice may be a more practical answer, helping the end user, the system integrator and the data protection authority to gain acceptance for their deployment [43]. In view of the diversity of applications of biometric methods, separate versions for government applications and commercial deployments may be required. Clearly a Europe-wide scheme would be preferable, and activities such as those in the BIOVISION project and proposals from the Biometric Working Group are directed towards building a consensus as to its form and content. Among the possibilities that are under consideration are:

- A statement of purpose of the installation, together with a rationale for use of a biometric over conventional means of authentication.
- Prominent display of the identity of the "controller" of the installation near each biometric terminal with a physical or web address to contact him or her (standards will be needed for positioning and size of any such notice). Electronic systems might be required to provide a link to these details, next to each instance of the use of a biometric.
- A maximum time-scale within which the controller will respond to any questions.
- A statement of compliance with the provisions of named local data protection laws, together with any exemptions or derogations that have been used.
- A statement in respect of "opt-in" or "opt-out" opportunities for end users, together with any rights afforded to the end users in respect of all personal data held on them.
- Stated retention periods for personal data.
- Any accesses permitted for third parties, including those permitted for lawful authorities.
- Wherever possible, the logic behind decision-making should be made available.
- Any Privacy Impact Assessment (PIA) that may have been made prior to deployment (a study that may include analysis of proportionality and finality, procedures in the event of failure to enrol, false rejection or false

acceptance etc.), together with any implementation plan based upon the recommendations of the PIA.
- Summary of the security measures employed to protect personal data relating to the end users, including details of ISO 17799 accreditation (if applicable), any external audit, maintenance procedures, etc.
- Specification of procedures to ensure secure disposal of the personal data in the event of withdrawal of the system.
- Specification of internal monitoring procedures to ensure compliance with the security policies applicable to the biometric and related systems.
- Details of the external audit of the system and whether this will be available to the public or end users.
- Review procedures and dates for re-examination of the operation of the system.

12.8 Social and Psychological Context of the Application of Biometric Methods

It is widely acknowledged that many individuals are fearful of the introduction of biometrics. Questionnaire studies have shown wide differences in the response to proposals such as the replacement of a PIN with a fingerprint or the use of an eye identification method for physical access. Often the questionnaire is produced after minimal user familiarization with a technology. Perhaps the developers of the questionnaires are unaware of the unspoken messages conveyed by the order and form of questions [44]. More subtle approaches will be needed if we are to uncover all of the concerns of potential end users of these innovative technologies. The shorthand of "concern with privacy" may hide a large number of other fears and uncertainties, some of which may be allayed by reassurance, targeted education, and additional time to familiarize the users with a radically new approach. The use of such new approaches to ascertain end user perception in the security field, e.g. semi-structured interviews in a focus group setting, has revealed many problems with existing authentication methods such as passwords [45]. The discussions have been analyzed using grounded theory approaches that search out repeated underlying themes. Proposals to extend this kind of in-depth analysis to biometric methods are under way.

If privacy concerns are just part of the story, what are the other components of the fear of biometrics? In an early paper, Simon Davies listed a number of concerns that were already being expressed in 1994 [46]. Since then, many other commentators have added to the list:

- Fear of "function creep" towards systems that will disadvantage the end user.
- The de-humanization of people by their reduction to bytes on a computer.

- The uniqueness promoted by some systems does not allow the representation of the multiple facets of identity that a person offers towards friends, family, work colleagues etc.
- The high integrity of identification reverses the "natural" relationship of government serving citizens and society.
- The symbolic representation of a "perfect" system in authority.
- Fear of society being increasingly driven by a technocracy, rather than a democratically elected government.
- Driving out criminal activity through the use of biometrics would displace it to more violent or socially disruptive activities.
- Exemptions and exceptions that would be made for the powerful in society.
- A system that would entrench fraud and criminality through technologically secure systems.
- The methods are the mechanism foretold in religious prophecies.

The impact of Hollywood's association of biometric methods with spies, advanced military hardware and science fiction may have increased these concerns, portraying these as perfect technologies in the service of powerful organizations.

Of course, these concerns could be viewed as positive features of the biometric approach. For every comment that interviewees make concerning the possible theft of fingerprints and reuse in an unrelated criminal case, there is another person who is encouraged by the use of a technology that has such a long history of integrity in the service of society. Both extremes make judgements based upon inadequate knowledge and understanding of technologies and institutions.

The first stage in addressing these concerns is to gather together these issues from all sections of the target population and to organize them in ways that allow further investigation. Only once the models for these concerns have been elucidated can we begin to structure questionnaires that will reveal the extent of these concerns and the depth or intensity with which they are held, as an initial step to addressing them in an effective manner.

The privacy component of such issues has been examined already, albeit in contexts other than biometrics. Studies in the way that end users of multimedia systems react to breaches in privacy offer models that could be used in the biometrics field, specifically in the application of automatic face recognition to CCTV systems. Victoria Belotti [47] has developed a model, based upon the need of users to (1) have feedback on, and (2) then exert control over, four elements in complex environments:

- Capture of personal information into the system.
- What happens once the information is in the system.
- Who and what processes will make use of the information in the system.
- For what purpose they will use that personal information.

Further research by Anne Adams [48] has centered even more on the end user's perspective on the transfer of personal information to organizations.

Contrast this stance (and that of Belotti) with the legal perspective of the data protection directives and national laws that emphasize the need for controls on the organization that collects and processes this information. Of course, the laws acknowledge rights for the data subject to agree to the collection and use of the personal data, to check that the data is correct and up-to-date etc. These laws aim to treat the parties as equals in the exchange of data for mutual benefit to the data subject and the organization. The situation for the end user may well be different, with lack of knowledge and understanding of the longer term consequences of her decisions. Adams acknowledges that every transaction will be different, the response to a request for personal data being determined situationally by the data subject. For each situation, the outcome will be colored not only by the individual's perspective, but by the norms and values of her social group. In this model, based upon the results of empirical research, the end user has three principal concerns:

- That the trust she places in the *receiver* of the personal data is not misplaced.
- That the risk–benefit analysis she makes of the *usage* to which that data is put is correctly assessed.
- That her judgement as to the *sensitivity* of the information is correctly made.

These concerns will be modified by comments from her peer group; they will respond to stories in the media and the experience of working in similar environments. However, if that experience leads the user to make unwarranted assumptions with outcomes that are unexpected or embarrassing for her, a dangerous overreaction can result. This overreaction and the sharing of that disappointment can set back the efforts of the proponents of a system, not just for the individual, but also for those in her immediate community.

One approach to ensuring that such issues are considered during the system design is to carry out a Privacy Impact Assessment (PIA) at an early stage of the process, preferably at the requirements capture stage. Although specific concerns will not emerge, the process of examining the system from the perspective of all the stakeholders should highlight the additional studies that are necessary and the time-scale over which these should be completed. Since security and privacy policies and solutions are often interrelated, an early focus on privacy aspects should ensure a correspondingly early attention to the security dimension of the system design.

Acknowledging the psychological and social contexts of the use of a biometric device may require more than just developing a model of these user concerns and making the appropriate interventions in standard design approaches. Complex systems based upon novel technological components will inevitably pose problems for the customer and end user if their introduction is accompanied by new and unfamiliar processes. Often, newly introduced systems that run counter to existing processes force end users into inefficient "workarounds" in an attempt to complete the required tasks (see [49], for example).. To avoid these inefficiencies in biometric-enabled systems, all of the stakeholders – the task, the end user and the customer – need to be in the forefront of a socio-technical design, from the inception of the project through to its deployment and operation [50]. All too often, however, the technology and its demands drive the design and the design methodology. Innovative trialing of new methodologies, that balance the technical and the human aspects, has been under way for many years, predominantly in Scandinavian countries [51]. We believe that future biometric-enabled systems will have a higher likelihood of success if they take account of such approaches.

12.9 Conclusions

The European perspective on the privacy implications of the application of biometrics is presented in a historical context. Many decades of debate have culminated in a framework for the protection of personal data that is both complex and, in some cases, difficult to apply to new technologies such as biometrics. Technical and organizational measures to secure the operation of the biometric-enabled system are a key part of the Eight Principles of Data Protection, and specific technical measures in the form of Privacy-Enhancing Biometric Technologies would provide a degree of reassurance to end users. Further protection of the end user's rights is offered by Article 8 of the European Convention on Human Rights that guarantees respect by the state for a citizen's private and family life, home and correspondence.

However, it is clear that users' views on the privacy implications of the application of biometrics may be only a part of a package of concerns. Situationally determined models of privacy are a first step towards a richer, more comprehensive model for the response of individuals to the introduction of biometrics. Such a model will be a valuable input to socio-technical designs of biometric systems that give equal weight to the social dimension and to issues of performance, standardization etc.

As biometric methods become more prevalent, we believe that familiarity with their operation should allay some of these fears. In the interim, a number of measures could help with their acceptance. Codes of Practice written in simpler terms than those used in the Personal Data Directive should clarify the rights and obligations of users and system owners. Targeted educational initiatives and installation of devices in high-profile limited deployments will begin the process of familiarization.

References

[1] Council of Europe, *Convention for the Protection of Human Rights and Fundamental Freedoms as amended by Protocol No. 11* (1950). Available online at: `http://conventions.coe.int/treaty/en/Treaties/Html/005.htm`.

[2] Directive 95/46/EC of the European Parliament and of the Council of 24 October 1995 on the protection of individuals with regard to the processing of personal data and on the free movement of such data. Available online at:

http://europa.eu.int/comm/internal_market/en/dataprot/law/index.htm.

[3] This has been the case with the 1997 "Telecommunications" directive, superseded by directive 2002/58/EC on Privacy and Electronic Communications: http://europa.eu.int/information_society/topics/telecoms/regulatory/new_rf/documents/l_20120020731en00370047.pdf (2002).

[4] For an introduction to the development of ideas of privacy from primitive societies to modern political thought see F. Schoeman (ed.), *Philosophical Dimensions of Privacy: an Anthology*, Cambridge University Press, 1984, and specifically the paper by Alan Westin, The origins of modern claims to privacy, reprinted from *Privacy and Freedom*, Association of the Bar of the City of New York, 1967.

[5] S. Shapiro, places and spaces: the historical interaction of technology, home and privacy. *The Information Society*, **14**, 275–284, 1998.

[6] I. About, *The Foundations of the National System of Police Identification in France (1893–1914). Anthropometry, Descriptions and Files, Geneses (54)*, http://www.iresco.fr/revues/geneses/summaries.htm, 2004.

[7] http://europa.eu.int/comm/internal_market/privacy/law/treaty_en.htm

[8] OECD, *Guidelines on the Protection of Privacy and Transborder Flows of Personal Data*, 1980.

[9] Council of Europe, *Convention for the Protection of Individuals with Regard to Automatic Processing of Personal Data*, 1981.

[10] http://europa.eu.int/comm/dgs/internal_market/mission_en.htm.

[11] *Directive 95/46/EC*, Reference 3.

[12] http://europa.eu.int/comm/internal_market/privacy/law/implementation_en.htm and Commission's First Report on the transposition of the Data Protection Directive (2003), http://europa.eu.int/comm/internal_market/privacy/lawreport/data-directive_en.htm

[13] *Directive 2002/58/EC of the European Parliament and of the Council of 12 July 2002 concerning the processing of personal data and the protection of privacy in the electronic communications sector*: http://europa.eu.int/comm/internal_market/privacy/law_en.htm.

[14] Report on data protection conference at http://europa.eu.int/comm/internal_market/privacy/lawreport/programme_en.htm (30 September –2 October 2002).

[15] http://www.dataprotection.gov.uk/.

[16] GUIDES Project, *E-business Guidelines on DPD 95/46/EC*. http://eprivacyforum.jrc.it/.

[17] http://www.collegebeschermingpersoonsgegevens.nl/en/download_audit/PrivacyAuditFramework.pdf

[18] Article 29 Data Protection Working Party: http://europa.eu.int/comm/internal_market/privacy/workingroup_en.htm.

[19] Dutch Data Protection Authority, *At face value – on biometrical identification and privacy*, 1999: http://www.cbpweb.nl/documenten/av_15_At_face_value.htm

[20] CNIL, *22ème rapport d'activité*, 2001: http://www.ladocumentationfrancaise.fr/brp/notices/024000377.shtml.

[21] *Opinion 10.2001 on the need for a balanced approach in the fight against terrorism* (14 December 2001).

[22] http://www.eubiometricforum.com/

[23] http://www.cesg.gov.uk/site/publications/index.cfm?displayPage=1
[24] http://pi.ijs.si/ProjectIntelligence.Exe?Cm=Project&Project=FINGER_CARD
[25] Data Protection Directive (95/46/EC) *Proposals for Amendment made by Austria, Finland, Sweden and the United Kingdom*: http://www.lcd.gov.uk/ccpd/dpdamend.htm (paragraph 6).
[26] J. Grijpink, Privacy law biometrics and privacy. *Computer Law and Security Report*, 17(3), 154–160, 2001.
[27] ISO/IEC, *Information technology – Code of practice for information security management.* http://www.iso.ch/iso/en/CatalogueDetailPage.CatalogueDetail?CSNUMBER=33441&ICS.
[28] The distinction between acquiescence and consent should be considered. In the UK, the *British Gas Trading Ltd (1998)* case provides one guideline.
[29] Commission decisions on the adequacy of the protection of personal data in third countries, http://europa.eu.int/comm/internal_market/privacy/adequacy_en.htm.
[30] I. van der Ploeg, Written on the body: biometrics and identity. *Computers and Society*, March, 37–44, 1999.
[31] Convention for the Protection of Human Rights and Fundamental Freedoms as amended by Protocol No. 11 (1950): http://conventions.coe.int/Treaty/en/Treaties/Html/005.htm
[32] J. Wadham and H. Mountfield, *Human Rights Act, 1998*, 2nd edn. Blackstone Press, 2000. A comprehensive compilation of cases relating to Article 8 is to be found in A. Mowbray, *Cases and Materials on the European Convention on Human Rights*. Butterworths, 2001.
[33] For example, S. Kenny and J. Borking, The value of privacy engineering. *The Journal of Information, Law and Technology (JILT)*, 2002(1) at http://elj.warwick.ac.uk/jilt/02-1/kenny.html, the PISA project at http://pet-pisa.openspace.nl/pisa_org/pisa/index.html, and the results of the RAPID roadmap project of research into privacy and identity management technologies, http://www.ra-pid.org/.
[34] ISO 15408 at http://www.commoncriteriaportal.org/
[35] Unofficial translation at Personal Data Protection Act (Wet bescherming persoonsgegevens), http://www.cbpweb.nl/en/index.htm.
[36] Federal Data Protection Commissioner, *Federal Data Protection Act (BDSG)*, http://www.bfd.bund.de/information/bdsg_eng.pdf (2002), although note that there is a common part applicable to both private and public operations. A debate is currently under way to reduce the complexity of laws on privacy in Germany.
[37] Data Protection Commissioner (Ireland), *Data Protection (Amendment) Bill, 2002*: http://www.dataprivacy.ie/images/dpbill2002.pdf (February 2002).
[38] Data Inspection Board, Sweden, *Personal Data Act – Section 12*: http://www.datainspektionen.se/in_english/legislation/data.shtml (1998: 204).
[39] Lord Chancellor's Department, *Data Protection Act 1998: Post-Implementation Appraisal. Summary of Responses to September 2000 Consultation*: http://www.lcd.gov.uk/ccpd/dparesp.htm (December 2001).
[40] Frits Bolkestein, Closing remarks at European Commission conference on "Data Protection", http://europa.eu.int/comm/commissioners/bolkestein/speeches_nl.htm (2002).

[41] Initiative on Privacy Standardisation in Europe – Final Report, http://www.cenorm.be/cenorm/businessdomains/businessdomains/isss/activity/ipsefinalreport.pdf (February 2002).

[42] L. Bergkamp, EU data protection policy: the privacy fallacy: adverse effects of Europe's data protection policy in an information-driven economy. *Computer Law and Security Report*, **18**(1), 31–47, 2002.

[43] Article 27 of the framework directive encourages the drawing up of such codes of practice. For example, in the UK, a *Code of practice for CCTV* installations, complying with the Data Protection Act, has been developed and is available at http://www.informationcommissioner.gov.uk/eventual.aspx?id=5739 (2000).

[44] J. Tanur, *Questions about Questions: Inquiries into the Cognitive Bases of Surveys*. Russell Sage Foundation, 1992.

[45] A. Sasse *et al.*, Transforming the "weakest link" – a human/computer interaction approach to usable and effective security. *BT Technology Journal*, **19**(3), 122–131, 2001.

[46] S. Davies, Touching Big Brother: how biometric technology will fuse flesh and machine. *Information Technology and People*, 7(4), 1994.

[47] V. Belotti, Design for privacy in multimedia computing and communications environments, in P. Agre and M. Rotenberg (eds), *Technology and Privacy: The New Landscape*. MIT Press, 1998.

[48] A. Adams, Users' perceptions of privacy in multimedia communications. *PhD Thesis*, University College London, 2001.

[49] J. Orr, *Talking about Machines: An Ethnography of a Modern Job*. Cornell University Press, 1996.

[50] M. Rejman-Greene, A framework for the development of biometric systems. In M. Lockie (ed.), *Biometrics Technology International*. Global Projects Group, 2003.

[51] An example of such methodologies comes from Jens Rasmudden's team in Denmark (summarised in Kim Vicente's survey *Cognitive Work Analysis*, Laurence Erlbaum Associates, Mahweh, NJ, 1999).

Index